Chemical Reactor Design

Wiley Series in Chemical Engineering

Bird, Armstrong and Hassager: DYNAMICS OF POLYMERIC LIQUIDS, Vol. I FLUID MECHANICS

Bird, Hassager, Armstrong and Curtiss: DYNAMICS OF POLYMERIC LIQUIDS, Vol. II KINETIC THEORY

Bird, Stewart and Lightfoot: TRANSPORT PHENOMENA

Brownell and Young: PROCESS EQUIPMENT DESIGN: VESSEL DESIGN

Davis: NUMERICAL METHODS AND MODELING FOR CHEMICAL ENGINEERS

Doraiswamy and Sharma: HETEROGENEOUS REACTIONS ANALYSIS, EXAMPLES AND REACTOR DESIGN, Vol. 1, Vol. 2

Felder and Rousseau: ELEMENTARY PRINCIPLES OF CHEMICAL PROCESSES, 2nd Edition

Foust, Wenzel, Clump, Maus and Andersen: PRINCIPLES OF UNIT OPERATIONS, 2nd Edition

Froment and Bischoff: CHEMICAL REACTOR ANALYSIS AND DESIGN

Franks: MODELING AND SIMULATION IN CHEMICAL ENGINEERING

Henley and Seader: EQUILIBRIUM–STAGE SEPARATION OPERATIONS IN CHEMICAL ENGINEERING

Hill: AN INTRODUCTION TO CHEMICAL ENGINEERING KINETICS AND REACTOR DESIGN

Jawad and Farr: STRUCTURAL ANALYSIS AND DESIGN OF PROCESS EQUIPMENT

Kellogg Company: DESIGN OF PIPING SYSTEMS, Revised 2nd Edition

Klein and Bischoff: CHEMICAL KINETICS AND REACTOR DESIGN, 2nd Edition

Levenspiel: CHEMICAL REACTION ENGINEERING, 2nd Edition

Nauman and Buffham: MIXING IN CONTINUOUS FLOW SYSTEMS

Nauman: CHEMICAL REACTOR DESIGN

Rase: CHEMICAL REACTOR DESIGN FOR PROCESS PLANTS PRINCIPLES AND TECHNIQUES, Vol. 1

Rase and Barrow: PIPING DESIGN FOR PROCESS PLANTS

Reklaitis: INTRODUCTION TO MATERIAL AND ENERGY BALANCES

Rudd, Fathi-Afshar, Trevino and Stadtherr: PETROCHEMICAL TECHNOLOGY ASSESSMENT

Rudd and Watson: STRATEGY OF PROCESS ENGINEERING

Sandler: CHEMICAL AND ENGINEERING THERMODYNAMICS

Seborg, Edgar and Mellichamp: PROCESS DYNAMICS AND CONTROL

Smith and Corripo: PRINCIPLES AND PRACTICE OF AUTOMATIC PROCESS CONTROL

Smith and Missen: CHEMICAL REACTION EQUILIBRIUM ANALYSIS

Ulrich: A GUIDE TO CHEMICAL ENGINEERING PROCESS DESIGN AND ECONOMICS

Welty, Wicks and Wilson: FUNDAMENTALS OF MOMENTUM, HEAT AND MASS TRANSFER, 3rd Edition

Chemical Reactor Design

E. B. Nauman
Rensselaer Polytechnic Institute
Troy, New York

John Wiley & Sons

New York   Chichester   Brisbane   Toronto   Singapore

Library of Congress Cataloging in Publication Data:

Nauman, E. B.
   Chemical reactor design.

   Includes bibliography and index.
   1. Chemical reactors--Design and construction.
I. Title.

TP157.N39   1987   660.2′83   86-19031
ISBN 0-471-84580-9

Printed in the United States of America

10 9 8 7 6 5 4 3 2 1

About The Author

E. Bruce Nauman is Professor of Chemical Engineering at Rensselaer Polytechnic Institute. Dr. Nauman has served as Department Chairman and now acts as Director of the Industrial Liaison Program in addition to leading a large and active research group in chemical and polymer reaction engineering. Prior to joining RPI, Dr. Nauman spent seventeen years at Union Carbibe and Xerox Corporation in a variety of research and general management positions. He has also held academic appointments at the University of Tennessee, the University of Rochester, and the State University of New York at Buffalo. He is an active consultant and is a member of several corporate boards in addition to a variety of academic and editorial advisory boards. Dr. Nauman has written a previous Wiley text, *Mixing in Continuous Flow Systems*, with B. A. Buffman, and has published more than fifty rersearch articles. The author of several patents, he has invented a new technique for microparticulation, which is being widely licensed in the polymer industry. Dr. Nauman founded the ongoing series of Engineering Foundation Conferences on Chemical Reaction Engineering and has chaired the Conference on Mixing.

Preface

Some areas of science and technology are led by academians. Others are paced by industrialists. Chemical reaction engineering has benefited from both inputs; and, at least in the research areas, is practiced at a high level of competence. However, undergraduate education has not yet reflected the remarkable changes that have occurred in academic and industrial reaction engineering over the last decade. The transport phenomena revolution of the 1950s and 1960s has fully permeated chemical engineering. Digital computation has become incredibly cheap in the 1970s and 1980s. This combination has led to a quantitative emphasis in chemical reactor design that far surpasses the standard undergraduate texts.

An emphasis throughout this text is numerical computation; it should be understood that this is not an absolute emphasis but one relative to other texts in the field. Physical insights remain of primary concern. The numerical methods used in this book are not sophisticated. Computers are cheap and becoming cheaper. Nowadays, a model too complex to be computable is probably too complex to be real. The student of chemical reactor design can come up with hard numbers rather than qualitative arguments. Obviously, the numbers will be no better than the physical insights that generated them. Horror stories can always be told of numerical precision disguising physical inaccuracy. However, at least one source of error is eliminated when the consequences of a model can be evaluated exactly. To me at least, this is a big step forward.

Simple methods which retain physical transparency are stressed. References are cited to more sophisticated techniques. These should be used for detailed design and optimization studies with complex models. In my judgment, the simplest possible numerical scheme is best at the beginning. Thus Euler's method is used for ordinary differential equations, fully explicit marching techniques are used for partial differential equations, and generous use is made of Richardson extrapolation. These methods are simple and robust, and they rarely mislead.

Two traditional topics, the analysis of rate data and chemical thermodynamics, have been postponed until fairly late in the text. They appear at the end of Chapter 4 which introduces heat balances. This postponement can be rationalized in several ways, but the real motivation was pedagogical. My personal feeling is that students prefer to start a new course by seeing what is distinctly different from previous courses. Should you disagree with this approach, Section 4.3 and Appendix 4.1 can be introduced earlier without significant difficulty.

A standard undergraduate course should include Chapters 1–4, 6 and 7. Chapter 5 constitutes a rather open-ended project where the student can get some feel for process integration and optimization. It, together with a distributed reactor problem from Chapters 6 or 7, is an integral part of RPI's one semester course in chemical reactor design. Also included in that one semester is Chapter 8 and either of Chapters 9 and 10. However, any of Chapters 8 through 12 can be used to round out the undergraduate course depending on the student's interests and abilities.

The book can also be used for a first graduate course, particularly if supplemented by a text, such as that by Aris,[1] which places greater stress on fundamentals. Most students will benefit from a quick tour of Chapters 1 through 4. Although they will have been exposed previously to kinetics and reactor design, few will be comfortable with the relatively heavy emphasis on computation that characterizes this text. Chapters 6 and 7 need to be covered in some detail, since few undergraduate programs give the necessary background in distributed parameter systems. However, it is still possible to cover all of Chapters 8 through 12 in a one-semester graduate course. These chapters include introductions to unsteady reactors and control systems, residence time and micromixing theory, heterogeneous catalysis, microbial processes, multiphase reactors, electronic device fabrication, and polymer reaction engineering. These introductions are intended to be just that. I have had the good fortune to work and consult in most of the mentioned areas. Hopefully, I have managed to distill the essence of these topics without succumbing to the temptation of writing far more than can be covered in an introductory course. Courses beyond this level, whether formal or informal, can be best served by the research literature and specialized monographs.

Many acknowledgments are in order. Excellent and extensive typing services were provided by Jean Mulson and Rosemary Primett. I was fortunate to have my own graduate students serve as teaching assistants for courses that used the manuscript: Ramesh Mallikarjun, Hugh McLaughlin, Keith Dackson, Tom Bergman, Muhammad Atiqullah, and Holly Haughney. It was they who suffered —almost as much as the students—from the early typos and unworkable

[1]R. Aris, Introduction to the Analysis of Chemical Reactors, Prentice-Hall, Englewood Cliffs, NJ, 1965.

problems. Colleagues who read and used the manuscript deserve credit for its strengths: Robert Adler, Alfred Engel, L. T. Fan, John Friedly, A. H. Johannes, and F. Shadman. They share none of the blame for its weaknesses. Also deserving credit for the book's strengths are the Wiley editors Don Ford and Christina Mediate. Occasionally, I had the good judgment to heed their advice.

E. B. Nauman

Troy, New York
August 1986

Contents

Chapter 2 Complex Reactions in Batch Reactors

Chapter 3 Complex Reactions in Ideal Flow Reactors

Chapter 4 Thermal Effects and Energy Balances   89

Chapter 5 Design and Optimization Studies 153

Chapter 6 Real Tubular Reactors in Laminar Flow 165

Chapter 7   Real Tubular Reactors in Turbulent Flow   203

Chapter 8   Unsteady Reactors   227

Chapter 9   Mixing in Continuous-Flow Systems   255

Chapter 10 Heterogeneous Catalysis

293

Chapter 11   Multiphase Reactors   331

Chapter 12   Polymer Reaction Engineering   377

Index

A Note on Notation

Occasionally, chemical reactions will be spelled out in detail, for example,

$$CH_3COOH + C_2H_5OH \rightleftharpoons CH_3COOC_2H_5 + H_2O \qquad (0.1)$$

and the concentrations (in moles per unit volume) of the various species will be represented using square brackets, for example,

$$K_{equil} = \frac{[CH_3COOC_2H_5][H_2O]}{[CH_3COOH][C_2H_5OH]} \qquad (0.2)$$

This notation quickly becomes cumbersome when the molecules are complicated. Thus we normally use shorthand:

$$A + B \rightleftharpoons C + D \qquad (0.3)$$

$$K_{equil} = \frac{[C][D]}{[A][B]} \qquad (0.4)$$

or

$$K_{equil} = \frac{C_C C_D}{C_A C_B} \qquad (0.5)$$

or

$$K_{equil} = \frac{cd}{ab} \qquad (0.6)$$

All the concentration notations in Equations 0.4 through 0.6 are found in the literature. We shall usually use the lowercase letters $a, b, c, \ldots$ to denote concentrations of components A, B, C, ... as illustrated in Equation 0.6. This form is easy to write and lends itself to subscripts: a_0, a_{in}, a_{out}.

When the capital letters A, B, C, ... appear as subscripts, they denote a particular component. Thus N_A is the number of moles of A present in the system and $\mathscr{R}_A$ is the rate of formation of A (in moles per unit volume per unit time). Roman numeral subscripts are used to distinguish between different reactions. Thus $\mathscr{R}_{A,II}$ is the rate of formation of component A by Reaction II.

Two types of averages are used in this text. The spatial average of a is denoted as $\hat{a}$ and the convected mean is denoted as $\bar{a}$.

List of Symbols

Roman Symbols

Symbol		Reference Equation
a	Concentration of component A	1.4
a_0	Initial concentration of A	1.30
$a(0+)$	Concentration of A immediately after reactor inlet	7.8
$\mathbf{a}$	Vector of concentrations	2.74
$\mathbf{a}_0$	Vector of initial concentrations	2.74
$\hat{a}$	Spatial average concentration	1.4
$\bar{a}$	Mixing cup or convected mean concentration	12.82
a_b	Concentration of A in the bubble phase	11.74
a_{batch}	Concentration of A after batch reaction	9.82
a_c	Specific surface area of catalyst	10.13
a_e	Concentration of A in the emulsion phase	11.73
a_g	Concentration of A in the gas phase	11.1
a_g^*	Concentration of A in gas phase at interface	11.1
a_i	Concentration of A inside a pore	10.4
a_{in}	Concentration of A at reactor inlet	1.3
a_j	Concentration of A at jth step of calculation	3.5
a_l	Concentration of A in liquid phase	11.1
a_l^*	Concentration of A in liquid phase at interface	11.1

a_{mix}	Concentration of A after mixing	3.85
a_n	Concentration of oligometer of length n	12.47
a_{out}	Concentration of A at reactor outlet	1.3, 8.4
a_r	Concentration of A inside a spherical pellet	11.99
a_s	Surface concentration of A	10.3
a_t	Concentration of A at time t	1.30
a_t^*	Estimate of $a(t)$	2.13
A	Subscript denoting reactive species A	1.3
A_b	Cross-sectional area of bubble phase	11.74
A_c	Cross-sectional area of reactor	3.7
A_e	Cross-sectional area of emulsion phase	11.73
A_{ext}	External surface area of reactor	4.23, 11.43
A'_{ext}	External surface area per unit length of reactor	4.26
A_g	Cross-sectional area of gas phase	11.46
A_i	Interfacial area per unit volume of reactor	11.1
A'_i	Interfacial area per unit height of column	11.45
A_l	Cross-sectional area of liquid phase	11.45
A_s	Cross-sectional area of solid phase	1.111
Av	Avogadro's number, molecules/mol	1.9
b	Concentration of component B	1.7
b_0	Initial concentration of B	1.45
b_{bulk}	Concentration of B in the bulk fluid	11.72
b_{in}	Concentration of B at reactor inlet	1.80
b_j	Concentration of B at jth step in calculation	3.6
b_n	Concentration of oligometer of length n	12.47
b_{out}	Concentration of B at reactor outlet	1.80
b_t	Concentration of B at time t	2.51
B	Subscript denoting reactive species B	2.9
c	Concentration of component C	1.18
c	Concentration of live cells	10.79
c_0	Initial concentration of C	2.53
c_{in}	Concentration of C at reactor inlet	3.32
c_n	Concentration of oligometer, possible cyclic, of length n	12.46, 12.47
c_{out}	Concentration of C at reactor outlet	3.32
C	Constant of integration	1.41
C	Subscript denoting component C	2.9
C	Concentration of nonreactive tracer	9.2
C_A	Capacity of ion exchange resin for reactant A	11.103
C_{AB}	Collision rate between A and B molecules	1.9
C_0	Initial concentration of tracer	9.2
C_p	Heat capacity	4.29
d	Concentration of component D	2.10

d_0	Initial concentration of D	2.10
d_{in}	Concentration of component D at reactor inlet	3.63
d_{out}	Concentration of component D at reactor outlet	3.63
d_p	Diameter of packing or of catalyst particle	3.14
d_t	Concentration of component D at time t	2.53
d_s	Diameter of unreacted zone	11.98
d	Diameter of tube	5.3
D	Subscript denoting reactive species D	2.9
D	Axial dispersion coefficient	7.5
D_r	Radial dispersion coefficient	11.95
D_z	Axial dispersion coefficient	11.95
$\mathscr{D}$	Molecular diffusivity	6.1
$\mathscr{D}_A$	Molecular diffusivity of component A	6.13
$\mathscr{D}_{eff}$	Effective diffusion coefficient inside catalyst particle	10.59
$\mathscr{D}_K$	Knudsen diffusivity	10.52
$\mathscr{D}_{net}$	Combined Knudsen and bulk diffusivity	10.53
$\mathscr{D}_P$	Molecular diffusivity of component P	10.9
e	Concentration of epoxy	8.40
e	Denotes the emulsion phase	11.73
E	Denotes an enzyme	Section 10.4.1
E	Activation energy	4.1
E	Axial dispersion coefficient for energy	7.26
$\mathscr{E}$	Enhancement factor in reactive gas absorption	11.71
E_I, E_{II}	Activation energies for Reactions I, II	Section 4.1.2
E_0	Concentration of reactive enzyme sites	10.64
E_f	Activation energy of forward reaction	4.6
E_r	Activation energy of reverse reaction	4.6
E_r	Radial dispersion coefficient for energy	7.46
$f(\ell)$	Frequency function for chain lengths	12.50
$f(t)$	Frequency function for residence times	9.10
f_G	Frequency function for growing chains	12.66
f_D	Frequency function for dead chains	12.67
$f_{\mathscr{R}}$	Collision efficiency factor	1.8
F	Arbitrary function	3.90
$F(t)$	Cumulative distribution of residence times	9.7
F_1, F_2	Arbitrary functions	3.51, 12.87
Fa	Fanning friction factor	3.13
F_C	Mole fraction of component C	3.67
F_j	Gas flow rate for jth tray	11.35
F_X, F_Y	Mole fractions of components X and Y	12.95
g	Denotes the gas phase	11.1

Symbol	Description	Reference
k_l	Gas-side mass transfer coefficient	11.3
k_g	Liquid-side mass transfer coefficient	11.3
k_M, k_P	Rate constants	11.14, 11.15
k_r	Rate constant for reverse reaction step	1.15, 11.49
k_H	Henry's law constant	11.2
k_I	Rate constant for initiator decomposition	12.17
k_n	Rate constant involving molecule of length n	12.1
k_{mn}	Rate constant for reaction between molecules of lengths m and n	12.2
k_p	Propagation rate constant	12.20
$k_{\mathscr{R}}$	Rate constant	10.37
$k_{\mathscr{R}}^+$	Forward rate constant for surface reaction	10.24
$k_{\mathscr{R}}^-$	Reverse rate constant for surface reaction	10.24
k_s	Mass transfer coefficient for catalyst particle	10.3
k_t	Termination rate constant	12.21
k_{XX}	Propagation rate constant for X–X homopolymerization	12.28
k_{XY}	Propagation rate constant for X–Y copolymerization	12.28
k_{YX}	Propagation rate constant for Y–X copolymerization	12.28
k_{YY}	Propagation rate constant for Y–Y homopolymerization	12.28
K	Equilibrium constant	4.69
K_I, K_{II}	Equilibrium constants for Reactions I and II	10.72
K_a	Adsorption equilibrium constant	10.34
K_d	Desorption equilibrium constant	10.31
K_{equil}	Equilibrium constant	1.16
K_P	Reciprocal desorption constant	10.38
$K_{\mathscr{R}}$	Equilibrium constant for surface reaction	10.30
l	Denotes liquid phase	11.1
ℓ	Distance down a pore	10.46
ℓ	Chain length	12.50
$\bar{\ell}, \bar{\ell}_n$	Number-average chain length	12.52
$\bar{\ell}_w$	Weight-average chain length	12.55
L	Length of tubular reactor	1.71
$\mathscr{L}$	Length of a pore	10.46
m	Constant in Arrhenius temperature dependence	4.1
m	Concentration of monoalkylate	11.15
m	Number of monomer units	12.2
m_A, m_B	Molecular masses, kg/molecule, of components A, B	1.9
$M \cdot$	Denotes free radical of monomer	12.1
$\overline{M}_n$	Number-average molecular weight	12.56

ρ_{min}	Density of bed at incipient fluidization	11.81
ρ_{molar}	Total molar density	3.65
ρ_{out}	Fluid density at outlet	4.12
σ^2	Dimensionless variance	9.36
σ_t^2	Variance of $f(t)$	9.34
τ	Scaled time	7.2
τ_p	Fractional tubularity	9.44
Υ	Heat flux	4.26
ϕ	Arbitrary function	2.69, 9.14
Φ	Vector of component fluxes	3.18
Φ_A	Molecular flux of component A	3.16
$(\Phi_A)_j$	Molecular flux of A at step j	3.19
ψ	Vector of random velocity components	7.1
Ψ	Concentration of active chains	12.64
ω	Proportionality factor in Flory distribution	12.65
Ω	Integral	6.74

Chemical
Reactor
Design

Simple Reactions in Ideal Reactors

1.1   The General Mass Balance

Consider any region of space having a finite volume and having prescribed boundaries which unambiguously separate the region from the rest of the universe. Such a region will be called a *control volume*, and the law of conservation of mass and energy may be applied to it. Here, we are concerned with nonnuclear processes so that there are separate conservation laws for mass and energy. For mass,

> Rate at which mass enters the volume
>
> = Rate at which mass leaves the volume   (1.1)
>
>   + Rate at which mass accumulates in the volume

where "entering" and "leaving" apply to the flow of material across the boundaries. See Figure 1.1. This equation applies to the total mass within a control volume, as measured in kilograms. In reactor design, we are interested in chemical reactions that transform one kind of mass into another. A mass balance can be written for each chemical component as follows:

> Rate at which component enters the volume
>
>   + Net rate at which component is formed by reaction
>
> = Rate at which component leaves the volume   (1.2)
>
>   + Rate at which the component accumulates

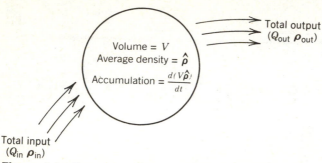

Figure 1.1   Control volume for overall mass balance.

See Figure 1.2. The *net rate of formation* accounts for chemical reaction between the various components. A component balance can always be expressed in terms of mass. Usually, by *component* we mean a distinct molecular species so that the balance can also be expressed in moles. When written in mass units, the sum of Equation 1.2 over all components must equal Equation 1.1 since the net rate at which total mass is formed must be zero. The similar summation does not apply in molar units since chemical reactions can change the number of moles.

Equation 1.2 can be given a more mathematical appearance by writing it as

$$Q_{in}a_{in} + \hat{\mathscr{R}}_A V = Q_{out}a_{out} + \frac{d(V\hat{a})}{dt}$$

(1.3)

where a (moles/volume) represents the concentration of some specified component A. In this formulation, Q_{in} and Q_{out} are volumetric flow rates (volume/time), $\hat{\mathscr{R}}_A$ is the volume rate of formation of the component (moles/time/volume), and V is the volume. Unless the volume is very well mixed, concentrations and reaction rates will vary from point to point within it. The material balance applies to the entire volume so that $\hat{\mathscr{R}}_A$ and $\hat{a}$ denote spatial averages. They may

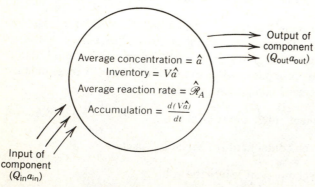

Figure 1.2   Control volume for component balance.

be related to local or point values by

$$\hat{a} = \frac{1}{V} \int \int \int_V a \, dV'$$

(1.4)

and

$$\hat{\mathcal{R}}_A = \frac{1}{V} \int \int \int_V \mathcal{R}_A \, dV'$$

(1.5)

Loosely speaking, the subject of *chemical reactor design* is concerned with evaluating the above integrals so that a_{out} can be found from Equation 1.3. This evaluation requires knowledge of the local reaction rate $\mathcal{R}_A$, which depends on local conditions such as concentration and temperature. Finding $\mathcal{R}_A$ is the province of *chemical kinetics*. This course stresses reactor design but will make occasional forays into the field of chemical kinetics to develop the necessary background. It is assumed that the reader has had previous contact with chemical kinetics and equilibrium thermodynamics. (Appendix 4.1 provides a brief review of thermodynamics.)

1.2 Elementary Reactions

Consider the reaction of two chemical species according to the stoichiometric equation

$$A + B \rightarrow P$$

(1.6)

This reaction is said to be *homogeneous* if it occurs within a single phase. For the time being we are concerned only with reactions that take place in the gas phase or in a single liquid phase. These reactions are said to be *elementary* if they result from a single interaction (i.e., collision) between the molecules appearing on the left-hand side of Equation 1.6. The rate at which collisions between A and B molecules occur should be proportional to their concentrations. Not all collisions cause a reaction, but at constant environmental conditions (e.g., temperature, electric field strength), some definite fraction should. Thus we expect

$$\mathcal{R} = k[A][B] = kab$$

(1.7)

where k is a constant of proportionality known as the *rate constant*.

Example 1.1 Use the kinetic theory of gases to rationalize the functional form of Equation 1.7.

Solution   We suppose that a collision between A and B molecules is necessary but not sufficient for reaction to occur. Thus we expect

$$\mathcal{R} = \frac{C_{AB} f_{\mathcal{R}}}{Av} \tag{1.8}$$

where C_{AB} is the collision rate (collisions per unit volume per unit time) and $f_{\mathcal{R}}$ is a reaction efficiency. Avogadro's number Av has been included in Equation 1.8 so that $\mathcal{R}$ will have normal units, $mol\,m^{-3}\,s^{-1}$, rather than units of molecules $m^{-3}\,s^{-1}$. By hypothesis, $0 \le f_{\mathcal{R}} \le 1$.

The molecules are treated as rigid spheres having radii r_A and r_B. They collide if they approach each other within a distance $r_A + r_B$. The kinetic theory of ideal gases gives[1]

$$C_{AB} = \left[\frac{8\pi R_g T(m_A + m_B)}{Av\, m_A m_B} \right]^{1/2} (r_A + r_B)^2 \, Av^2 \, ab \tag{1.9}$$

where R_g is the gas constant, T is the absolute temperature, and m_A and m_B are the molecular masses. This result agrees with Equation 1.7 when

$$k = \left[\frac{8\pi R_g T(m_A + m_B)}{Av\, m_A m_B} \right]^{1/2} (r_A + r_B)^2 \, Av \, f_{\mathcal{R}} \tag{1.10}$$

Collision theory is mute about the value of $f_{\mathcal{R}}$. Typically, $f_{\mathcal{R}} \ll 1$ so that the number of molecules colliding is much greater than the number reacting. See Problem 1.2. Not all collisions have enough energy to produce a reaction. Steric effects may also be important. As will be discussed in Chapter 4, $f_{\mathcal{R}}$ is strongly dependent on temperature. This dependence usually overwhelms the $T^{1/2}$ dependence predicted for the collision rate.

Note that the rate constant k is positive so that $\mathcal{R}$ is positive. We have written this "for the reaction." The rate at which A and B are formed is negative. Thus

$$\mathcal{R}_A = \mathcal{R}_B = -kab \tag{1.11}$$

while P is formed at a positive rate:

$$\mathcal{R}_P = +kab \tag{1.12}$$

The sign convention we have adopted is that the rate of the reaction is always positive. The *rate of formation of a component* is positive when the

---

[1]See any good book on physical chemistry. Note that $Av\, m_A m_B/(m_A + m_B) = M_A M_B/(M_A + M_B)$, where m_A and m_B are in kg/molecule while M_A and M_B are in kg/mol $\equiv$ kg/gm-mole.

component is formed by the reaction and is negative when the component is consumed.

A general expression for any single reaction is

$$0_M \rightarrow \nu_A A + \nu_B B + \cdots + \nu_R R + \nu_S S + \cdots \tag{1.13}$$

As an example, the reaction $2H_2 + O_2 \rightarrow 2H_2O$ can be written as $0_M \rightarrow -2H_2 - O_2 + 2H_2O$. This form is obtained by setting all participating species, whether products or reactants, on the right-hand side of the stoichiometric equation. The remaining term on the left is the *zero molecule*, which is denoted by 0_M to avoid confusion with atomic oxygen. The $\nu_A, \nu_B, \ldots$ terms are the *stoichiometric coefficients*. They are positive for products and negative for reactants. Using them, the relationship between the *rate of the reaction* and the *rate of formation of component A* is given in general by

$$\boxed{\mathscr{R}_A = \nu_A \mathscr{R}} \tag{1.14}$$

If the reaction of Equation 1.13 were reversible and elementary, its rate would be

$$\mathscr{R} = k_f [A]^{-\nu_A} [B]^{-\nu_B} \cdots - k_r [R]^{\nu_R} [S]^{\nu_S} \tag{1.15}$$

and it would have equilibrium constant

$$K_{equil} = \frac{k_f}{k_r} = [R]^{\nu_R} [S]^{\nu_S} \ldots [A]^{\nu_A} [B]^{\nu_B} = \frac{[R]^{\nu_R} [S]^{\nu_S} \ldots}{[A]^{-\nu_A} [B]^{-\nu_B} \ldots} \tag{1.16}$$

where $A, B, \ldots$ are reactants and $R, S, \ldots$ are products.

As a shorthand notation for indicating that a reaction is elementary, we shall include the rate constants in the stoichiometric equation. Thus the reaction

$$A + B \underset{k_r}{\overset{k_f}{\rightleftharpoons}} 2C \tag{1.17}$$

is elementary and reversible and has for a rate expression

$$\mathscr{R} = k_f ab - k_r c^2 \tag{1.18}$$

We deal with many reactions that are not elementary. Most industrially important reactions go through a complex kinetic mechanism before the final products are reached. These mechanisms may give rate expressions far different from Equation 1.15 even though they involve only short-lived intermediates that never appear in conventional chemical analyses. Elementary reactions are generally limited to the following types.

First Order, Unimolecular

$$A \xrightarrow{k} \text{Products} \qquad \mathscr{R} = ka$$

(1.19)

The best example of a truly first-order reaction is radioactive decay, for example,

$$U^{238} \rightarrow Th^{234} + He^4$$

since it occurs spontaneously as a single-body event. Among strictly chemical reactions, thermal decompositions such as

$$CH_3OCH_3 \rightarrow CH_4 + CO + H_2$$

follow first-order kinetics at normal gas densities. However, collisions between the reactant and other molecules are necessary to provide energy for the reaction. Thus a second order dependence is predicted for the pure gas at very low densities.

Second Order, One Reactant

$$2A \xrightarrow{k} \text{Products} \qquad \mathscr{R} = ka^2$$

(1.20)

(Note that $\mathscr{R}_A = -2ka^2$ according to the convention we have adopted.) A gas phase reaction believed to be elementary and second order is

$$2HI \rightarrow H_2 + I_2$$

Here, collisions between two HI molecules supply energy and also supply the reactants needed to satisfy the observed stoichiometry.

Second Order, Two Reactants

$$A + B \xrightarrow{k} \text{Products} \qquad \mathscr{R} = kab$$

(1.21)

Liquid phase esterifications such as

$$
\begin{array}{cc}
\quad\quad O & \quad\quad O \\
\quad\quad \| & \quad\quad \| \\
C_2H_5OH + CH_3COH & \rightarrow C_2H_5OCCH_3 + H_2O
\end{array}
$$

usually follow second order kinetics.

As suggested by these examples, the **order** of a reaction is the sum of the exponents $m, n, \ldots$ in

$$\mathscr{R} = ka^m b^n \ldots$$

(1.22)

This definition is meaningful only for rate expressions that have the simple form of Equation 1.22. For elementary reactions, $m = -\nu_A$, $n = -\nu_B$, and so on, where the stoichiometric coefficients are all integers. Complex reactions can give rise to rate expressions that have the form of Equation 1.22 but with fractional exponents.

For homogeneous systems, complex reactions can be broken into a number of series and parallel elementary reactions, possibly involving short-lived intermediates such as free radicals. All the individual reactions are treated as elementary. They are typically second order, although third order elementary reactions are occasionally reported. The number of such reactions may be surprisingly large. For example, a good model for

$$CH_4 + 2O_2 \rightarrow CO_2 + 2H_2O \tag{1.23}$$

will involve 20 to 30 elementary reactions, even assuming that the indicated products are the only ones formed in significant quantities.

Any elementary reaction must have Equation 1.15 as its rate expression. A complex reaction may closely approximate Equation 1.15 without being elementary. Consider (abstractly, not concretely) the nitration of toluene:

or, in shorthand,

$$A + 3B \rightarrow C + 3D$$

This reaction cannot be elementary. We can hardly expect three nitric acid molecules to react at all three toluene sites in a glorious, four-body collision. Thus the reaction rate

$$\mathscr{R}_I = kab^3 \tag{1.24}$$

is implausible.

An elementary reaction might be

since this would require only a two-body collision. Then the rate for this elementary step would be

$$\mathscr{R}_{II} = kab \tag{1.25}$$

The overall Reaction I might consist of seven elementary steps (two reactions leading to mononitrotoluenes, three leading to dinitrotoluenes, and two leading to TNT itself). Each step would require only a two-body collision and would be governed by a second order rate equation similar to Equation 1.25. In Chapter 2 we shall see how rate expressions for the elementary steps can be combined to give an expression for the overall reaction. The result will not resemble Equation 1.24. It could be as simple as

$$\mathscr{R}_{\mathrm{I}} = ka \tag{1.26}$$

which would result from using a large excess of nitric acid.

1.3 Ideal, Isothermal Reactors

1.3.1 Batch Reactors

A batch reactor has no input or output of mass. The amounts of individual components may change due to reaction but not due to flow into or out of the system. The general mass balance, Equation 1.3, becomes

$$\hat{\mathscr{R}}_{\mathrm{A}}V = \frac{d(V\hat{a})}{dt} \tag{1.27}$$

where $\hat{a}$ and $\hat{\mathscr{R}}_{\mathrm{A}}$ are defined by Equations 1.4 and 1.5. An ideal batch reactor has no temperature or concentration gradients within the system volume. The concentration will change with time due to the reaction, but at any time it is everywhere uniform. The reaction rate is also uniform and is everywhere equal to the system average rate. Thus $a = \hat{a}$ and $\hat{\mathscr{R}}_{\mathrm{A}} = \mathscr{R}_{\mathrm{A}}$:

$$\mathscr{R}_{\mathrm{A}}V = \frac{d(Va)}{dt} = V\frac{da}{dt} + a\frac{dV}{dt} \tag{1.28}$$

This section treats only the constant-volume case, $dV/dt = 0$. This corresponds to assuming constant fluid density, an assumption that is usually quite good for liquid phase reactions. Industrial gas phase reactions are normally conducted in flow systems rather than batch systems. Under the assumption of constant volume,

$$\mathscr{R}_{\mathrm{A}} = \frac{da}{dt} \tag{1.29}$$

This equation formally integrates as

$$t = \int_0^t d\alpha = \int_{a_0}^{a_t} \frac{da}{\mathscr{R}_{\mathrm{A}}} \tag{1.30}$$

where a_t is the concentration of component A after a batch reaction that lasted t time units. We use a_0 to denote the initial concentration of A. Any symbol could be used for α since it is a dummy variable of integration. However, we prefer to identify it as the *age* of the batch and will henceforth write Equation 1.29 as

$$\mathscr{R}_A = \frac{da}{d\alpha} \qquad (1.31)$$

for an isothermal batch reactor. Instead of using definite integrals, we can regard Equation 1.31 as a differential equation (albeit a very simple one) that is subject to the initial condition

$$a = a_0 \qquad \text{at} \qquad \alpha = 0 \qquad (1.32)$$

Example 1.2   The First Order Reaction

This can be written

$$A \xrightarrow{k} \text{Products} \qquad \mathscr{R} = ka \qquad (1.33)$$

$$\mathscr{R}_A = \nu_A \mathscr{R} = -ka \qquad (1.34)$$

Substituting into Equation 1.31 gives

$$\frac{da}{d\alpha} + ka = 0 \qquad (1.35)$$

Solving the ordinary differential equation and applying the initial condition gives

$$a = a_0 e^{-k\alpha} \qquad (1.36)$$

Evaluating this result at $\alpha = t$ gives the final concentration, a_t. It is usually convenient to divide through by a_0 to obtain dimensionless concentrations. Thus

$$\frac{a_t}{a_0} = e^{-kt} \qquad (1.37)$$

gives the *fraction unreacted* and

$$X_A \equiv 1 - \frac{a_t}{a_0} = 1 - e^{-kt} \qquad (1.38)$$

gives the *conversion*.

Example 1.3   The Second Order Reaction, One Reactant

We may choose to write the stoichiometric equation as

$$2A \xrightarrow{\frac{k}{2}} \text{Products} \qquad \mathscr{R} = \left(\frac{k}{2}\right)a^2 \tag{1.39}$$

so that

$$\mathscr{R}_A = -2\mathscr{R} = -ka^2 \tag{1.40}$$

Equation 1.31 integrates to

$$-a^{-1} + C = -kt \tag{1.41}$$

where C is a constant. Applying the initial condition gives $C = (a_0)^{-1}$ and

$$\frac{a_t}{a_0} = \frac{1}{1 + a_0 kt} \tag{1.42}$$

It is worth observing that $a_0 k$ has units of reciprocal time so that $a_0 kt$ is dimensionless.

Example 1.4   The Second Order Reaction, Two Reactants

Now we consider

$$A + B \xrightarrow{k} \text{Products} \qquad \mathscr{R} = kab \tag{1.43}$$

$$\mathscr{R}_A = -kab \tag{1.44}$$

Direct integration of Equation 1.31 is impossible since $\mathscr{R}_A$ now depends on the concentration of B as well as A. Two approaches are possible. The first is to observe that A and B are consumed in equal quantities. Thus

$$a_0 - a = b_0 - b \tag{1.45}$$

from which

$$\mathscr{R}_A = -kab = -k\left[a^2 + a(b_0 - a_0)\right] \tag{1.46}$$

Equation 1.31 can now be integrated and the initial condition applied in the usual manner to give

$$\frac{a_t}{a_0} = \frac{b_0 - a_0}{b_0 \exp\left[(b_0 - a_0)kt\right] - a_0} \tag{1.47}$$

An alternative approach is to apply Equations 1.31 and 1.32 to component B as well as component A. This gives a set of simultaneous ODEs:

$$\frac{da}{d\alpha} = -kab \qquad \frac{db}{d\alpha} = -kab \qquad (1.48)$$

subject to $a(0) = a_0$, $b(0) = b_0$. We shall see in subsequent sections that numerical integration is often the only feasible method of solving such a set. However, this particular set of ODEs can be solved analytically. From Equation 1.48 we have

$$\frac{da}{d\alpha} = \frac{db}{d\alpha} \qquad (1.49)$$

or

$$a(\alpha) = b(\alpha) + C \qquad (1.50)$$

where C is a constant. Application of the initial condition gives Equation 1.45 without explicit appeal to the reaction stoichiometry.

The above examples have shown that the conversion in a constant-volume isothermal reactor is relatively easy to determine provided $\mathscr{R}_A$ is a function of a alone. Usually, analytical results are possible. When $\mathscr{R}_A = \mathscr{R}_A(a, b, \ldots)$, the stoichiometry of the reaction can be used to eliminate all but one of the concentrations. The general stoichiometric relationships for a single reaction are

$$\boxed{\frac{N_A - (N_A)_0}{\nu_A} = \frac{N_B - (N_B)_0}{\nu_B} = \cdots} \qquad (1.51)$$

where N_A is the number of moles of species A present at any time. Expressed this way, Equation 1.51 is valid for any batch reaction. If the volume is constant, one simply divides through by V to obtain

$$\frac{a - a_0}{\nu_A} = \frac{b - b_0}{\nu_B} = \cdots \qquad (1.52)$$

Example 1.5 Consider the reaction

$$A + 3B \rightarrow \text{Products} \qquad (1.53)$$

and suppose

$$\mathscr{R} = kab^{1/2} \qquad (1.54)$$

This is another example where the reaction kinetics disagree with Equation 1.15. For a batch reactor,

$$\frac{da}{d\alpha} = -kab^{1/2} = \mathscr{R}_A(a, b)$$

(1.55)

so that $\mathscr{R}_A$ involves both a and b.

Assume constant density (or, equivalently, constant volume since the system mass is constant in a simple batch reactor). Then Equation 1.52 gives

$$\frac{a - a_0}{-1} = \frac{b - b_0}{-3}$$

(1.56)

or

$$b = 3(a - a_0) + b_0$$

(1.57)

Substitution into Equation 1.55 gives

$$\frac{da}{d\alpha} = -ka\sqrt{3(a - a_0) + b_0}$$

(1.58)

which now has the form

$$\frac{da}{d\alpha} = \mathscr{R}_A(a)$$

(1.59)

so that $\mathscr{R}_A$ has been obtained as a function of a alone.

Any reactor where Equation 1.51 is satisfied at each point is said to *preserve local stoichiometry*. Ideal batch reactors preserve stoichiometry as do the ideal flow reactors (piston flow and perfect mixers). Real reactors may or may not preserve local stoichiometry. Stoichiometry will be lost in real flow systems having significant levels of molecular diffusion. It will also be lost whenever there are competitive reactions. A general approach, valid for all situations, is to write a material balance for each component. If stoichiometric constraints such as Equation 1.51 exist, this approach will automatically satisfy them. When they do not exist, a separate material balance for each component is the only way to determine the composition of the system.

Example 1.6 Returning to the Example 1.5 and Reaction 1.53, separate component balances give

$$\frac{da}{d\alpha} = -kab^{1/2} \qquad \frac{db}{d\alpha} = -3kab^{1/2}$$

(1.60)

Just using algebra,

$$\frac{db}{d\alpha} = 3\frac{da}{d\alpha} \tag{1.61}$$

or

$$\frac{db}{da} = 3 \tag{1.62}$$

This has the solution

$$b - b_0 = 3(a - a_0) \tag{1.63}$$

which leads to Equation 1.59 without explicit appeal to the stoichiometry. Of course, Equation 1.63 is equivalent to the stoichiometric constraint.

Write the material balances correctly, and the stoichiometry takes care of itself.

1.3.2 Piston Flow Reactors

A piston flow reactor is usually visualized as a long tubular reactor such as that illustrated in Figure 1.3. For concreteness, suppose the tube has length L and circular cross section of radius R. Position in the tube is defined using a cylindrical coordinate system (z, r, θ) so that the concentration at a point is denoted as $a(z, r, \theta)$. For the tube to be a *piston flow reactor* (also called plug flow, slug flow, or ideal tubular), three conditions must be satisfied:

(i) The axial velocity profile is flat, $v_z = \bar{u}$.

(ii) There is complete mixing across the tube, for example, $a(z, r, \theta) = a(z)$ is a function of z alone.

(iii) There is no mixing in the axial direction.

Application of the general mass balance gives

$$Q_{in}a_{in} + \hat{\mathscr{R}}_A V = Q_{out}a_{out} \tag{1.64}$$

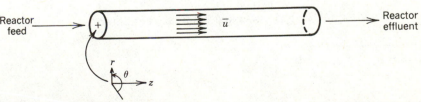

Figure 1.3 Piston flow reactor.

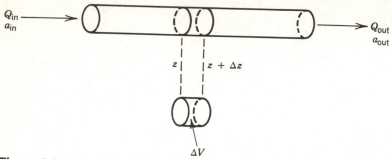

Figure 1.4   Piston flow reactor differential element.

In writing this equation we have assumed steady state operation so that the accumulation term vanishes. For the moment we also assume constant fluid density so that $Q_{in} = Q_{out} = Q$, independent of z. However, this still leaves us with an intractable result. The concentration of reactants will obviously vary down the tube so that $\mathscr{R}_A$ will be a function of z, and there is no immediately apparent way of calculating the average $\hat{\mathscr{R}}_A$.

Piston flow reactors have spatial variations in concentration, $a = a(z)$. Such systems are called **distributed**, and analysis of their steady state performance requires solution of a differential equation. The derivation of Equation 1.64 used the entire reactor as the control volume and consequently produced an expression containing the average reaction rate $\hat{\mathscr{R}}_A$. We now use a control volume of differential size, over which a and $\mathscr{R}_A$ can be regarded as approximately constant. Applying a steady state mass balance to the differential element of length Δz as shown in Figure 1.4 gives

$$Q\,a(z) + \mathscr{R}_A\,\Delta V = Q\left[a(z) + \frac{da}{dz}\,\Delta z\right]$$

(1.65)

where $Q = \pi R^2 \bar{u}$ and $\Delta V = \pi R^2\,\Delta z$. Simplifying,

$$\boxed{\mathscr{R}_A = \bar{u}\,\frac{da}{dz}}$$

(1.66)

which is the required ODE. The associated boundary condition is

$$\boxed{a = a_{in} \quad \text{at } z = 0}$$

(1.67)

The outlet concentration a_{out} is found by evaluating the solution at $z = L$.

Example 1.7   First Order Reaction

Substituting $\mathcal{R}_A = -ka$ into Equation 1.66 gives

$$\bar{u}\frac{da}{dz} + ka = 0 \qquad (1.68)$$

Solution and application of the boundary condition gives

$$\frac{a_{out}}{a_{in}} = e^{-kL/\bar{u}} \qquad (1.69)$$

The term $L/\bar{u}$ in Equation 1.69 is the time $\bar{t}$ needed for fluid to flow through the tube. In piston flow, all portions of the fluid take the same time since they have the same velocity $\bar{u}$ and travel the same distance L. In more complex flow situations there will be a distribution of times spent in the system and an average time that is the **mean residence time $\bar{t}$**. For constant-density systems

$$\boxed{\bar{t} = \frac{V}{Q}} \qquad (1.70)$$

of which $\bar{t} = L/\bar{u}$ is a special case applicable to tubular reactors.

In piston flow, the time a molecule has been in the system is directly related to that molecule's position. Define

$$\boxed{\alpha = \left(\frac{z}{L}\right)\bar{t}} \qquad (1.71)$$

Then α is the age of the molecule in a similar sense to that used in Equation 1.31. Furthermore, substituting Equation 1.71 in 1.66 gives

$$\mathcal{R}_A = \frac{da}{d\alpha} \qquad (1.72)$$

which is identical to Equation 1.31. Even the initial condition is identical if we set $a_{in} = a_0$. There is thus a one-to-one correspondence between batch and piston flow reactors. The composition at time α in a batch reactor is identical to the composition at position $z = (\alpha/\bar{t})L$ in a piston flow reactor. This correspondence—which extends beyond the isothermal, constant-density case—is detailed in Table 1.1.

Table 1.1   Relationship Between Batch and Piston Flow Reactors

Batch	Piston Flow
Initial concentration, a_0	Inlet concentration, a_{in}
Composition uniform at any time	Composition uniform at any axial position
Governing equation, $\dfrac{da}{d\alpha} = \mathscr{R}_A$	Governing equation, $\bar{u}\dfrac{da}{dz} = \mathscr{R}_A$
Final concentration, $a(t) = a_t$	Final concentration, $a(L) = a_{out}$
Variable density, $\rho = \rho(\alpha)$	Variable density, $\rho = \rho(z)$
Variable temperature, $T = T(\alpha)$	Variable temperature, $T = T(z)$
Age equivalent to position in a piston flow reactor, $\alpha = \left(\dfrac{z}{L}\right)\bar{t}$	Position equivalent to age in a batch reactor, $z = \left(\dfrac{\alpha}{\bar{t}}\right)L$

Example 1.8   Second Order Reaction, One Reactant

Evaluating the batch result, Equation 1.42, for a reaction time $\bar{t} = L/\bar{u}$ gives

$$\frac{a_{out}}{a_{in}} = \frac{1}{1 + a_{in}k(L/\bar{u})} \tag{1.73}$$

for a piston flow reactor.

1.3.3   Perfectly Mixed Reactors

Figure 1.5 illustrates a flow reactor in which the contents are mechanically agitated. If mixing caused by the agitator is sufficiently fast,[2] the entering feed will be quickly dispersed throughout the vessel and the composition at any point will approximate the average composition $\hat{a}$. Thus the reaction rate at any point will be approximately the same, $\mathscr{R}_A(\hat{a}, \hat{b}, \dots) = \hat{\mathscr{R}}_A$. Also, the outlet concentration will be identical to the interior composition, $a_{out} = \hat{a}$. The general mass balance then becomes

$$Qa_{in} + \mathscr{R}_A(a_{out}, b_{out}, \dots)V = Qa_{out} \tag{1.74}$$

Note that we have set $Q_{in} = Q_{out} = Q$, which assumes constant density. A

---

[2]More specifically, the vessel will have a characteristic throughput time $\bar{t}$, the reaction will have a characteristic time $t_{1/2}$ (see Problem 1.11), and there will be a characteristic time for mixing, t_{mix}. We now assume t_{mix} to be much smaller than either $\bar{t}$ or $t_{1/2}$. This concept is explored at greater length in Chapter 9.

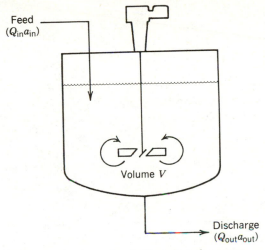

Feed
$(Q_{in}a_{in})$

Volume V

Discharge
$(Q_{out}a_{out})$

Figure 1.5 A mechanically agitated reactor.

reactor with performance governed by Equation 1.74 is known as a constant-density **perfect mixer**. The chemical engineering literature frequently refers to such a device as a CSTR (for Continuous-Flow Stirred Tank Reactor), as an ideal mixer, or just as a mixed flow reactor. This terminology is ambiguous in light of micromixing theory, Chapter 9, but is well entrenched. Unless otherwise qualified, we accept all these to mean that the reactor is a perfect mixer. The term perfect mixer is unambiguous. It denotes instantaneous and complete mixing on the molecular scale. Obviously, no real reactor can achieve this ideal just as no tubular reactor can achieve true piston flow. However, it is often possible to design reactors that very closely approach these limits.

Dividing Equation 1.74 by Q gives

$$a_{in} + \mathscr{R}_A \bar{t} = a_{out} \qquad (1.75)$$

Writing $\mathscr{R}_A$ as an explicit function of a_{out} and then solving for a_{out} gives the outlet concentration for a perfect mixer.

Example 1.9   First Order Reaction

Here,

$$\mathscr{R}_A = -ka_{out}$$

and Equation 1.75 becomes

$$a_{in} - ka_{out}\bar{t} = a_{out} \qquad (1.76)$$

Solution for a_{out} gives

$$\frac{a_{out}}{a_{in}} = \frac{1}{1 + k\bar{t}}$$

(1.77)

Example 1.10 Second Order Reaction, One Reactant

For the case $\mathcal{R}_A = -ka_{out}^2$, Equation 1.75 gives

$$\frac{a_{out}}{a_{in}} = \frac{-1 + \sqrt{1 + 4a_{in}k\bar{t}}}{2a_{in}k\bar{t}}$$

(1.78)

Note that the negative square root has been rejected since $a_{out} \geq 0$.

Example 1.11 Second Order Reaction, Two Reactants

Here,

$$\mathcal{R}_A = \mathcal{R}_B = -ka_{out}b_{out}$$

and we write Equation 1.75 twice:

$$a_{in} - k\bar{t}a_{out}b_{out} = a_{out}$$

(1.79)

and

$$b_{in} - k\bar{t}a_{out}b_{out} = b_{out}$$

(1.80)

Solution of these simultaneous equations gives two roots for a_{out}, but only one of the roots is positive:

$$\frac{a_{out}}{a_{in}} = \frac{-1 - (b_{in} - a_{in})k\bar{t} + \sqrt{[1 + (b_{in} - a_{in})k\bar{t}]^2 + 4a_{in}k\bar{t}}}{2a_{in}k\bar{t}}$$

(1.81)

The solution also gives

$$b_{out} - b_{in} = a_{out} - a_{in}$$

(1.82)

from which it is perhaps evident that perfect mixers preserve stoichiometry just as batch and piston flow reactors do.

Perfect mixers have no spatial distribution of compositions (or of temperatures). Such systems are called **lumped**. The steady state performance of lumped systems is determined by algebraic equations rather than the differential equations needed for distributed systems such as piston flow reactors.

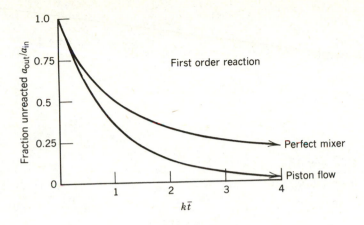

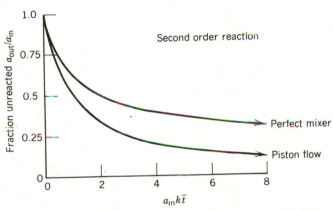

Figure 1.6 Fraction unreacted for perfect mixer versus piston flow.

Figure 1.6 displays the conversion behavior for first and second order reactions in a perfect mixer and contrasts this behavior to that of a piston flow reactor. It is apparent that piston flow is substantially better than perfect mixing for obtaining high conversions. The comparison is even more dramatic when made in terms of the volume needed to achieve a given conversion, Figure 1.7. The generalization that

Conversion for piston flow > Conversion for perfect mixing

is true for most kinetic schemes. The major exceptions to this rule, autocatalytic reactions, are discussed in Chapter 2.

For producing high-volume chemicals, flow reactors are usually preferred to batch reactors. Flow reactors are operated continuously; that is, at steady state

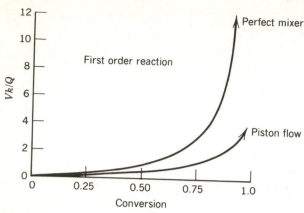

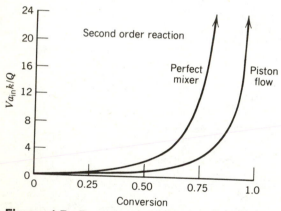

Figure 1.7 Reactor comparison: volume required to achieve a desired conversion.

with reactants continuously entering the vessel and with products continuously leaving. Batch reactors are operated discontinuously. The reaction cycle has periods for charging, reaction, and discharging. The continuous nature of a flow reactor lends itself to larger productivities and greater economies of scale than the cyclic operation of a batch reactor. The volume productivity (moles of product per unit volume of reactor) for batch systems is identical to that of piston flow reactors and is higher than most real flow reactors. However, this volume productivity is achieved only when the reaction is actually occurring and not when the reactor is being charged or discharged, being cleaned, and so on. Within the class of flow reactors, piston flow is usually desired for the reasons illustrated in Figure 1.6. However, there are many instances where a close approach to piston flow is infeasible or where a superior product results from the special reaction environment possible in stirred tanks.

For small-volume chemicals, the economics usually favor batch reactors. This is particularly true when general-purpose equipment can be shared among several products. Batch reactors are used for a greater number of products—some estimates say 90 percent—than flow reactors. However, flow reactors produce the overwhelmingly larger volume as measured in tons.

Suggestions for Further Reading

There are many good texts on chemical engineering kinetics, and the reader may wish to browse through several of them to see how they introduce the subject. Three popular books are:

O. Levenspiel, *Chemical Reaction Engineering*, 2nd ed., Wiley, New York, 1972.
C. G. Hill, Jr., *An Introduction to Chemical Engineering Kinetics and Reactor Design*, Wiley, New York, 1977.
J. M. Smith, *Chemical Engineering Kinetics*, 3rd ed., McGraw-Hill, New York, 1983.

An extended treatment of material balance equations, with substantial emphasis on component balances in reacting systems, is given in:

G. V. Reklaitis and D. R. Schneider, *Introduction to Material and Energy Balances*, Wiley, New York, 1983.

Problems

1.1   (a) Write the overall and component mass balances for an unsteady, perfectly mixed reactor.
  (b) Simplify for the case of constant reactor volume and for constant density, time-independent flow streams.
  (c) Suppose no reaction is occurring but that the input concentration of some key component varies with time according to $C_{in} = C_0, t < 0; C_{in} = 0, t > 0$. Find $C_{out}(t)$.
  (d) Repeat (c) for the case where the key component is consumed by a first order reaction.

1.2   The homogeneous gas phase reaction

$$NO + NO_2Cl \rightarrow NO_2 + NOCl$$

is believed to be elementary with rate $\mathcal{R} = k[NO][NO_2Cl]$. Use the kinetic theory of gases to estimate $f_{\mathcal{R}}$ at 300 K. Assume $r_A + r_B = 3.5 \times 10^{-10}$ m. The experimentally observed rate constant at 300 K is $k = 8$ m^3 mol^{-1} s^{-1}.

1.3   Rate expressions for gas phase reactions are sometimes based on reactant partial pressures rather than concentrations. A literature source[3] gives $k = 1.1 \times$

[3] J. M. Smith, *Chemical Engineering Kinetics*, 1st ed., McGraw-Hill, New York, 1956, p. 131.

10^{-3} mol cm^{-3} atm^{-2} hr^{-1} for the reaction of gaseous sulfur with methane at 873 K:

$$CH_4 + 2S_2 \rightarrow CS_2 + 2H_2S$$

where $\mathscr{R} = kP_{CH_4}P_{S_2}$ mol cm^{-3} hr^{-1}. Determine k when the rate is based on concentrations: $\mathscr{R} = k[CH_4][S_2]$. Give k in SI units.

1.4 Find the spatial distribution of concentration, $a(z)$, in a piston flow reactor for a component that is consumed by a first order reaction. Use this result to show that Equation 1.69 is consistent with the general mass balance, Equation 1.3.

1.5 Repeat Problem 1.4 for a second order reaction, $2A \overset{k/2}{\rightarrow} B$.

1.6 Suppose that the following reactions are elementary. Write rate expressions for the reaction and for each of the components:

(a) $2A \underset{k_r}{\overset{k_f}{\rightleftharpoons}} B + C$

(b) $2A \underset{k_r}{\overset{k_{f/2}}{\rightleftharpoons}} B + C$

(c) $B + C \underset{k_f}{\overset{k_r}{\rightleftharpoons}} 2A$

(d) $2A \overset{k_I}{\rightarrow} B + C$

   $B + C \overset{k_{II}}{\rightarrow} 2A$

1.7 Consider the reaction

$$A + B \overset{k}{\rightarrow} P$$

with $k = 1$ m^3 mol^{-1} s^{-1}. Suppose $b_{in} = 10$ mol/m^3. It is desired to achieve $b_{out} = 0.01$ mol/m^3.

(a) Find the mean residence time needed to achieve this, assuming piston flow and $a_{in} = b_{in}$.

(b) Repeat (a) assuming perfect mixing.

(c) Repeat (a) and (b) assuming $a_{in} = 10b_{in}$.

1.8 The esterification reaction

$$RCOOH + R'OH \underset{k_r}{\overset{k_f}{\rightleftharpoons}} RCOOR' + H_2O$$

can be driven to completion by removing the water of condensation. This might be done continuously in a stirred tank reactor or in a horizontally compartmented, progressive flow reactor. This type of reactor gives a reasonable approximation to piston flow in the liquid phase while providing a vapor space for the removal of the by-product water. Suppose it is desired to obtain an ester product containing not more than 1% (by mole) residual alcohol and 0.01% residual acid.

    (a) What are the limits on initial stoichiometry if the product specifications are to be achieved?

    (b) What value of $a_0 k\bar{t}$ is needed in a perfect mixer?

    (c) What value of $a_0 k\bar{t}$ is needed in the progressive reactor?

    (d) Discuss the suitability of a batch reactor for this situation.

1.9   Will an irreversible elementary reaction go to completion in a reactor having a finite $\bar{t}$?

1.10 Write a plausible reaction mechanism, including appropriate rate expressions, for the toluene nitration example of Section 1.2.

1.11 The *half-life* $t_{1/2}$ is defined for a batch reaction as the time when $a_t/a_0 = 0.5$, where A is the limiting reactant.

    (a) Suppose a first order reaction and $k = 1 \text{ s}^{-1}$. Find $t_{1/2}$.

    (b) Suppose a second order reaction, $\mathcal{R}_A = -ka^2$, $ka_0 = 1 \text{ s}^{-1}$. Find $t_{1/2}$.

1.12 The reaction of trimethylamine with *n*-propyl bromide gives a quaternary ammonium salt:

$$N(CH_3)_3 + C_3H_7Br \rightarrow (CH_3)_3(C_3H_7)NBr$$

Suppose laboratory results at 110°C using toluene as a solvent show the reaction to be second order with rate constant $k = 5.6 \times 10^{-7} \text{ m}^3 \text{ mol}^{-1} \text{ s}^{-1}$. Suppose $[N(CH_3)_3]_0 = [C_3H_7Br]_0 = 80 \text{ mol/m}^3$.

    (a) Estimate the time required to achieve 99% conversion in a batch reactor.

    (b) Estimate the volume required in a perfect mixer to achieve 99% conversion if a production rate of 100 kg/hr of the salt is desired.

    (c) Suggest means for increasing the productivity, that is, reducing the batch reaction time or the volume of the perfect mixer.

1.13 It is always possible to divide Equation 1.51 by V even though the batch reaction volume may vary with time, $V = V(\alpha)$. Why isn't Equation 1.52 valid for the variable-volume case?

1.14 Suppose an elementary, third order reaction really exists:

$$3A \xrightarrow{k} P$$

Determine the fraction of A unreacted in a batch reactor after time t.

1.15 Ethyl acetate can be formed from dilute solutions of ethanol and acetic acid according to the reversible reaction

$$C_2H_5OH + CH_3COOH \underset{k_r}{\overset{k_f}{\rightleftharpoons}} C_2H_5OOCCH_3 + H_2O$$

Ethyl acetate is somewhat easier to separate from water than either ethanol or acetic acid. For example, the relatively large acetate molecule has much lower permeability through a membrane ultrafilter. Thus esterification is sometimes proposed as an economical approach for recovering dilute fermentation products.

Suppose fermentation effluents are available as separate streams containing 3% by weight acetic acid and 5% by weight ethanol. Devise a reaction scheme for generating ethyl acetate using the reactants in stoichiometric ratio. After reaction, the ethyl acetate concentration is increased first to 25% by weight using ultrafiltration and then to 99% by weight using distillation. The reactants must ultimately be heated for the distillation step. Thus we can suppose both the esterification and membrane separation to be conducted at 100°C. At this temperature,

$$k_f = 8.0 \times 10^{-9} \ \mathrm{m^3 \ mol^{-1} \ s^{-1}}$$

$$k_r = 2.7 \times 10^{-9} \ \mathrm{m^3 \ mol^{-1} \ s^{-1}}$$

Determine $\bar{t}$ for a perfect mixer that will approach equilibrium within 5%, that is,

$$\frac{a_{out} - a_{equil}}{a_{in} - a_{equil}} = 0.05$$

What is a_{out} from this reactor?

2

Complex Reactions in Batch Reactors

2.1 Multiple and Nonelementary Reactions

Chapter 1 treated single reactions in the three types of ideal reactors: batch, piston flow, and perfectly mixed. Chapters 2 and 3 now extend this treatment to multiple and complex reactions. Multiple reactions involve two or more stoichiometric equations, each with its own rate expression. They are often classified as *consecutive* as in

$$A + B \xrightarrow{k_I} C \qquad \mathscr{R}_I = k_I ab$$

$$C + D \xrightarrow{k_{II}} E \qquad \mathscr{R}_{II} = k_{II} cd$$

(2.1)

or *competitive* as in

$$A + B \xrightarrow{k_I} C \qquad \mathscr{R}_I = k_I ab$$

$$A + D \xrightarrow{k_{II}} E \qquad \mathscr{R}_{II} = k_{II} ad$$

(2.2)

or *independent* as in

$$A \xrightarrow{k_I} B \qquad \mathscr{R}_I = k_I a$$

$$C + D \xrightarrow{k_{II}} E \qquad \mathscr{R}_{II} = k_{II} cd$$

(2.3)

Even *reversible* reactions can be regarded as multiple:

$$A + B \overset{k_I}{\to} C \qquad \mathscr{R}_I = k_I ab$$

$$C \overset{k_{II}}{\to} A + B \qquad \mathscr{R}_{II} = k_{II} c$$

(2.4)

Note that the Roman numeral subscripts refer to numbered reactions and have nothing to do with iodine. All these examples have involved elementary reactions. Multiple reactions and apparently single but nonelementary reactions are called *complex*. Complex reactions, even when apparently single, consist of a number of elementary steps. These steps, some of which may be quite fast, constitute the *mechanism* of the observed, complex reaction. As an example, suppose

$$A \overset{k_I}{\to} B + C$$

$$B \overset{k_{II}}{\to} D$$

(2.5)

where $k_{II} \gg k_I$. Then the observed reaction will be

$$A \to C + D \qquad \mathscr{R} = ka$$

(2.6)

This reaction is complex even though it has a stoichiometric equation and rate expression which could correspond to an elementary reaction.

To solve a problem in reactor design, knowledge of the reaction mechanism may not be critical to success but it is always desirable. Two reasons are:

1. Knowledge of the mechanism will allow fitting experimental data to a theoretical rate expression. This will presumably be more reliable on extrapolation than an empirical fit.

2. Knowing the mechanism may suggest chemical modifications and optimization possibilities for the final design that would otherwise be missed.

The best way to find a reaction mechanism is to find a good chemist. Chemical insight can be used to hypothesize a mechanism, and the hypothesized mechanism can then be tested against experimental data. If inconsistent, the mechanism must be rejected. This is seldom the case. More typically, there are several mechanisms that will fit the data equally well. A truly definitive study of reaction mechanisms requires direct observation of all chemical species including intermediates which may have low concentrations and short lives. Such studies are not always feasible. Working hypotheses for the reaction mechanisms must then be selected based on general chemical principles and on analogous systems that have been studied in detail. There is no substitute for understanding the chemistry or at least for having faith in the chemist.

2.2 Complex Kinetics in Batch Systems

2.2.1 Multiple Reactions

For multiple reactions in a batch system, the mass balance, Equation 1.31, still holds for each component. However, the net rate of formation of the component may be due to several different reactions. Thus

$$\mathscr{R}_A = \nu_{A,I}\mathscr{R}_I + \nu_{A,II}\mathscr{R}_{II} + \nu_{A,III}\mathscr{R}_{III} + \cdots \qquad (2.7)$$

Here we envision component A being formed by Reactions I, II, III, ..., each of which has a stoichiometric coefficient with respect to the component. Equivalent to Equation 2.7 we can write

$$\boxed{\mathscr{R}_A = \sum_{\text{Reactions}} \nu_{A,I}\mathscr{R}_I} \qquad (2.8)$$

Obviously, $\nu_{A,I} = 0$ if component A does not appear in Reaction I.

Example 2.1 Consecutive First Order Reactions

The reaction network of Equation 2.5 gives $\nu_{A,I} = -1$, $\nu_{A,II} = 0$, $\nu_{B,I} = +1$, $\nu_{B,II} = -1$, $\nu_{C,I} = +1$, $\nu_{C,II} = 0$, $\nu_{D,I} = 0$, $\nu_{D,II} = +1$. Applying Equations 1.31 and 2.8 to each component gives

$$\frac{da}{d\alpha} = \mathscr{R}_A = -k_I a$$

$$\frac{db}{d\alpha} = \mathscr{R}_B = +k_I a - k_{II} b$$

$$\frac{dc}{d\alpha} = \mathscr{R}_C = +k_I a \qquad (2.9)$$

$$\frac{dd}{d\alpha} = \mathscr{R}_D = +k_{II} b$$

This set of ODEs is readily solvable:

$$a = a_0 e^{-k_I \alpha}$$

$$b = \frac{k_I a_0 e^{-k_I \alpha} + \left[b_0(k_{II} - k_I) - k_I a_0 \right] e^{-k_{II} \alpha}}{k_{II} - k_I}$$

$$c = c_0 + a_0 - a \qquad (2.10)$$

$$d = d_0 + a_0 + b_0 - b - a$$

where we have used the initial conditions that $a = a_0$, $b = b_0$, $c = c_0$, and $d = d_0$ at $\alpha = 0$. In this particular case, the term "readily solvable" means that it should not take more than an hour the first time you try a problem of this sort. Things get much worse if any of the reaction steps are other than first order.

Example 2.2 Consecutive Reactions

$$2A \overset{k_{\mathrm{I}}}{\rightarrow} B \qquad \mathscr{R}_{\mathrm{I}} = k_{\mathrm{I}} a^2$$

$$B + C \overset{k_{\mathrm{II}}}{\rightarrow} P \qquad \mathscr{R}_{\mathrm{II}} = k_{\mathrm{II}} bc \qquad\qquad \textbf{(2.11)}$$

The various stoichiometric coefficients are $\nu_{A,\mathrm{I}} = -2$, $\nu_{A,\mathrm{II}} = 0$, $\nu_{B,\mathrm{I}} = +1$, $\nu_{B,\mathrm{II}} = -1$, $\nu_{C,\mathrm{I}} = 0$, $\nu_{C,\mathrm{II}} = -1$, $\nu_{P,\mathrm{I}} = 0$, $\nu_{P,\mathrm{II}} = +1$. The rates of formations are $\mathscr{R}_A = -2\mathscr{R}_{\mathrm{I}}$, $\mathscr{R}_B = +\mathscr{R}_{\mathrm{I}} - \mathscr{R}_{\mathrm{II}}$, $\mathscr{R}_C = -\mathscr{R}_{\mathrm{II}}$, $\mathscr{R}_P = +\mathscr{R}_{\mathrm{II}}$. For a batch system of constant volume, Equation 1.31 can be applied to each component. This gives

$$\frac{da}{d\alpha} = \mathscr{R}_A = -2k_{\mathrm{I}} a^2$$

$$\frac{db}{d\alpha} = \mathscr{R}_B = k_{\mathrm{I}} a^2 - k_{\mathrm{II}} bc$$

$$\frac{dc}{d\alpha} = \mathscr{R}_C = -k_{\mathrm{II}} bc \qquad\qquad \textbf{(2.12)}$$

$$\frac{dp}{d\alpha} = \mathscr{R}_P = +k_{\mathrm{II}} bc$$

The initial conditions are $a = a_0$ at $t = 0$, and so on for the other components. This set of equations must be solved simultaneously to find the concentrations a, b, c, p as a function of the batch reaction time. The reader is challenged to find an analytical solution. The attempt, even if successful, should be a convincing argument for the use of numerical methods.

2.2.2 Numerical Methods for Distributed Systems

Industrially realistic kinetic schemes soon become intractable from the viewpoint of obtaining analytical solutions. This is true even for idealized situations such as the isothermal, constant-volume batch reactors considered in the previous section. It becomes manifestly true for real reactors with spatial variations in composition and temperature. Thus the design engineer must frequently resort to numerical techniques. For batch and piston flow reactors, it is necessary to solve sets of simultaneous, first order, ordinary differential equations. Usually, the equations are nonlinear. Sophisticated and computationally efficient methods have been developed for solving such sets of equations. One popular method, called Runge–Kutta, is described in Appendix 2.1. This or even more sophisticated techniques should be used if the cost of computation becomes significant. However, computer costs will often be inconsequential compared to the costs of the engineer's personal time. Then the use of a simple technique can save time and money by allowing the engineer to focus on the physics and chemistry of the problem rather than on the numerical mathematics.

The simplest possible method for solving a set of ordinary differential equations—subject to given initial values—is called *marching ahead*. It is also known as Euler's method. We suppose that all concentrations are known at time $\alpha = 0$. This allows the initial reaction rates to be calculated, one for each component. Choose some time increment $\Delta\alpha$, which is so small that, given the calculated reaction rates, the concentrations will change very little during the time increment. Calculate these small changes in concentration assuming the reaction rates are constant. Use the new concentrations to revise the reaction rates. Pick another time increment and repeat the calculations. Continue until the various $\Delta\alpha$ sum to the specified batch reaction time.

Example 2.3   Solve Equations 2.12 for $t = 1$ h given $a_0 = c_0 = 30$ moles/m^3, $b_0 = p_0 = 0$, $k_I = 0.01$ m^3 mole^{-1} hr^{-1}, $k_{II} = 0.02$ m^3 mole^{-1} hr^{-1}. Use $\Delta\alpha = 0.125$ hr.

α	a	b	c	p	$\Delta a = \mathcal{R}_A \Delta\alpha$	$\Delta b = \mathcal{R}_B \Delta\alpha$	$\Delta c = \mathcal{R}_C \Delta\alpha$	$\Delta p = \mathcal{R}_p \Delta\alpha$
0	30.00	0	30.00	0				
					-2.25	$+1.13$	0	0
0.125	27.75	1.13	30.00	0				
					-1.93	$+0.87$	-0.08	$+0.08$
0.250	25.82	2.00	29.92	0.08				
					-1.66	$+0.69$	-0.15	$+0.15$
0.375	24.16	2.69	29.77	0.23				
					-1.46	$+0.53$	-0.20	$+0.20$
0.500	22.70	3.22	29.57	0.43				
					-1.29	$+0.40$	-0.24	$+0.24$
0.625	21.41	3.62	29.33	0.67				
					-1.15	$+0.31$	-0.27	$+0.27$
0.750	20.26	3.93	29.06	0.94				
					-1.02	$+0.23$	-0.28	$+0.28$
0.875	19.24	4.16	28.78	1.22				
					-0.93	$+0.16$	-0.30	$+0.30$
1.000	18.31	4.32	28.48	1.52				

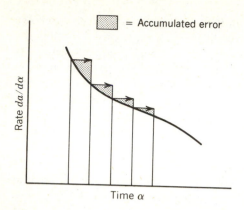

☒ = Accumulated error

Rate $da/d\alpha$

Time α

Figure 2.1 Marching-ahead systematic error.

The true value for a_t is 18.75 moles/m³ rather than the above value of 18.31. The marching-ahead technique systematically overestimates $\mathscr{R}_A$ since the rate is calculated using concentrations at the beginning of each time increment. This creates a systematic error similar to the numerical integration error shown in Figure 2.1. Such error can be dramatically reduced by use of more sophisticated numerical techniques. It can also be reduced by the simple expedient of using a smaller $\Delta\alpha$.

Example 2.4   Repeat the above calculations as a function of $\Delta\alpha$.

$\Delta\alpha$	a_t	b_t	c_t	p_t	Error in a_t
1	12.000	9.000	30.000	0.000	−6.750
$\frac{1}{2}$	16.590	5.355	28.650	1.350	−2.160
$\frac{1}{4}$	17.819	4.609	28.518	1.482	−0.931
$\frac{1}{8}$	18.313	4.321	28.478	1.522	−0.437
$\frac{1}{16}$	18.538	4.193	28.462	1.538	−0.212
$\frac{1}{32}$	18.645	4.132	28.454	1.546	−0.105
$\frac{1}{64}$	18.698	4.102	28.451	1.549	−0.052
⋮	⋮	⋮	⋮	⋮	⋮
0	18.750	4.072	28.448	1.552	0

Note that the errors are relative to the fully converged value, $a_t = 18.750$.

The error in the marching-ahead method is approximately proportional to $\Delta\alpha$. Proportionality holds exactly as $\Delta\alpha$ becomes small, at least until round-off error becomes significant in the calculations. The marching-ahead method is said to *converge 0(Δα)*. Other integration schemes converge $0(\Delta\alpha^2)$, $0(\Delta\alpha^3)$, and even faster. Also, the absolute error in the marching-ahead method (i.e., the error at some predetermined value of $\Delta\alpha$) tends to be larger than for more sophisticated methods. However, as is apparent from the above examples, the marching-ahead method is quite satisfactory for many design purposes. Also the systematic way in which the error is reduced as $\Delta\alpha$ becomes small lends itself to extrapolation. If the step size is halved, the estimates for $a_t, b_t, \ldots$ will change by amounts $\Delta a^*, \Delta b^*, \ldots$. For convergence $0(\Delta\alpha)$, another halving will cause just half the previous change, and so on for additional halvings.

Let $a_t^*(\Delta\alpha)$ and $a_t^*(\Delta\alpha/2)$ be two successive estimates, the first using a step size of $\Delta\alpha$ and the second using a step size of $\Delta\alpha/2$. Define the difference between successive estimates as

$$\Delta a_t^* = a_t^*\left(\frac{\Delta\alpha}{2}\right) - a_t^*(\Delta\alpha) \tag{2.13}$$

This difference will become smaller as $\Delta\alpha$ becomes smaller. The next Δa_t^* will be one-half of the one that just occurred. The Δa_t^* after that will be one-fourth of the one that just occurred. The Δa_t^* yet to come form a series $\frac{1}{2} + \frac{1}{4} + \frac{1}{8} + \cdots = 1$. In words, *the total change yet to come is equal to the change that just occurred*. In mathematics,

$$\lim_{N \to \infty} a_t^*\left(\frac{\Delta\alpha}{2^N}\right) = a_t^*\left(\frac{\Delta\alpha}{2}\right) + \Delta a_t^* \tag{2.14}$$

This equation can be used to extrapolate the results in Example 2.4. Begin with $\Delta\alpha = \frac{1}{32}$, which gives $a_t^* = 18.645$. Halve the step size to $\Delta\alpha = \frac{1}{64}$, which gives $a_t^* = 18.698$. The difference, $\Delta a_t^* = 0.053$, is the change that just occurred. The total change yet to come is $0.053/2 + 0.053/4 + \cdots = 0.053$. Thus the extrapolated value for a_t is $18.698 + 0.053 = 18.751$.

In practice, one calculates the various Δa_t^* and extrapolates when it is clear that the results are converging $0(\Delta\alpha)$. This scheme gives the following results for Example 2.4:

$\Delta\alpha$	a_t^*	Δa_t^*	b_t^*	Δb_t^*	c_t	Δc_t^*
$\frac{1}{16}$	18.538		4.193		28.462	
		+0.107		−0.061		−0.008
$\frac{1}{32}$	18.645		4.132		28.454	
		+0.053		−0.030		−0.003
$\frac{1}{64}$	18.698		4.102		28.451	
		+0.053		−0.030		−0.003
Extrapolated to $\Delta\alpha = 0$	18.751		4.072		28.448	

The extrapolated results are accurate to one unit in the third place. Comparable accuracy without extrapolation requires $\Delta\alpha = 2^{-12}$.

The marching-ahead technique is also known as *forward differencing*. It is unstable if too large a step size is chosen. Thus it is only *conditionally stable*. For sufficiently small $\Delta\alpha$, the method is stable and will converge. When it converges, it converges to the true, analytical solution. These statements can be proved for the case of linear differential equations. They cannot be proved, and are indeed untrue, for the general case of nonlinear equations. Fortunately, the marching-ahead method behaves quite well for most nonlinear systems of engineering importance. Practical problems do arise in *stiff* sets of differential equations where some members of the set have characteristic times much smaller than other members. This occurs in reaction kinetics when some reactions are much faster than others. In free-radical kinetics, reaction rates may differ by three orders of magnitude. With a simple marching-ahead approach, the allowable size for $\Delta\alpha$ is determined by the fastest reaction and may be too small even for modern computers. Special numerical methods have been devised to deal with stiff sets of differential equations. In free-radical processes, it is also possible to avoid stiff sets of equations through use of the *quasi-steady state hypothesis*, which is discussed in the next section.

2.3 Analytically Tractable Examples

2.3.1 The *n*th Order Reaction

Relatively few kinetic schemes admit analytical solutions. The present section is concerned with those special cases that do and also with some cases where preliminary analytical work will ease the subsequent numerical studies. We begin with the *n*th order reaction

$$A \rightarrow \text{Products} \qquad \mathscr{R} = ka^n \tag{2.15}$$

This occurs in elementary reactions with $n = 1$, 2, or possibly 3. Nonintegral values for n are often found when fitting rate data empirically, sometimes for sound kinetic reasons as we shall shortly see. For the batch, isothermal, constant-volume reactor,

$$\frac{da}{d\alpha} = \mathscr{R}_A = -ka^n \tag{2.16}$$

For $n \neq 1$, this integrates to

$$\left(\frac{a_t}{a_0}\right)^{n-1} = \frac{1}{1 + (n-1)a_0^{n-1}kt} \tag{2.17}$$

For $n > 1$, this result is well behaved with the dimensionless concentration a_t/a_0, starting at 1 and gradually decreasing toward zero as $t \to \infty$. For $n < 1$, an equivalent form for Equation 2.17 is preferred:

$$\left(\frac{a_t}{a_0}\right)^{1-n} = 1 - (1 - n)\frac{kt}{a_0^{1-n}} \tag{2.18}$$

which goes to zero at

$$t = \frac{a_0^{1-n}}{(1-n)k} \tag{2.19}$$

If the apparent reaction order did not change, reactions with $n < 1$ would go to completion in finite time. In principle, the reaction order must change to $n \geq 1$ in the limit of low concentrations. However, Equation 2.16 sometimes remains useful as an approximation down to quite low concentrations. Mathematically, Equation 2.16 should be supplemented with $\mathscr{R}_A = 0$ if $a \leq 0$. Otherwise, the mathematics would predict $a_t < 0$ for times greater than that given by Equation 2.19. Indeed, Equation 2.18 will give negative concentrations if used at inappropriately long times.

2.3.2 The Steady State Hypothesis

Many reactions involve short-lived intermediates. The intermediates may be so reactive that they never accumulate in large quantities and are difficult to detect. However, their presence is important in the reaction mechanism and may dictate the functional form of the rate equation. Consider the following reaction:

$$A \underset{k_{-I}}{\overset{k_I}{\rightleftarrows}} B \overset{k_{II}}{\to} C \tag{2.20}$$

This system contains only first order steps so that a rigorous but somewhat cumbersome analytical solution is available. Assuming $b_0 = 0$,

$$a_t = \frac{k_I a_0}{S_I - S_{II}}\left[\left(1 - \frac{k_{II}}{S_I}\right)e^{-S_I t} - \left(1 - \frac{k_{II}}{S_{II}}\right)e^{-S_{II} t}\right] \tag{2.21}$$

$$b_t = \frac{k_I a_0}{S_I - S_{II}}\left(e^{-S_{II} t} - e^{-S_I t}\right) \tag{2.22}$$

where $S_I, S_{II} = \frac{1}{2}\left[k_I + k_{-I} + k_{II} \pm \sqrt{(k_I + k_{-I} + k_{II})^2 - 4k_I k_{II}}\right]$. Suppose that B is highly reactive; when formed, it rapidly reverts back to A or decomposes into C. The *quasi-steady state hypothesis* assumes that an "equilibrium" concentration of B is formed early in the reaction and changes only slowly thereafter.[1] Alternately stated, the rate of formation of B equals its rate of disappearance. Either statement requires $db/d\alpha = 0$, giving

$$\frac{db}{d\alpha} = +k_{+I}a - (k_{-I} + k_{II})b = 0 \tag{2.23}$$

Thus

$$b = \frac{k_I a}{k_{-I} + k_{II}} \tag{2.24}$$

and

$$a_t = a_0 \exp\left(\frac{-k_I k_{II} t}{k_{-I} + k_{II}}\right) \tag{2.25}$$

which is a convenient simplification of Equation 2.21.

A more typical use of the steady state hypothesis arises in *chain reactions* propagated by free radicals. Free radicals are molecules or atoms having an unpaired electron. Many common organic reactions such as thermal cracking and vinyl polymerization occur by free-radical processes. The following mechanism has been postulated for the gas phase decomposition of acetaldehyde:

Initiation

$$CH_3CHO \overset{k_I}{\rightarrow} CH_3 \cdot + \cdot CHO \tag{2.26}$$

This spontaneous or thermal initiation generates two free radicals by breaking a covalent bond. The aldehyde radical is long-lived and does not markedly influence the subsequent chemistry. The methane radical is highly reactive; but rather than disappearing, most reactions regenerate it.

Propagation

$$CH_3 \cdot + CH_3CHO \overset{k_{II}}{\rightarrow} CH_4 + CH_3 \cdot CO \tag{2.27}$$

$$CH_3 \cdot CO \overset{k_{III}}{\rightarrow} CH_3 \cdot + CO \tag{2.28}$$

---

[1] The concentration of B is not at equilibrium in the thermodynamic sense. We mean only that its rate of formation equals its rate of disappearance over an extended period of time.

The propagation reactions use a methane radical but also generate one. There is no net consumption, so a single initiation reaction can cause an indefinite number of propagation reactions. Assuming the methane radicals do not accumulate, the overall stoichiometry is given by the net sum of the propagation steps:

$$CH_3CHO \rightarrow CH_4 + CO \tag{2.29}$$

The methane radicals do not accumulate because of termination reactions. The major one postulated for the acetaldehyde decomposition is:

Termination

$$2CH_3 \cdot \overset{k_{IV}}{\rightarrow} C_2H_6 \tag{2.30}$$

Applying the steady state hypothesis to the $CH_3 \cdot$ and $CH_3 \cdot CO$ radicals gives

$$\frac{d[CH_3 \cdot]}{d\alpha} = k_I[CH_3CHO] - k_{II}[CH_3CHO][CH_3 \cdot]$$

$$+ k_{III}[CH_3 \cdot CO] - 2k_{IV}[CH_3 \cdot]^2 = 0 \tag{2.31}$$

$$\frac{d[CH_3 \cdot CO]}{d\alpha} = +k_{II}[CH_3CHO][CH_3 \cdot] - k_{III}[CH_3 \cdot CO] = 0 \tag{2.32}$$

and thus

$$k_I[CH_3CHO] = 2k_{IV}[CH_3 \cdot]^2 \tag{2.33}$$

The consumption of acetaldehyde is given by

$$\frac{-d[CH_3CHO]}{dt} = k_I[CH_3CHO] + k_{II}[CH_3CHO][CH_3 \cdot] \tag{2.34}$$

$$= k_I[CH_3CHO] + \left(\frac{k_{II}^2 k_I}{2k_{IV}}\right)^{1/2} [CH_3CHO]^{3/2}$$

The rate of Reaction 2.26 is presumably slow compared to the propagation steps. Thus the second term dominates Equation 2.34, and the overall reaction has the form

$$A \rightarrow Products \qquad \mathcal{R}_A = -ka^{3/2} \tag{2.35}$$

This result agrees with experimental findings.[2] It is typical of free-radical mechanisms in that it generates half-integer kinetics.

[2] A. Boyer, M. Niclause, and M. Letort, *J. chim. phys.*, **49**, 345 (1952).

A second example is the monochlorination of hydrocarbons:

Initiation

$$Cl_2 \xrightarrow{k_I} 2Cl \cdot$$

(2.36)

Propagation

$$Cl \cdot + RH \xrightarrow{k_{II}} R \cdot + HCl$$

(2.37)

$$R \cdot + Cl_2 \xrightarrow{k_{III}} RCl + Cl \cdot$$

(2.38)

Termination

$$2Cl \cdot \xrightarrow{k_{IV}} Cl_2$$

(2.39)

or

$$Cl \cdot + R \cdot \xrightarrow{k_{IV}} RCl$$

(2.40)

or

$$2R \cdot \xrightarrow{k_{IV}} R_2$$

(2.41)

The steady state hypothesis is applied to the chlorine and hydrocarbon radicals. The resulting rate expression for the overall reaction is

$$Cl_2 + RH \rightarrow RCl + HCl$$

(2.42)

Depending on which termination mechanism is assumed,

$$\mathcal{R} = k[Cl_2]^{1/2}[RH] \qquad \text{using Equation 2.39}$$

(2.43)

$$\mathcal{R} = k[Cl_2][RH]^{1/2} \qquad \text{using Equation 2.40}$$

(2.44)

$$\mathcal{R} = k[Cl_2]^{3/2} \qquad \text{using Equation 2.41}$$

(2.45)

If two or three termination reactions are simultaneously important, an analytical solution for $\mathcal{R}$ is possible but complex. Laboratory results in such situations could probably be approximated as

$$\mathscr{R} = k[Cl_2]^m[RH]^n \tag{2.46}$$

where

$$\tfrac{1}{2} < m < \tfrac{3}{2} \quad \text{and} \quad 0 < n < 1$$

Our treatment of chain reactions has been confined to relatively simple situations where the number of participating species and their possible reactions have been sharply bounded. Most free-radical reactions of industrial importance involve many more species. The set of possible reactions is unbounded in polymerizations, and it is bounded but very large in processes such as naptha cracking and combustion. Analytical treatments are sometimes possible but numerical treatments are more likely. Often, of course, one can postulate a set of elementary reactions but does not know the rate constants. Reactor designs must then be based on empirical rate equations for the overall reaction.

2.3.3 Sets of First Order Reactions

Many reactions behave as though they were first order even though they are not unimolecular. The usually cited example of a *pseudo-first order reaction* is

$$A + B \xrightarrow{k} \text{Products} \qquad \mathscr{R} = kab \tag{2.47}$$

where component B is present in great excess. Typical examples are hydrations done in water and slow oxidations done in air. The concentration of B does not change appreciably during the course of the reaction. Thus b is approximately constant and $\mathscr{R} = k'a$, where $k' = kb$.

In principle, analytical solutions can always be found for networks of first order reactions since they give rise to linear differential equations of the form

$$\frac{dp}{d\alpha} = -kp + k_I a + k_{II} b + k_{III} c + \cdots \tag{2.48}$$

where p is formed from any of $A, B, C, \ldots$ and itself reacts according to first order kinetics. Such a set can always be solved although the algebra may be difficult. Reaction 2.20 is a linear network. So is the consecutive reaction sequence

$$A \xrightarrow{k_I} B \xrightarrow{k_{II}} C \xrightarrow{k_{III}} D \tag{2.49}$$

A solution is

$$a_t = a_0 e^{-k_I t}$$

(2.50)

$$b_t = \left[b_0 - \frac{a_0 k_I}{k_{II} - k_I} \right] e^{-k_{II} t} + \left[\frac{a_0 k_I}{k_{II} - k_I} \right] e^{-k_I t}$$

(2.51)

$$c_t = \left[c_0 - \frac{b_0 k_{II}}{k_{III} - k_{II}} + \frac{a_0 k_I k_{II}}{(k_{III} - k_I)(k_{III} - k_{II})} \right] e^{-k_{III} t}$$

$$+ \left[\frac{b_0 k_{II}}{k_{III} - k_{II}} - \frac{a_0 k_I k_{II}}{(k_{III} - k_{II})(k_{II} - k_I)} \right] e^{-k_{II} t}$$

(2.52)

$$+ \left[\frac{a_0 k_I k_{II}}{(k_{III} - k_I)(k_{II} - k_I)} \right] e^{-k_I t}$$

$$d_t = d_0 + (a_0 - a_t) + (b_0 - b_t) + (c_0 - c_t)$$

(2.53)

These results assume that all the rate constants are different. Special forms apply when some of the k values are identical, but the qualitative behavior of the reaction set remains the same. Figure 2.2 illustrates this qualitative behavior for a case with $b_0 = c_0 = d_0 = 0$. The concentrations of B and C start at zero, increase to maximums, and then decline back to zero. Typically, component B or C is the desired product whereas the others are undesired. If, say, B is desired, the batch reaction time can be picked to maximize its concentration. Setting $db/dt = 0$ gives

$$t_{max} = \frac{\ln(k_{II}/k_I)}{k_{II} - k_I}$$

(2.54)

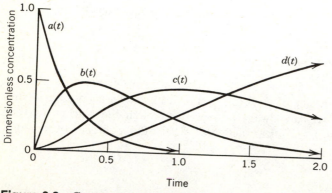

Figure 2.2 Consecutive reaction sequence.

for the case of $b_0 = 0$. Selection of the optimal time for the production of C requires a numerical solution but remains straightforward conceptually. This reactor optimization problem becomes more interesting when reaction temperature is a design variable. Such optimization problems are treated in Chapter 4.

2.3.4 Autocatalytic Reactions

As suggested by the name, the products of an autocatalytic reaction accelerate the rate of the forward reaction. A model reaction frequently used to represent autocatalytic behavior is

$$A \rightarrow B + C \tag{2.55}$$

with an assumed mechanism of

$$A + B \xrightarrow{k} 2B + C \tag{2.56}$$

For a batch system,

$$\frac{da}{d\alpha} = -kab = -ka(b_0 + a_0 - a) \tag{2.57}$$

which has the solution

$$\frac{a_t}{a_0} = \frac{[1 + (b_0/a_0)] \exp\{-[1 + (b_0/a_0)]a_0kt\}}{(b_0/a_0) + \exp\{-[1 + (b_0/a_0)]a_0kt\}} \tag{2.58}$$

Figure 2.3 illustrates the course of the reaction for various values of b_0/a_0. The S-shaped curve is typical of autocatalytic behavior. The reaction rate is initially low because the concentration of the catalyst B is low. Indeed, no reaction ever occurs if $b_0 = 0$. As B is formed, the rate accelerates and continues to increase so long as the term ab in Equation 2.57 is growing. Eventually, however, this term must decrease as component A is depleted even though the concentration of B continues to increase. At very high conversions, $a \rightarrow 0$ and the reaction rate again becomes small.

Autocatalytic reactions often show higher conversions in a stirred tank than in a piston flow reactor having the same $\bar{t}$. The catalyst B is present throughout the entire working volume of the stirred tank but may be quite low in concentration during the early portions of a batch or piston flow reaction.

The qualitative behavior shown in Figure 2.3 is characteristic of many systems, particularly biological ones, even though the reaction mechanism may not agree exactly with Equation 2.56. Polymerizations of vinyl monomers such as methylmethacrylate and styrene also show apparently autocatalytic behavior when the undiluted monomers are reacted by free-radical mechanisms. Here, the

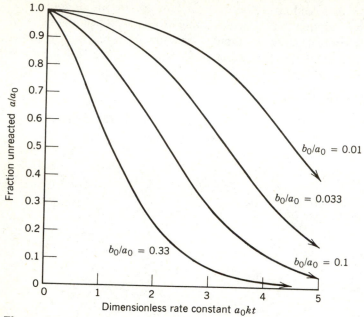

Figure 2.3 Course of an autocatalytic reaction.

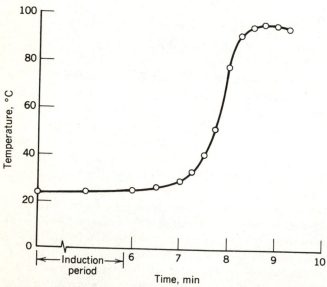

Figure 2.4 Reaction exotherm for a methylmethacrylate casting system.

reason for the autoacceleration is quite different from that illustrated by Equation 2.56. The free-radical initiation and propagation steps for the polymerization involve at least one small molecule, and the corresponding reaction rates are more or less independent of the extent of polymerization. The termination reaction involves two large molecules, and the rate for such reactions is retarded by chain entanglements and thus by high concentrations of polymer. Free radicals continue to be initiated, but the rate constant for the termination step is greatly reduced. This leads to a net increase in the concentration of growing chains in the system and consequently in the rate of propagation. Figure 2.4 gives an example for the free-radial polymerization of methylmethacrylate. This casting system is approximately adiabatic so that the reaction exotherm provides a good measure of the extent of polymerization. Another example of autocatalytic behavior, again with complex underlying kinetics, is the batch fermentation of sugars. Cell mass (e.g., yeast cells) and product yields (e.g., ethanol) both exhibit the S-shaped growth curves typical of autocatalysis.

Autoacceleration also occurs in branching chain reactions where a single chain propagating species can generate more than one new propagating species. Such reactions are obviously important in nuclear fission and fusion. They also occur in combustion processes. For example, the elementary reactions

$$H \cdot + O_2 \rightarrow HO \cdot + O \cdot$$

$$O \cdot + H_2 \rightarrow HO \cdot + H \cdot$$

are believed important in the burning of hydrogen.

2.4   Variable-Volume Batch Reactors

2.4.1   Systems with Constant Mass

We consider now batch reactors in which the feed is all charged at once and the products are all removed together, the mass in the system being held constant during the reaction step. Such reactors usually operate at nearly constant volume. The reason for this is that most batch reactors operate in the liquid phase, and liquid densities tend to be insensitive to composition. When the density and thus the volume do change, it is more convenient to express the mass balance in terms of moles than concentrations. Equation 1.28 can be written as

$$\mathscr{R}_A V = \frac{d(Va)}{d\alpha} = \frac{dN_A}{d\alpha} \qquad N_A = (N_A)_0 \text{ at } \alpha = 0 \qquad (2.59)$$

We suppose $\mathscr{R}_A$ is given as a function of concentration but this too can be converted to moles since $a = N_A/V$, $b = N_B/V$, and so on. For a first order reaction,

$$\mathscr{R}_A V = -kaV = -kN_A = \frac{dN_A}{d\alpha} \tag{2.60}$$

which has the solution

$$N_A = (N_A)_0 e^{-k\alpha} \tag{2.61}$$

Thus for first order reactions, the number of moles of the reactive component decreases exponentially with time. If it happens that the fluid density is constant, the concentration of the reactive component also decreases exponentially as in Equation 1.36. For reactions of order other than first, things are not so convenient. For a second order reaction. $\mathscr{R}_A = -2ka^2$, and

$$\frac{-2kN_A^2}{V} = \frac{dN_A}{d\alpha} \tag{2.62}$$

Solution requires that V be known as a function of N_A (or of time α). This is often more of a problem in thermodynamics than in chemical kinetics.

Example 2.5   Suppose $2A \overset{k}{\to} B$ in the gas phase isothermally and at constant pressure. Find the composition as a function of batch reaction time assuming ideal gas behavior.

For an ideal gas

$$PV = (N_A + N_B + N_I)R_g T \tag{2.63}$$

where N_I is the number of moles of inerts.

For the given reaction stoichiometry, Equation 1.51 becomes

$$\frac{N_A - (N_A)_0}{-2} = \frac{N_B - (N_B)_0}{+1} \tag{2.64}$$

This gives

$$PV = \left[\frac{N_A + (N_A)_0}{2} + (N_B)_0 + N_I \right] R_g T \tag{2.65}$$

so that Equation 2.62 becomes

$$\frac{-4PkN_A^2}{[(N_A)_0 - N_A + 2(N_B)_0 + 2N_I] R_g T} = \frac{dN_A}{d\alpha} \tag{2.66}$$

Analytical solution is possible in the above example and other cases with simple kinetics and ideal gas behavior. In more complex situations, numerical solutions are needed. These numerical solutions require an equation of state for the system. Consider a variable-volume reaction involving components A, B, ... The mass balance equations can be written as

$$\frac{dN_A}{d\alpha} = V\mathscr{R}_A(N_A, N_B, \ldots, V) \tag{2.67}$$

$$\frac{dN_B}{d\alpha} = V\mathscr{R}_B(N_A, N_B, \ldots, V) \tag{2.68}$$

The equation of state can be written as

$$\phi(N_A, N_B, \ldots, P, V, T) = 0 \tag{2.69}$$

This set of equations is solved simultaneously using any suitable method such as marching ahead. The marching-ahead technique can be formalized as

$$(N_A)_{\alpha + \Delta\alpha} = (N_A)_\alpha + \Delta\alpha V_\alpha \mathscr{R}_A\left[(N_A)_\alpha, (N_B)_\alpha, \ldots, V_\alpha\right] \tag{2.70}$$

$$(N_B)_{\alpha + \Delta\alpha} = (N_B)_\alpha + \Delta\alpha V_\alpha \mathscr{R}_B\left[(N_A)_\alpha, (N_B)_\alpha, \ldots, V_\alpha\right] \tag{2.71}$$

where V_α is found from

$$\phi\left[(N_A)_\alpha, (N_B)_\alpha, \ldots, P_\alpha, V_\alpha, T_\alpha\right] = 0 \tag{2.72}$$

It should be apparent that the above formalism also applies to variable-temperature reactors given the temperature dependence of the various rate constants and given an auxiliary equation from which temperature can be calculated as a function of α and/or composition. We shall see in Chapter 4 that the auxiliary equation is obtained from a heat balance and simply adds another differential equation to the above set.

2.4.2 Semibatch Systems

Many industrial reactors operate in the semibatch mode. In semibatch operation, reactants may be charged to the system at different times, or some of the products may be removed at different times. Occasionally, a "heel" of material from a previous batch is retained to start the new batch.

There are a variety of reasons for operating in a semibatch mode. Some typical ones are:

1. A starting material is subjected to several different reactions, one after the other. It is merely convenient to use the same vessel, and each reaction is essentially independent.

2. Reaction starts as soon as the reactants come into contact during the charging process. The initial reaction environment will differ depending on whether the reactants are charged sequentially or simultaneously.

3. One reactant is charged to the reactor in small increments to control the composition distribution of the products. Partial nitrations or vinyl copolymerizations are typical examples.

4. A by-product comes out of solution or is intentionally removed to avoid an equilibrium limitation.

5. One reactant is sparingly soluble in the reaction phase and would be depeleted were it not added continuously.

All but the first of these has chemical reactions occurring simultaneously with mixing or mass transfer operations. A general treatment requires the combination of transport equations with the chemical kinetics, and it becomes necessary to solve sets of partial differential equations rather than ordinary differential equations. Although this approach is becoming common in continuous-flow systems, it remains very difficult in batch systems. The central difficulty is in developing good equations for the mixing and mass transfer steps.

Semibatch reactors can be treated fairly easily when the mixing and mass transfer steps are fast compared to the reaction steps. Compositions and reaction rates will then be spatially uniform, and a flow term is simply added to the mass balance. Instead of Equation 2.67, we write

$$\frac{dN_A}{d\alpha} = (Qa)_{\text{flow}} + V\mathcal{R}_A(N_A, N_B, \ldots, V)$$

(2.73)

where the term $(Qa)_{\text{flow}}$ represents the net molar flow rate of A into or out of the reactor. It is positive for inflow and negative for outflows.

Many semibatch reactions involve more than one phase and are thus classified as **heterogeneous**. Examples are aerobic fermentations, where oxygen is supplied continuously to a liquid substrate, and chemical vapor deposition reactors, where gaseous reactants are supplied continuously to a solid substrate. Typically, the overall reaction rate will be limited by the rate of interphase mass transfer. Such systems are treated using the methods of Chapters 10 and 11. Occasionally, the reaction will be kinetically limited so that the transferred component saturates the reaction phase. The system can then be treated as a batch reaction, with the concentration of the transferred component being dictated by its solubility. The early stages of a batch fermentation will behave in this fashion but will shift to a mass transfer limitation as the cell mass and thus the oxygen demand increase.

2.5   Reaction Coordinates

A batch reaction involving N components undergoing M reactions can be described with a set of N ordinary differential equations, one for each component. This set of equations can be written as

$$\frac{d(aV)}{d\alpha} = \underset{\approx}{\nu}\mathscr{R}V \qquad \text{where } \mathbf{a} = \mathbf{a}_0 \text{ at } \alpha = 0 \qquad (2.74)$$

Here, $\mathbf{a}$ is a vector ($N \times 1$ matrix) of the component concentrations, $\underset{\approx}{\nu}$ is an $N \times M$ matrix of stoichiometric coefficients, and $\mathscr{R}$ is a vector ($M \times 1$ matrix) of reaction rates.

Example 2.6 Consider a constant-volume batch reactor with the following set of reactions:

$$\begin{aligned}
\text{A} + 2\text{B} &\rightarrow \text{C} \qquad \mathscr{R}_{\text{I}} = k_{\text{I}}a \\
\text{A} + \text{C} &\rightarrow \text{D} \qquad \mathscr{R}_{\text{II}} = k_{\text{II}}ac \qquad (2.75) \\
\text{B} + \text{C} &\rightarrow \text{E} \qquad \mathscr{R}_{\text{III}} = k_{\text{III}}c
\end{aligned}$$

(These rate equations would be plausible if B were present in great excess, say, water in an aqueous-phase reaction.)

Equation 2.74 becomes

$$\frac{d}{d\alpha}\begin{pmatrix} a \\ b \\ c \\ d \\ e \end{pmatrix} = \begin{pmatrix} -1 & -1 & 0 \\ -2 & 0 & -1 \\ +1 & -1 & -1 \\ 0 & +1 & 0 \\ 0 & 0 & +1 \end{pmatrix}\begin{pmatrix} k_{\text{I}}a \\ k_{\text{II}}ac \\ k_{\text{III}}c \end{pmatrix} \qquad (2.76)$$

Note that the matrix of stoichiometric coefficients devotes a row to each component and a column to each reaction. Multiplying out gives the expected set of ordinary differential equations:

$$\frac{da}{d\alpha} = -k_{\text{I}}a - k_{\text{II}}ac$$

$$\frac{db}{d\alpha} = -2k_{\text{I}}a \qquad - k_{\text{III}}c$$

$$\frac{dc}{d\alpha} = +k_{\text{I}}a - k_{\text{II}}ac - k_{\text{III}}c \qquad (2.77)$$

$$\frac{dd}{d\alpha} = \qquad + k_{\text{II}}ac$$

$$\frac{de}{d\alpha} = \qquad\qquad + k_{\text{III}}c$$

This is a set of five equations and five unknowns which can be solved for **a**. Note however that the first three equations contain only a, b, and c. This subset can be solved independently so that the effective dimensionality is only three.

Equation 2.74 generates a set of N equations which can be solved simultaneously to give the N compositional unknowns. However, as the example shows, the dimensionality of the equation set can sometimes be smaller than N. In fact, by choosing the correct variables to substitute for the component compositions, the dimensionality can always be reduced to M. There need be no more equations than there are independent chemical reactions. A set of reactions is independent if no member of the set can be obtained by adding or subtracting multiples of the other members. A set will be independent if every reaction contains one species not present in the other reactions.[3]

Define ε, the *extent of reaction*, or *reaction coordinate*, as

$$\boxed{\mathbf{N} - \mathbf{N}_0 = \underset{\approx}{\nu}\varepsilon}$$

(2.78)

where $\mathbf{N}$ and $\mathbf{N}_0$ are vectors ($N \times 1$ matrices) giving the final and initial number of moles of each component, $\underset{\approx}{\nu}$ is the matrix of stoichiometric coefficients, and ε is the reaction coordinate vector ($M \times 1$ matrix). In more explicit form,

$$\begin{pmatrix} N_A \\ N_B \\ \vdots \end{pmatrix} - \begin{pmatrix} N_A \\ N_B \\ \vdots \end{pmatrix}_0 = \begin{pmatrix} \nu_{A,I} & \nu_{A,II} & \cdots \\ \nu_{B,I} & \nu_{B,II} & \\ \vdots & & \end{pmatrix} \begin{pmatrix} \varepsilon_I \\ \varepsilon_{II} \\ \varepsilon_{III} \\ \vdots \end{pmatrix}$$

(2.79)

Example 2.7 For any single reaction, Equation 2.79 becomes

$$\begin{pmatrix} N_A \\ N_B \\ \vdots \end{pmatrix} - \begin{pmatrix} N_A \\ N_B \\ \vdots \end{pmatrix}_0 = \begin{pmatrix} \nu_A \\ \nu_B \\ \vdots \end{pmatrix}(\varepsilon)$$

(2.80)

Equating each component of the vector gives

$$\frac{N_A - (N_A)_0}{\nu_A} = \frac{N_B - (N_B)_0}{\nu_B} = \cdots = \varepsilon$$

(2.81)

---

which is Equation 1.51.

[3] See pp. 150–165 of G. V. Reklaitis and D. R. Schneider, *Introduction to Material and Energy Balances*, Wiley, New York, 1983, for a detailed discussion of independence. See also Appendix 4.1.

Equation 2.78 represents a generalization to M reactions of the stoichiometric constraints of Equation 1.51. If the vector $\boldsymbol{\varepsilon}$ is known, Equation 2.78 allows the composition in the entire system to be calculated. What is now needed is some means for calculating $\boldsymbol{\varepsilon}$. To do this, it is useful to consider some component, H, which is formed only by Reaction I, which does not appear in the feed, and which has a stoichiometric coefficient of $\nu_{H,I} = 1$. Then

$$\varepsilon_I = N_H = Vh \qquad (2.82)$$

There may be a real component that satisfies the requirements on H, but H need not exist. It can be hypothetical. Some may prefer to consider H as a kind of pixie dust formed by the reaction.[4] Note that each different reaction has its own H.

Consider a batch reaction that generates H. A component material balance on H gives

$$\frac{d(Vh)}{d\alpha} = V\mathcal{R}_H = V\nu_H\mathcal{R} = V\mathcal{R} \qquad (2.83)$$

where we have used the fact that $\nu_H = +1$. Substituting $\varepsilon_I = Vh$ gives

$$\frac{d\varepsilon_I}{d\alpha} = V\mathcal{R} \qquad (2.84)$$

The initial condition associated with Equation 2.84 is $\varepsilon_I = 0$ at $\alpha = 0$ since $h = 0$ at $\alpha = 0$. We now consider a different H for each of the M reactions, giving

$$\boxed{\frac{d\boldsymbol{\varepsilon}}{d\alpha} = V\boldsymbol{\mathcal{R}} \qquad \text{where } \boldsymbol{\varepsilon} = 0 \text{ at } \alpha = 0} \qquad (2.85)$$

---

[4] Since stoichiometric equations can be multiplied by an arbitrary constant, it is always possible to arrange that $\nu_P = +1$ for some real product P. For example, the decomposition of ozone,

$$2O_3 \rightarrow 3O_2$$

can be rewritten as

$$\tfrac{2}{3}O_3 \rightarrow O_2$$

In this form, the **numerical** value of ε can be regarded as the **fractional extent of completion** for the reaction.

In a more explicit form,

$$\frac{d}{d\alpha}\begin{pmatrix} \varepsilon_I \\ \varepsilon_{II} \\ \vdots \end{pmatrix} = V\begin{pmatrix} \mathscr{R}_I \\ \mathscr{R}_{II} \\ \vdots \end{pmatrix}$$

(2.86)

This is the set of M equations needed to find $\boldsymbol{\varepsilon}$. The reaction rate vector $\mathscr{R}$ will be expressed in terms of component concentrations rather than in terms of the ε. However, concentrations can be related to the ε using Equation 2.78. Thus there are M simultaneous equations which can be written in terms of the M unknown components of $\boldsymbol{\varepsilon}$.

Example 2.8   For the reactions in Example 2.6,

$$\frac{d}{d\alpha}\begin{pmatrix} \varepsilon_I \\ \varepsilon_{II} \\ \varepsilon_{III} \end{pmatrix} = V\begin{pmatrix} k_I a \\ k_{II} ac \\ k_{III} c \end{pmatrix} = \begin{pmatrix} k_I N_A \\ k_{II} N_A N_C / V \\ k_{III} N_C \end{pmatrix}$$

(2.87)

Writing Equation 2.78 for this reaction system gives

$$N_A - (N_A)_0 = -\varepsilon_I - \varepsilon_{II}$$

$$N_B - (N_B)_0 = -2\varepsilon_I \quad - \varepsilon_{III}$$

$$N_C - (N_C)_0 = +\varepsilon_I - \varepsilon_{II} - \varepsilon_{III}$$

(2.88)

$$N_D - (N_D)_0 = \quad + \varepsilon_{II}$$

$$N_E - (N_E)_0 = \quad + \varepsilon_{III}$$

These equations are used to eliminate N_A, N_B, and N_C from Equation 2.87. The result is

$$\frac{d\varepsilon_I}{d\alpha} = k_I\big[(N_A)_0 - \varepsilon_I - \varepsilon_{II}\big]$$

$$\frac{d\varepsilon_{II}}{d\alpha} = \frac{k_{II}}{V}\big[(N_A)_0 - \varepsilon_I - \varepsilon_{II}\big]\big[(N_C)_0 + \varepsilon_I - \varepsilon_{II} - \varepsilon_{III}\big]$$

(2.89)

$$\frac{d\varepsilon_{III}}{d\alpha} = k_{III}\big[(N_C)_0 + \varepsilon_I - \varepsilon_{II} - \varepsilon_{III}\big]$$

In a formal sense, Equations 2.78 and 2.85 represent the "solution" to all batch reaction problems. These equations are perfectly general and do not require the assumption of constant volume. However, if V is variable, an auxiliary equation—an equation of state—will be needed:

$$V = f(a, b, \ldots) = f\left(\frac{N_A}{V}, \frac{N_B}{V}, \ldots\right) = f(\varepsilon_I, \varepsilon_{II}, \ldots) \tag{2.90}$$

In variable-volume problems there will be M ordinary differential equations and one algebraic equation in $M + 1$ unknowns.

Suggestions for Further Reading

Most undergraduate texts on physical chemistry provide a survey of chemical kinetics and reaction mechanisms. A comprehensive treatment is provided in:

S. W. Benson, *Foundations of Chemical Kinetics*, McGraw-Hill, New York, 1960.

A briefer and more recent description is:

J. H. Espenson, *Chemical Kinetics and Reaction Mechanisms*, McGraw-Hill, New York, 1981.

An account of the reaction coordinate method as applied to chemical equilibrium is given in Chapter 9 of:

J. M. Smith and H. C. Van Ness, *Introduction to Chemical Engineering Thermodynamics*, 4th ed., McGraw-Hill, 1986.

Many techniques for integrating ordinary differential equations are discussed in:

M. E. Davis, *Numerical Methods and Modeling for Chemical Engineers*, Wiley, New York, 1984.

Another excellent reference on numerical methods is:

B. A. Finlayson, *Nonlinear Analysis in Chemical Engineering*, McGraw-Hill, New York, 1980.

Problems

2.1   Solve Equations 2.12 analytically. Compare your results with the numerical example of Section 2.2.2.

2.2   For the reaction network of Equation 2.20, find the value of the batch reaction time that maximizes the concentration of B. Compare this maximum value for b to the value for b obtained using the quasi-steady state hypothesis.

2.3   The following reactions are occurring in a constant-volume, isothermal batch reactor:

$$A + B \xrightarrow{k_1} C$$

$$C + B \xrightarrow{k_2} D$$

The following are known: $a_0 = b_0 = 10$ mol/m^3, $c_0 = d_0 = 0$, $k_1 = k_2 = 0.01$ m^3/mol^{-1} hr^{-1}, $t_{batch} = 4$ hr.

(a) Find the concentration of C at the end of the batch cycle.

(b) Find a general relationship between the concentrations of A and C when that of C is at a maximum.

2.4 The following kinetic scheme is postulated for a batch reaction:

$$A + B \rightarrow C \qquad \mathcal{R} = k_I a^{1/2} b$$

$$C + B \rightarrow D \qquad \mathcal{R} = k_{II} c^{1/2} b$$

Determine a, b, c, and d as functions of time. Continue your calculations until the limiting reagent is 90% consumed given $a_0 = 10$ mol/m^3, $b_0 = 2$ mol/m^3, $c_0 = d_0 = 0$, $k_I = k_{II} = 0.02$ m$^{3/2}$ mol$^{-1/2}$ s^{-1}.

2.5 Dimethyl ether thermally decomposes at temperatures above 450°C. The predominant reaction is

$$CH_3OCH_3 \rightarrow CH_4 + H_2 + CO$$

Suppose a homogeneous, gas phase reaction occurs in a constant-volume batch reactor. Assume ideal gas behavior.

(a) Show how the reaction rate can be determined from pressure measurements. Specifically, relate $\mathcal{R}$ to $dP/d\alpha$.

(b) Determine $P(t)$ assuming the decomposition is first order.

2.6 Consider the sequential reaction of Equation 2.49 and suppose $b_0 = c_0 = d_0 = 0$, $k_I = 3$ hr^{-1}, $k_{II} = 2$ hr^{-1}, $k_{III} = 4$ hr^{-1}. Determine the ratios a_t/a_0, b_t/a_0, c_t/a_0, and d_t/a_0, where t is the batch reaction time chosen such that

(a) The final concentration of A is maximized.

(b) The final concentration of B is maximized.

(c) The final concentration of C is maximized.

(d) The final concentration of D is maximized.

2.7 The bromine–hydrogen reaction

$$Br_2 + H_2 \rightarrow 2HBr$$

is believed to proceed by the following elementary reactions:

$$Br_2 + M \underset{k_{-1}}{\overset{k_1}{\rightleftharpoons}} 2Br \cdot + M \qquad\qquad \text{(I)}$$

$$Br \cdot + H_2 \underset{k_{-2}}{\overset{k_2}{\rightleftharpoons}} HBr + H \cdot \qquad\qquad \text{(II)}$$

$$H \cdot + Br_2 \overset{k_3}{\rightarrow} HBr + Br \cdot \qquad\qquad \text{(III)}$$

The initiation step, Reaction I, represents the thermal dissociation of bromine, which is brought about by collision with another molecule denoted by M. The only termination reaction is the reverse of the initiation step and is third order. Apply the steady state hypothesis to [Br $\cdot$] and [H $\cdot$] to obtain

$$\mathscr{R} = \frac{k[H_2][Br_2]^{3/2}}{[Br_2] + k'[HBr]}$$

2.8 A proposed mechanism for the thermal cracking of ethane is

$$C_2H_6 + M \xrightarrow{k_I} 2CH_3 \cdot + M \qquad (I)$$

$$CH_3 \cdot + C_2H_6 \xrightarrow{k_{II}} CH_4 + C_2H_5 \cdot \qquad (II)$$

$$C_2H_5 \cdot \xrightarrow{k_{III}} C_2H_4 + H \cdot \qquad (III)$$

$$H \cdot + C_2H_6 \xrightarrow{k_{IV}} H_2 + C_2H_5 \cdot \qquad (IV)$$

$$2C_2H_5 \cdot \xrightarrow{k_V} C_4H_{10} \qquad (V)$$

The overall reaction has variable stoichiometry:

$$C_2H_6 \rightarrow \nu_B C_2H_4 + \nu_C C_4H_{10} + (2 - 2\nu_B - 4\nu_C)CH_4$$
$$+ (-1 + 2\nu_B + 3\nu_C)H_2$$

where we have assumed the free radical concentration is negligible.

(a) Apply the steady state hypothesis to obtain an expression for the disappearance of ethane.

(b) What does the steady state hypothesis predict for ν_B and ν_C?

2.9 Plot a_t/a_0 as a function of the dimensionless rate constant $a_0 kt$ for the autocatalytic batch reaction of Equation 2.58 with $b_0/a_0 = 0.01$. Repeat for a perfect mixer.

2.10 It is proposed to study the chlorination of ethylene

$$C_2H_4 + Cl_2 \rightarrow C_2H_4Cl_2$$

in a constant-pressure, gas phase batch reactor. Derive an expression for the reactor volume as a function of time assuming second order kinetics, ideal gas behavior, perfect stoichiometry, and 50% inerts by volume at $\alpha = 0$.

2.11 Consider the liquid phase reaction of a diacid with a diol, the first reaction step being

$$HO-R-OH + HOOC-R'-COOH \rightarrow HO-ROOCR'-COOH + H_2O$$

Suppose the desired product is the single-step mixed acidol. A large excess of the diol is used, and batch reactions are conducted to experimentally determine the reaction time t_{max}, which maximizes the yield of acidol. Devise a kinetic model for the system and explain how the parameters in this model can be fit to the experimental data.

2.12 A numerical integration scheme has produced the following results:

Δx	Integral
1.0	0.23749
0.5	0.20108
0.25	0.19298
0.125	0.19104
0.0625	0.19056

(a) What is the apparent order of convergence?

(b) Extrapolate the results to $\Delta x = 0$. (*Note:* Such extrapolation should not be done unless the integration scheme has a theoretical order of convergence that agrees with the apparent order. Assume it does.)

(c) What value for the integral would you expect at $\Delta x = \frac{1}{32}$?

2.13 Suppose that ethylene oxide is decomposing in the gas phase at constant pressure:

$$\underset{\diagdown \ \ \diagup}{\underset{O}{CH_2 \ CH_2}} \xrightarrow{k} CH_4 + CO$$

The reaction is batch and isothermal. Ideal gas behavior may be assumed. The initial mixture contains 90 mole-% inerts. The initial volume is V_0. Find $V(\alpha)$.

2.14 Determine the maximum batch reactor yield of B for a reversible, first order reaction:

$$A \underset{k_r}{\overset{k_f}{\rightleftharpoons}} B$$

Do not assume $b_0 = 0$.

2.15 The photoinitiated chlorination of sulfolane (tetrahydrothiophene 1,1-dioxide) is believed[5] to follow the sequence:

[5]S. W. Suh, M. Richard, and M. Lenzl, Paper presented at 1985 Annual AIChE Meeting, Chicago.

It has been established that substitution of hydrogen by chlorine occurs only at the β-position as indicated above. Suppose that each of the above reactions is pseudo-first order with respect to the precursor organic.

(a) Write the set of ODEs governing the above reaction sequence in a batch reaction. Assume that only the unsubstituted sulfone is present initially.

(b) Denote the rate constant for the first monosubstitution reaction as k_1. What ratio to k_1 would you expect for the other rate constants assuming equal reactivity of the various β-hydrogens?

(c) Plot the various species concentrations versus dimensionless time $k_1 t$ for the equal reactivity case considered in part (b) above.

(d) The structure we have denoted as

actually admits two possibilities, the cis and trans isomers. Suh, Richard, and Lenzl[5] did not detect *cis*-dichlorosulfolane. Use this fact to modify the equal reactivity assumption and the ratios of rate constants determined in part (b). Revise the graph in part (c) to reflect the new kinetics.

Appendix 2.1

Numerical Solution of Ordinary Differential Equations

Chapter 2 described the marching-ahead method for solving sets of ordinary differential equations. The method is extremely simple from a conceptual and programming viewpoint. It is computationally inefficient in the sense that a great many arithmetic operations are necessary to produce accurate solutions. More efficient techniques should be used when the same set of equations is to be solved many times as in optimization studies. One such technique, **fourth order Runge–Kutta**, has proved very popular and can be generally recommended for all but very stiff sets of first order ordinary differential equations. The set of equations to be solved is

$$\frac{da}{d\alpha} = \mathscr{R}_A(a, b, \ldots, \alpha)$$

$$\frac{db}{d\alpha} = \mathscr{R}_B(a, b, \ldots, \alpha) \qquad (2.91)$$

$$\vdots \qquad \vdots$$

A value of $\Delta\alpha$ is selected, and values for $\Delta a, \Delta b, \ldots$ are estimated by evaluating the functions $\mathscr{R}_A, \mathscr{R}_B, \ldots$. In the marching-ahead method, this evaluation is done at the initial point $(a_0, b_0, \ldots, t_0)$ so that the estimate for Δa is just $\Delta \alpha \mathscr{R}_A(a_0, b_0, \ldots, t_0) = \Delta \alpha (\mathscr{R}_A)_0$. In fourth order Runge–Kutta, the evaluation is done at four points and the estimates for $\Delta a, \Delta b, \ldots$ are based on weighted averages of the $\mathscr{R}_A, \mathscr{R}_B, \ldots$ at these four points:

$$\Delta a = \Delta \alpha \frac{(\mathscr{R}_A)_0 + 2(\mathscr{R}_A)_1 + 2(\mathscr{R}_A)_2 + (\mathscr{R}_A)_3}{6}$$

$$\Delta b = \Delta \alpha \frac{(\mathscr{R}_B)_0 + 2(\mathscr{R}_B)_1 + 2(\mathscr{R}_B)_2 + (\mathscr{R}_B)_3}{6} \tag{2.92}$$

$$\vdots \qquad \vdots$$

where

$$(\mathscr{R}_A)_0 = \mathscr{R}_A(a_0, b_0, \ldots, \alpha_0)$$

$$(\mathscr{R}_A)_1 = \mathscr{R}_A(a_1, b_1, \ldots, \alpha_1)$$

$$(\mathscr{R}_A)_2 = \mathscr{R}_A(a_2, b_2, \ldots, \alpha_2) \tag{2.93}$$

$$(\mathscr{R}_A)_3 = \mathscr{R}_A(a_3, b_3, \ldots, \alpha_3)$$

with similar equations applying for $B, C, \ldots$. The values for $a_1, a_2, \ldots$ are determined in a sequential fashion:

$$a_1 = a_0 + \tfrac{1}{2} \Delta \alpha (\mathscr{R}_A)_0$$

$$a_2 = a_0 + \tfrac{1}{2} \Delta \alpha (\mathscr{R}_A)_1 \tag{2.94}$$

$$a_3 = a_0 + \Delta \alpha (\mathscr{R}_A)_2$$

with similar equations holding for $b_1, b_2, \ldots$. Time rarely appears explicitly in the $\mathscr{R}$, but should it appear,

$$\alpha_1 = \alpha_0 + \tfrac{1}{2} \Delta \alpha$$

$$\alpha_2 = \alpha_1 \tag{2.95}$$

$$\alpha_3 = \alpha_0 + \Delta \alpha$$

Example 2.9   Solve Equations 2.12 for $t = 1$ hr given $a_0 = c_0 = 30$ moles/m³, $b_0 = p_0 = 0$, $k_I = 0.01$, $k_{II} = 0.02$. Use $\Delta \alpha = 1.0$ and 0.5.

Solution For $\Delta\alpha = 1.0$:

i	a_i	b_i	c_i	$(\mathscr{R}_A)_i$	$(\mathscr{R}_B)_i$	$(\mathscr{R}_C)_i$
0	30.000	0	30.000	-18.000	9.000	0
1	21.000	4.500	30.000	-8.820	1.710	-2.700
2	25.590	0.855	28.650	-13.097	6.059	-0.490
3	16.903	6.059	29.510	-5.714	-0.719	-3.576
Δ's	-11.258	3.970	-1.659			
a_t	18.742	3.970	28.341			

For $\Delta\alpha = 0.5$:

i	a_i	b_i	c_i	$(\mathscr{R}_A)_i$	$(\mathscr{R}_B)_i$	$(\mathscr{R}_C)_i$
0	30.000	0	30.000	-18.000	9.000	0
1	25.500	2.250	30.000	-13.005	5.153	-1.350
2	26.749	1.288	29.663	-14.310	6.391	-0.764
3	22.845	3.195	29.236	-10.438	3.351	-1.868
Δ's	-6.992	2.953	-0.508			
$a_{t/2}$	23.078	2.953	29.492			
0	23.078	2.953	29.492			
1	20.415	3.849	29.057			
2	20.993	3.436	28.933			
3	18.670	4.162	28.498			
Δ's	-4.328	1.116	-1.047			
a_t	18.750	4.069	28.445			

The fourth order Runge–Kutta method converges $0(\Delta\alpha^5)$. Thus halving the step size decreases the error by a factor of 32. With this in mind, Runge–Kutta calculations can be extrapolated. Since $32^{-1} + 32^{-2} + 32^{-3} + \cdots = 0.03226$, the extrapolation formula analogous to Equation 2.14 is

$$\lim_{N \to \infty} a_t^*(\Delta\alpha\, 2^{-N}) = a_t^*\left(\frac{\Delta\alpha}{2}\right) + 0.03226\,\Delta a^* \tag{2.96}$$

Application to the above example gives $a_t = 18.750$, $b_t = 4.072$, $c_t = 28.448$, which are correct to three decimal places. However, such extrapolation should not really be done without confirming that $\Delta\alpha$ is small enough to be giving convergence $0(\Delta\alpha^5)$. This requires another halving of the time step, which, in the present example, gives results identical to the extrapolated values.

Runge–Kutta is a powerful integration technique that remains fairly easy to implement. It will be unstable if the step size is too large. In a set of equations, the Runge–Kutta algorithm uses the same step size for each member of the set. This obviously causes problems if the set is stiff. Most alternatives to Runge–Kutta suffer similar problems, but a variety of fairly complex, "canned" computer programs is now available for stiff sets. See Davis (1984)[6] for a summary and comparison of available routines.

[6]M. E. Davis, *Numerical Methods and Modeling for Chemical Engineers*, Wiley, New York, 1984.

Complex Reactions in Ideal Flow Reactors

3.1 Piston Flow Reactors

This chapter begins the treatment of complex reactions in flow systems. The simplest case is that of a piston flow reactor having constant cross-sectional area and operating with a constant-density fluid.

3.1.1 Complex Kinetics

Provided the fluid density remains constant, piston flow reactors are treated identically to constant-volume batch reactors except that the time coordinate α is replaced by the distance coordinate z:

$$d\alpha = \frac{dz}{\bar{u}} \tag{3.1}$$

The general material balance for some reactive component A becomes

$$\bar{u}\,\frac{da}{dz} = \mathscr{R}_A \tag{3.2}$$

A set of simultaneous reactions occurring in the piston flow reactor can be described by a set of ordinary differential equations, each based on Equation 3.2. These can be solved analytically when the reactions are all first order, whether reversible or irreversible. More typically, the equations must be solved numerically using methods such as marching ahead or Runge–Kutta as discussed in

Section 2.2.2. A typical set of equations is

$$\bar{u}\,\frac{da}{dz} = \mathscr{R}_{\mathrm{A}}$$

$$\bar{u}\,\frac{db}{dz} = \mathscr{R}_{\mathrm{B}} \tag{3.3}$$

$$\vdots \qquad \vdots$$

where $\mathscr{R}_{\mathrm{A}}, \mathscr{R}_{\mathrm{B}}, \ldots$ are identical to the rate expressions used for batch reactions. Suppose the reactor is divided into a number of increments in the axial direction, each of length Δz. Then the spatial derivatives in Equations 3.3 can be approximated as

$$\frac{da}{dz} \approx \frac{a(z + \Delta z) - a(z)}{\Delta z} \tag{3.4}$$

or, in a different notation,

$$\frac{da}{dz} \approx \frac{a_{j+1} - a_j}{\Delta z} \qquad j = 0, 1, \ldots, J - 1 \tag{3.5}$$

where $J = L/\Delta z$. Equation 3.4 holds exactly in the limit of small Δz. Indeed, this is just the definition of a derivative. Equation 3.5 is a restatement of this fact assuming the reactor length has been divided into an integral number of equal size increments. A step-by-step solution to Equations 3.3 can be written as

$$a_{j+1} = a_j + \frac{\mathscr{R}_{\mathrm{A}}(a_j, b_j, \ldots)}{\bar{u}}\,\Delta z$$

$$b_{j+1} = b_j + \frac{\mathscr{R}_{\mathrm{B}}(a_j, b_j, \ldots)}{\bar{u}}\,\Delta z \tag{3.6}$$

$$\vdots \qquad \vdots$$

where $a_0, b_0, \ldots$ are the initial, or tube inlet, conditions and $a_J, b_J, \ldots$ are the final, or outlet, conditions.

We have used a first order approximation for the derivatives such as da/dz. This means that the marching-ahead calculations will converge $0(\Delta z)$. Obviously, the numerical properties of this solution technique are identical to those found for the batch reactor example of Section 2.2.2. The marching-ahead method is conceptually straightforward. It is useful for the initial formulation of reactor design problems. More elaborate solution techniques, which in essence are based on higher order approximations for the derivatives, should be used if computer time becomes a critical factor, as it can in design optimization.

3.1.2 Variable-Density Reactors

Gas phase tubular reactors may have appreciable density differences between the inlet and outlet. The mass density can vary at constant pressure if there is a change in the number of moles upon reaction. The density can also vary when there is a pressure drop down the reactor due to skin friction. To account for such cases, we must return to the general material balance and allow the volumetric flow rate as well as the concentration to vary from point to point within the system. Figure 3.1 shows the system and indicates the nomenclature. A material balance for a reactive component A gives

$$\frac{d(Qa)}{dV} = \frac{1}{A_c}\frac{d(Qa)}{dz} = \frac{1}{A_c}\frac{d(A_c\bar{u}a)}{dz} = \mathcal{R}_A \qquad (3.7)$$

The derivative can be expanded into three separate terms:

$$\mathcal{R}_A = \bar{u}\frac{da}{dz} + a\frac{d\bar{u}}{dz} + \frac{\bar{u}}{A_c}a\frac{dA_c}{dz} \qquad (3.8)$$

The first of these terms must always be retained since A is a reactive component and thus varies in the z-direction. The second term must be retained if either the mass density or the reactor cross-sectional area varies with z. The last term is needed only for reactors with variable cross sections such as the annular flow reactor of Problem 3.9.

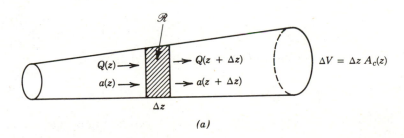

(a)

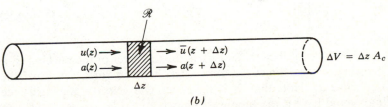

(b)

Figure 3.1 Piston flow reactors (a) Variable cross section. (b) Constant cross section.

The usual but not universal case is for the cross-sectional area A_c to be constant. Then

$$\frac{d(\bar{u}a)}{dz} = \mathscr{R}_A \tag{3.9}$$

This naturally reduces to Equation 3.2 when $\bar{u}$ is constant.

We shall treat only the case of constant mass flow down the tube.[1] The volumetric flow rate and the average velocity are related to the mass density:

$$Q\rho = A_c\bar{u}\rho = Q_{in}\rho_{in} = const \tag{3.10}$$

An equation of state is required to relate density to composition. We will use the ideal gas law for simplicity's sake although the conceptual and computational framework does not force this restriction.

When skin friction is important, as it is in most industrial reactors, a fluid dynamical equation is needed to calculate the pressure drop down the tube. For laminar flow in a circular tube, the pressure drop is given by the Poiseuille equation:

$$\frac{dP}{dz} = -\frac{8\bar{u}\mu}{R^2} \tag{3.11}$$

For turbulent flow, the pressure drop is calculated from

$$\frac{dP}{dz} = -\frac{\mathrm{Fa}\,\rho\bar{u}^2}{R} \tag{3.12}$$

where the Fanning friction factor Fa can be approximated as

$$\mathrm{Fa} = 0.079\,\mathrm{Re}^{-1/4} \tag{3.13}$$

For packed beds in either turbulent or laminar flow, the Ergun equation is often satisfactory:

$$\frac{dP}{dz} = -\frac{\rho\bar{u}_s^2}{d_p}\frac{1-\varepsilon}{\varepsilon^3}\left[\frac{150(1-\varepsilon)\mu}{d_p\,\rho\bar{u}_s} + 1.75\right] \tag{3.14}$$

where ε is the void fraction of the bed and d_p is the diameter of the packing. For nonspherical packing, use six times the ratio of volume to surface area of the packing as an effective d_p. Note that $\bar{u}_s$ is the superficial velocity, this being the velocity the fluid would have if the tube were empty.

---

[1] Transpired wall reactors exist and violate this assumption. See Problem 3.12.

Equations 3.11 through 3.14 give expressions for the pressure gradient. These expressions have not been integrated to give the total pressure drop since the fluid properties may vary from point to point. The pressure gradient may be approximated by a forward difference

$$\frac{dP}{dz} \approx \frac{P_{j+1} - P_j}{\Delta z} \qquad (3.15)$$

similar to that used for the concentration. This allows the pressure profile to be calculated as a function of axial position using the same numerical techniques as for concentration.

For most problems with variable mass density or variable cross section, it is easier to formulate the solution in terms of flux rather than concentration:

$$\Phi_A = \bar{u}a \qquad (3.16)$$

(The student of mass transfer will note that a diffusion term, $-D\,da/dz$, is usually included in the flux. This term is zero for piston flow.) When formulated in terms of flux, Equation 3.7 becomes

$$\frac{1}{A_c} \frac{d(A_c\Phi_A)}{dz} = \mathscr{R}_A \qquad A_c\Phi_A = (A_c\Phi_A)_{in} \quad \text{at} \quad z = 0 \qquad (3.17)$$

A version of Equation 3.17 can be written for each chemical species $A, B, \ldots$. These equations are usually coupled through the concentration dependence of rate, $\mathscr{R}_A = \mathscr{R}_A(a, b, \ldots)$, $\mathscr{R}_B = \mathscr{R}_B(a, b, \ldots)$, and so on, so that the set must be solved simultaneously. A compact way of writing the entire set of equations uses the vector and matrix notation of Section 2.5:

$$\frac{1}{A_c} \frac{d(A_c\Phi)}{dz} = \nu\mathscr{R} \qquad \Phi = \Phi_{in} \quad \text{at} \quad z = 0 \qquad (3.18)$$

where Φ is the vector ($N \times 1$ matrix) of component fluxes.

There are several possible ways of applying finite difference approximations to Equation 3.17. A simple, marching-ahead form is

$$(\Phi_A)_{j+1} = (\Phi_A)_j + \Delta z\,(\mathscr{R}_A)_j - \frac{(\Phi_A)_j}{(A_c)_j}\left(\frac{dA_c}{dz}\right)_j \Delta z \qquad (3.19)$$

The reactor geometry is assumed known. Then the flux may be calculated provided $\mathscr{R}_A(a_j, b_j, \ldots) = \mathscr{R}_A((\Phi_A)_j/\bar{u}_j, (\Phi_B)_j/\bar{u}_j, \ldots)$ is known. This in turn requires that $\bar{u}_j$ be known. A relationship between $\bar{u}_j$ and the fluxes at position j

can be found from an equation of state. For ideal gases,

$$\frac{P_j}{R_g T_j} = (\rho_{\text{molar}})_j = a_j + b_j + \cdots = \frac{(\Phi_A + \Phi_B + \cdots)_j}{\bar{u}_j} \tag{3.20}$$

The various fluxes will be known at position j. The velocity and hence the concentrations at position j can be calculated from Equation 3.20 provided that P_j and T_j are known. Temperature is considered constant in this chapter, but in Chapter 4 it will be obtained by a heat balance. Pressure must be obtained by a momentum balance, one of Equations 3.11, 3.12, or 3.14. Suppose it is obtained from Equation 3.12. In finite difference form,

$$P_{j+1} = P_j - \frac{\Delta z \, (\text{Fa})_j \, \rho_j \bar{u}_j^2}{R_j} \tag{3.21}$$

This equation can be marched ahead provided $(\text{Fa})_j$ and ρ_j are known. The mass density can be found from

$$\rho_j = M_A a_j + M_B b_j + \cdots \tag{3.22}$$

where M_A, M_B, ... are molecular weights. The friction factor Fa requires a correlation such as Equation 3.13. This correlation may in turn introduce new variables such as the viscosity, but such new variables can all eventually be found from the state variables of temperature, pressure, and composition, which are determined in the marching-ahead procedure.

Example 3.1   Ethylbenzene is catalytically dehydrogenated to styrene in a tubular, packed-bed reactor. A large excess of steam, 9 to 1 molar, is used to maintain an approximately isothermal reaction at 700°C. The heterogeneous reaction can be treated using pseudohomogeneous kinetics:

$$C_8 H_{10} \rightarrow C_8 H_8 + H_2 \qquad \mathscr{R} = k[C_8 H_{10}]$$

or

$$\tag{3.23}$$

$$A \rightarrow B + C \qquad \mathscr{R} = ka$$

with rate constant $k = 3.752$ s^{-1}. The tube length is 1 m. The inlet pressure is set at 1 atm and the downstream pressure is adjusted to give a superficial velocity of 4 m/s at the tube inlet. The catalyst pellets can be approximated as 3-mm spheres. The void fraction is 0.5 and the particle Reynolds number is 100 based on inlet conditions. Find the concentration, conversion, pressure, and velocity at the reactor exit.

Solution   This is a variable-velocity problem with $\bar{u}$ changing because of the reaction stoichiometry and the pressure drop. The flux marching equations for the various components are

$$(\Phi_A)_{j+1} = (\Phi_A)_j - ka\,\Delta z = (\Phi_A)_j - \frac{k(\Phi_A)_j\Delta z}{\bar{u}_j}$$

$$(\Phi_B)_{j+1} = (\Phi_B)_j + ka\,\Delta z = (\Phi_B)_j + \frac{k(\Phi_A)_j\Delta z}{\bar{u}_j}$$

(3.24)

$$(\Phi_C)_{j+1} = (\Phi_C)_j + ka\,\Delta z = (\Phi_C)_j + \frac{k(\Phi_A)_j\Delta z}{\bar{u}_j}$$

$$(\Phi_D)_{j+1} = (\Phi_D)_j$$

where D represents the inert steam. There is one equation for each component. It is perfectly feasible to retain each of these equations and to solve them simultaneously. Indeed, this is good practice if there is a complex reaction network or if molecular diffusion destroys local stoichiometry (see Chapter 6). For the current example, however, there is a simple stoichiometric constraint that may as well be used. At any step j,

$$\Phi_C = \Phi_B = (\Phi_A)_{in} - \Phi_A \tag{3.25}$$

Thus we need retain only the flux marching equation for component A.
The ideal gas law is used to relate $\bar{u}$ to the flux:

$$\bar{u}_j = \frac{R_gT}{P_j}(\Phi_A + \Phi_B + \Phi_C + \Phi_D)_j = \frac{R_gT}{P_j}\left[2(\Phi_A)_{in} - \Phi_A + \Phi_D\right]_j \tag{3.26}$$

The Ergun equation is used to find the pressure:

$$P_{j+1} = P_j - \Delta z\left[\frac{\rho_j\bar{u}_j^2}{d_p}\frac{1-\varepsilon}{\varepsilon^3}\right]\left[\frac{150(1-\varepsilon)}{(\mathrm{Re})_p} + 1.75\right] \tag{3.27}$$

where $(\mathrm{Re})_p = d_p\rho\bar{u}_s/\mu$ is approximately constant since $\rho\bar{u}_s$ is constant and since μ is a function of temperature alone for low-density gases. The inlet conditions are $a_{in} = 1.25$ mol/m^3, $b_{in} = c_{in} = 0$, $d_{in} = 11.3$ mol/m^3, $\rho_{in}\bar{u}_{in} = 1.33$ kg m^{-2} s^{-1}, and $(\Phi_A)_{in} = 5.0$ mol m^{-2} s^{-1}. Substituting known values and being careful with the units gives

$$(\Phi_A)_{j+1} = (\Phi_A)_j \left[1 - \frac{3.752 \, \Delta z}{\bar{u}_j} \right]$$

$$P_{j+1} = P_j - 0.0439 \, \Delta z \, \bar{u}_j \qquad\qquad (3.28)$$

$$\bar{u}_{j+1} = \frac{0.0798}{P_j} \left[55.13 - (\Phi_A)_j \right]$$

where P is in atm. Solutions for various step sizes are:

Δz	$(\Phi_A)_{out}$	P_{out}	$\bar{u}_{out}$
1	0.31	.824	4.00
0.5	1.41	.824	4.59
0.25	1.82	.815	4.87
0.125	2.01	.808	5.06
0.0625	2.11	.803	5.17
0.03125	2.15	.801	5.23
0	2.20	.798	5.29

The outlet conversion of ethylbenzene is

$$X = 1 - \frac{(\Phi_A)_{out}}{(\Phi_A)_{in}} = 1 - \frac{2.20}{5.01} = 0.56$$

The above problem captures the spirit of a typical heterogeneous catalytic reactor. Industrial design calculations would use a carefully determined and more elaborate rate expression. Other reactions such as the dealkylation of ethylbenzene to benzene and ethylene might be included in the calculations. Adiabatic rather than isothermal operation would probably be assumed, and measured pressure drops would be used to supplement or modify the Ergun equation. The heat balances needed for nonisothermal operation and rate expressions that are possibly more realistic for heterogeneous catalysis will be considered in subsequent chapters. However, these elaborations do not change the essential nature of the calculations. Nor does the fact that the downstream rather than upstream pressure will usually be specified. Specifying the downstream pressure forces an iterative solution (with P_{in} guessed and P_{out} calculated), but the calculations for each iteration follow the above scheme.

Also to be considered in subsequent chapters are the effects of nonideal flow behavior in the gas phase. Real systems will have velocity profiles that are not completely flat; there will be incomplete radial mixing; and there will also be some mixing in the axial direction due to molecular or eddy diffusion. However, these effects tend to be small in packed beds. The assumption of piston flow usually turns out to be quite good for the flow (inside tubes) of gases or turbulent liquids.

3.2 Perfectly Mixed Reactors

Perfect mixers show no spatial variations in composition or physical properties within the reactor or in the exit from it. Everything inside the system is uniform except at the very entrance. Molecules experience a step change in environment immediately upon entering. A perfect mixer has only two environments, one at the inlet and one within the system and at the outlet. These environments are specified by a set of compositions and operating conditions which can take on one of two possible states: either $a_{in}, b_{in}, \ldots, P_{in}, T_{in}, \ldots$ or $a_{out}, b_{out}, \ldots, P_{out}, T_{out}, \ldots$. For the case of a steady state perfect mixer, the inlet and outlet states are related by a series of algebraic equations. The other ideal reactors, and all real reactors, show a more gradual change from inlet to outlet, and the inlet and outlet states are related by a set of differential equations. Perhaps surprisingly, the differential equations are sometimes easier to solve than the algebraic ones.

3.2.1 Complex Kinetics and Reaction Coordinates

This section develops the minimum set of equations that must be solved for a steady state perfect mixer. The component material balances take the form

$$Q_{in}a_{in} + \mathscr{R}_A(a_{out}, b_{out} \ldots, P_{out}, T_{out}, \ldots)V = Q_{out}a_{out}$$

$$Q_{in}b_{in} + \mathscr{R}_B(a_{out}, b_{out} \ldots, P_{out}, T_{out}, \ldots)V = Q_{out}b_{out} \qquad (3.29)$$

$$\vdots \qquad\qquad \vdots \qquad\qquad \vdots$$

This set of equations can be summarized as

$$\boxed{Q_{in}\mathbf{a}_{in} + \underset{\approx}{\nu}\mathscr{R}V = Q_{out}\mathbf{a}_{out}} \qquad (3.30)$$

For now, we assume that all operating conditions are known. Suppose also that the inlet concentrations $a_{in}, b_{in}, \ldots$, volumetric flow rate Q_{in}, and working volume V are all known. Then Equations 3.29 or 3.30 are a set of N simultaneous equations in $N + 1$ unknowns, the unknowns being the N outlet concentrations $a_{out}, b_{out}, \ldots$ and the one volumetric flow rate Q_{out}. If the mass density of the fluid is constant, as is approximately true for liquid systems, then $Q_{in} = Q_{out}$. This allows Equations 3.29 to be solved for the outlet compositions. If Q_{out} is unknown, Equations 3.29 must be supplemented by an equation of state for the system.

Example 3.2 A liquid phase perfect mixer is used for the following reaction scheme:

$$A \xrightarrow{k_I} B$$

$$B \xrightarrow{k_{II}} C \tag{3.31}$$

Determine all outlet concentrations assuming constant density.

Solution With density constant, $Q_{in} = Q_{out} = Q$ and $\bar{t} = V/Q$. Equations 3.29 become

$$a_{in} - k_I \bar{t} a_{out} = a_{out}$$

$$b_{in} + k_I \bar{t} a_{out} - k_{II} \bar{t} b_{out} = b_{out} \tag{3.32}$$

$$c_{in} + k_{II} \bar{t} b_{out} = c_{out}$$

These can be solved sequentially to give

$$a_{out} = \frac{a_{in}}{1 + k_I \bar{t}}$$

$$b_{out} = \frac{b_{in} + k_I \bar{t} a_{in}/(1 + k_I \bar{t})}{1 + k_{II} \bar{t}} = \frac{b_{in} + k_I \bar{t}(a_{in} + b_{in})}{(1 + k_I \bar{t})(1 + k_{II} \bar{t})} \tag{3.33}$$

$$c_{out} = c_{in} + \frac{k_{II} \bar{t} [b_{in} + k_I \bar{t}(a_{in} + b_{in})]}{(1 + k_I \bar{t})(1 + k_{II} \bar{t})}$$

Example 3.3 Repeat the above example for the case where the density change upon reaction should not be ignored.

Solution An overall mass balance gives

$$\rho_{in} Q_{in} = \rho_{out} Q_{out} \tag{3.34}$$

Writing Equation 3.29 for component A gives

$$a_{in} - k_I(\bar{t})_{in} a_{out} = \left(\frac{\rho_{in}}{\rho_{out}}\right) a_{out} \tag{3.35}$$

where $(\bar{t})_{in} = V/Q_{in}$. An equation of state giving density as a function of composition is now needed. Assuming an ideal mixture is one possibility:

$$\rho = \frac{\rho_A a + \rho_B b + \cdots}{a + b + \cdots} = \frac{\Sigma \rho_A a}{\Sigma a} \tag{3.36}$$

The complete set of equations to be solved is thus

$$a_{in} - k_I \bar{t}_{in} a_{out} = a_{out} \frac{\Sigma (\rho_A a)_{in}}{\Sigma (\rho_A a)_{out}} \frac{\Sigma a_{out}}{\Sigma a_{in}}$$

$$b_{in} + k_I \bar{t}_{in} a_{out} - k_{II} \bar{t}_{in} b_{out} = b_{out} \frac{\Sigma (\rho_A a)_{in}}{\Sigma (\rho_A a)_{out}} \frac{\Sigma a_{out}}{\Sigma a_{in}} \tag{3.37}$$

$$c_{in} + k_{II} \bar{t}_{in} b_{out} = c_{out} \frac{\Sigma (\rho_A a)_{in}}{\Sigma (\rho_A a)_{out}} \frac{\Sigma a_{out}}{\Sigma a_{in}}$$

Since all inlet conditions are assumed known, Equations 3.37 contain only a_{out}, b_{out}, and c_{out} as unknowns. The solution, however, is appreciably more difficult than for Equations 3.32 of the previous example. A numerical solution is needed for all but the simplest situations.

Writing the system material balance as a set of N simultaneous equations in N compositional unknowns is theoretically correct but causes practical difficulties when N is large. The equations will be linear only if all the reactions are first order. They are usually nonlinear even in compositional variables and will certainly be nonlinear when operating variables such as P_{out} and T_{out} are also unknown. There is no truly robust method for solving a set of N nonlinear algebraic equations; and, in fact, a solution can be quite difficult to obtain when N is large. It is therefore important to reduce Equation 3.30 to the lowest possible dimensionality. As for batch reactions (see Section 2.5), this lowest dimensionality is one unknown for each independent reaction.

For a flow system, the reaction coordinate vector ε' is defined by

$$\boxed{Q\mathbf{a} - Q_{in}\mathbf{a}_{in} = \underset{\approx}{\nu}\varepsilon'} \tag{3.38}$$

The components of ε' have dimensions of moles per unit time. By way of comparison, the components of the batch extent of reaction, ε, have dimensions of moles. A component of ε', say, ε'_I, can be regarded as the rate of formation (moles per unit time) of a possibly hypothetical compound having a stoichiometric coefficient of $+1$ for Reaction I and of zero for all other reactions.

Example 3.4   Suppose

$$2A + B \rightarrow 3D \tag{I}$$

Then

$$-2\varepsilon'_I = \text{Rate of formation of A by Reaction I}$$

$$-\varepsilon'_I = \text{Rate of formation of B by Reaction I}$$

$$+3\varepsilon'_I = \text{Rate of formation of D by Reaction I}$$

In this example, there is no real component H such that $\nu_{H,I} = 1$, but $\nu_{H,II} = \nu_{H,III} = \cdots = 0$. However, this fact does not prevent ε'_I from being used to calculate the extents of reaction or formation for the components that do exist.

Given $\varepsilon'_I, \varepsilon'_{II}, \ldots$, it is possible to find the change in the number of moles for any component. For component A, for example,

$$Q_{out} a_{out} - Q_{in} a_{in} = \nu_{A,I}\varepsilon'_I + \nu_{A,II}\varepsilon'_{II} + \cdots = \sum_{\text{Reactions}} \nu_{A,I}\varepsilon'_I \tag{3.39}$$

Recall that the net rate of formation of a component due to several reactions is given by Equation 2.8. Then the component material balance becomes

$$0 = Q_{out} a_{out} - Q_{in} a_{in} - \sum_{\text{reactions}} \nu_{A,I}\mathscr{R}_I V \tag{3.40}$$

and substitution of Equation 3.39 gives

$$\sum_{\text{reactions}} \nu_{A,I}[\varepsilon'_I - \mathscr{R}_I V] = 0 \tag{3.41}$$

Equation 3.41 can be written for each component, and thus there are N versions of it. However, the number of those that are independent is never greater than the number of (independent) reactions. This independent subset of Equations 3.41 can be picked by writing Equation 3.41 for components that appear in different independent reactions. They are most easily picked by choosing that component which has a stoichiometric coefficient of $+1$ for a given reaction and of 0 for all other reactions. This component need not actually exist. Consider a reaction which we choose to denote as Reaction III, for example. Suppose there exists for this reaction a component H such that $\nu_{H,III} = 1$, $\nu_{H,I} = \nu_{H,II} = \nu_{H,IV} = \cdots = 0$. Then write Equation 3.41 for H and note that all terms in the summation vanish except for

$$\varepsilon_{III} - \mathscr{R}_{III} V = 0 \tag{3.42}$$

Equation 3.42 represents a steady state material balance for the possibly hypothetical component H. In essence, this equation serves as a material balance for the entire reaction, Reaction III. If we can solve Equation 3.42 for ε_{III}, we can immediately determine how many moles of the real components reacted or were formed by this particular reaction. Applying the same procedure to Reaction I, Reaction II, and so on gives

$$\boxed{\varepsilon' = \mathscr{R}V} \tag{3.43}$$

or, more explicitly,

$$\varepsilon'_I = \mathscr{R}_I V$$

$$\varepsilon'_{II} = \mathscr{R}_{II} V \tag{3.44}$$

$$\vdots \quad \vdots$$

as the design equations for a perfect mixer at steady state. These equations contain the ε' as unknowns. The various reaction rates will normally be given as functions of concentration, $\mathscr{R}_I = \mathscr{R}_I(a, b, \ldots)$, and so on for the various reactions. For a steady state perfect mixer, the reaction rate is evaluated at the outlet conditions. Thus we need $\mathscr{R}(a_{out}, b_{out}, \ldots)$ for the various reactions. Equation 3.39 allows the outlet concentrations to be replaced by reaction coordinates. This gives $\mathscr{R}(\varepsilon'_I, \varepsilon'_{II}, \ldots)$ so that Equations 3.44 contain only the ε' as compositional unknowns. The dimensionality of Equations 3.44 is equal to the number of independent reactions.

Example 3.5   The consecutive-competitive reactions

$$A + B \overset{k_I}{\rightarrow} R \tag{I}$$

$$R + B \overset{k_{II}}{\rightarrow} S \tag{II}$$

have the following stoichiometry coefficients: $\nu_{A,I} = -1$, $\nu_{A,II} = 0$, $\nu_{B,I} = -1$, $\nu_{B,II} = -1$, $\nu_{R,I} = +1$, $\nu_{R,II} = -1$, $\nu_{S,I} = 0$, $\nu_{S,II} = +1$. The component material balance, Equation 3.41, gives

$$-1[\varepsilon'_I - \mathscr{R}_I V] = 0 \qquad \text{for A}$$

$$-1[\varepsilon'_I - \mathscr{R}_I V] - 1[\varepsilon'_{II} - \mathscr{R}_{II} V] = 0 \qquad \text{for B}$$

$$+1[\varepsilon'_I - \mathscr{R}_I V] - 1[\varepsilon'_{II} - \mathscr{R}_{II} V] = 0 \qquad \text{for R} \tag{3.45}$$

$$+1[\varepsilon'_{II} - \mathscr{R}_{II} V] = 0 \qquad \text{for S}$$

which, of course, are consistent with Equations 3.44. The two independent equations that must be solved are

$$\varepsilon'_I = \mathcal{R}_I V = k_I a_{out} b_{out} V$$

$$\varepsilon'_{II} = \mathcal{R}_{II} V = k_{II} r_{out} b_{out} V \tag{3.46}$$

where the unknown compositions can be expressed in terms of the reaction coordinates by using Equations 3.39.

This gives

$$a_{out} = \frac{Q_{in} a_{in} - \varepsilon'_I}{Q_{out}}$$

$$b_{out} = \frac{Q_{in} b_{in} - \varepsilon'_I - \varepsilon'_{II}}{Q_{out}} \tag{3.47}$$

$$r_{out} = \frac{Q_{in} r_{in} + \varepsilon'_I - \varepsilon'_{II}}{Q_{out}}$$

Substitution into Equation 3.46 gives a pair of simultaneous quadratic equations to solve for ε'_I and ε'_{II}:

$$\varepsilon'_I = \frac{k_I V}{Q_{out}^2} [Q_{in} a_{in} - \varepsilon'_I][Q_{in} b_{in} - \varepsilon'_I - \varepsilon'_{II}]$$

$$\varepsilon'_{II} = \frac{k_{II} V}{Q_{out}^2} [Q_{in} r_{in} + \varepsilon'_I - \varepsilon'_{II}][Q_{in} b_{in} - \varepsilon'_I - \varepsilon'_{II}] \tag{3.48}$$

These may be expressed in dimensionless form as

$$x = k_I \bar{t} b_{in} \left[\frac{Q_{in} a_{in}}{Q_{out} b_{in}} - x \right] \left[\frac{Q_{in}}{Q_{out}} - x - y \right]$$

$$y = k_{II} \bar{t} b_{in} \left[\frac{Q_{in} r_{in}}{Q_{out} b_{in}} + x - y \right] \left[\frac{Q_{in}}{Q_{out}} - x - y \right] \tag{3.49}$$

where $x = \varepsilon'_I / Q_{out} b_{in}$, $y = \varepsilon'_{II} / Q_{out} b_{in}$, and $\bar{t} = V / Q_{out}$.

The reaction coordinate method has been used to reduce the dimensionality of the design equations for a steady state perfect mixer. The same approach can be used for batch and piston flow reactors if a reduction in dimensionality is required. Equation 2.85 is the design equation for a batch reactor. For piston flow, Equation 3.38 can be differentiated and then substituted into Equation 3.18

to give

$$
\boxed{\frac{1}{A_c}\frac{d\varepsilon'}{dz} = \mathcal{R} \qquad \varepsilon' = 0 \quad \text{at} \quad z = 0}
$$

(3.50)

This result is quite general, being subject only to the restriction of constant mass flow down the tube. Equation 3.50 represents a set of M ordinary differential equations. However, in this situation the reduction in dimensionality gains little computational advantage. Replacement of concentrations with reaction coordinates tends to disguise the chemistry and physics of the reactor design problem. This can be a substantial disadvantage in formulating problems with variable physical properties.

3.2.2 Numerical Methods

Example 3.5 was for a relatively simple kinetic scheme yet it illustrates the awkwardness and potential impossibility of analytical solutions. Industrial reactor designs should allow for realistically complex kinetic schemes and for variable physical properties. This means that numerical methods must be used to solve the design equations. We present here a relatively robust method for solving a set of algebraic equations that are implicit in a single variable.

Consider the set of equations

$$
F_1(x, y) = 0
$$
$$
F_2(x, y) = 0
$$

(3.51)

As written, these are implicit in both x and y. However, for typical kinetic schemes, it is often possible to solve explicitly for one of the variables. Thus we might be able to rewrite Equations 3.51 as

$$
F_1(x, y) = 0
$$
$$
y = F_3(x)
$$

(3.52)

or, equivalently, as

$$
F_1(x, F_3(x)) = 0
$$

(3.53)

which is implicit in the single variable x.

It is next necessary to bound the range on x for which a solution is possible. In reactor design problems, there may be physical limits on the values that x can assume. For example, the dimensionless variables x and y in Equation 3.49 are bounded by 0 and 1, and all physically meaningful solutions must lie within this interval. Given the bounds, calculate $F(x_{min})$ and $F(x_{max})$. These must differ in sign if there is an odd number of solutions (preferably one) within $x_{min} < x < x_{max}$. Now calculate F at the midpoint of the interval, that is, at $x = (x_{min} + x_{max})/2$. The sign of F will be the same as at one of the endpoints. Discard that endpoint and replace it with the midpoint. The signs of F at the two new endpoints will differ, so that the range in which the solution must lie has been halved. This procedure can obviously be repeated i times to reduce the range in which a solution must lie to 2^{-i} of the original range. Given an original range of $0 < x < 1$, $i = 7$ will produce an answer accurate to two decimals, $i = 11$ will give three decimals, and $i = 14$ will give four decimals. This method is known as a **binary search**. It is crude but effective and simple to program. For hand calculations and simple $F(x)$, however, you will probably prefer trial and error. A gradient method, useful when the set of equations cannot be reduced to a single, implicit variable, is outlined in Appendix 3.1.

Example 3.6 Determine the outlet composition for the consecutive-competitive reactions in Example 3.5. Assume $a_{in} = b_{in}$ (perfect initial stoichiometry), $r_{in} = b_{in}$, $Q_{in} = Q_{out}$ (constant mass density), $k_I t b_{in} = 4$ and $k_{II} t b_{in} = 1$. Equations 3.49 become

$$x = 4(1 - x)(1 - x - y)$$

$$y = 1(1 + x - y)(1 - x - y) \tag{3.54}$$

The second equation is easily solved for y giving

$$F(x, y) = 4(1 - x)(1 - x - y) - x$$

$$y = \frac{3 - \sqrt{5 + 4x^2}}{2} \tag{3.55}$$

where the positive square root has been eliminated from knowledge that both x and y are bounded by the interval $(0, 1)$. The binary search proceeds as follows:

i	x	$y(x)$	$F(x)$	Discard	
0	0	0.3820	+		
0	1	0	−		
1	0.5000	0.2753	−	1.000	
2	0.2500	0.3544	+	0	
3	0.3750	0.3208	+	0.2500	
4	0.4375	0.2994	+	0.3750	
5	0.4688	0.2877	+	0.4375	
6	0.4844	0.2815	−	0.5000	
7	0.4766	0.2846	+	0.4688	Ans. = 0.48
8	0.4805	0.2831	+	0.4766	
9	0.4825	0.2823	+	0.4805	
10	0.4834	0.2819	+	0.4825	
11	0.4839	0.2817	−	0.4844	Ans. = 0.484
12	0.4837	0.2818	+	0.4834	
13	0.4838	0.2818	+	0.4837	
14	0.4839	0.2817			Ans. = 0.4839

Completing the problem gives

$$\frac{a_{out}}{b_{in}} = 0.5161$$

$$\frac{b_{out}}{b_{in}} = 0.2344$$

(3.56)

$$\frac{r_{out}}{b_{in}} = 1.2022$$

$$\frac{s_{out}}{b_{in}} = 0.2817$$

Note that the second of Equations 3.54 is also easily solved for x. Do this; repeat the binary search; and confirm Equations 3.56.

As mentioned previously, it is often easier to solve a set of ordinary differential equations than a set of algebraic equations. Perfect mixers operating in the unsteady state are governed by ordinary differential equations. Thus it is sometimes reasonable to find a steady state solution by simulating a reactor startup and allowing the outlet concentrations to approach their steady state values. This approach has a side benefit. It shows that the steady state solution will indeed be achievable using the selected startup strategy. Chapter 8 gives the methodology for this approach.

3.2.3   Variable Physical Properties

The design equations for a perfect mixer do not require that the reacting mixture have constant physical properties or that operating conditions such as temperature and pressure be the same for the inlet and outlet environments. It is required, however, that these variables all be known. Operating conditions such as temperature and pressure can sometimes be specified independently of the extent of reaction. This is typically possible in laboratory equipment because of excellent heat transfer at the small scale. It is sometimes possible in industrial-scale reactors. When the operating conditions or physical properties are linked to the extent of reaction, independent specification of the variables is no longer possible. Additional equations, one for each unknown state variable, must be joined to Equations 3.44 and the entire set solved simultaneously. For temperature, the auxiliary equation is obtained from a heat balance which will be discussed in Chapter 4. Physical property variables are determined from correlations or an equation of state. Gas phase perfect mixers may show differences in mass density and thus in volumetric flow rate even when temperature and pressure are held constant. An equation of state is then necessary for the reactor calculations.

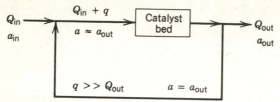

Figure 3.2   Recycle reactor for catalyst evaluation.

A form of stirred tank reactor is commonly used for laboratory studies with heterogeneous catalysts and gas phase reactants (see Figure 3.2). Gas is circulated at high rates through a small quantity of packed catalyst. Most of the gas exiting from the catalyst bed is recycled back through it, and the net throughput is relatively small. The high flow rates ensure good heat transfer to the catalyst and eliminate gas phase resistances to mass transfer. The high level of recirculation means that the per-pass conversion is very low so that the system can be analyzed as though it were a perfect mixer. High overall conversions can be obtained so that the problems of analytical chemistry are eased compared to a differential reactor with $a_{out} \approx a_{in}$.

Example 3.7   Suppose the reaction is

$$SO_2 + \tfrac{1}{2}O_2 \rightarrow SO_3 \tag{3.57}$$

and that studies on similar catalysts have suggested a rate expression of the form

$$\mathscr{R} = \frac{k[SO_2][O_2]}{1 + k'[SO_3]} = \frac{kab}{1 + k'c} \tag{3.58}$$

where $a = [SO_2]$, $b = [O_2]$, and $c = [SO_3]$. There is only one reaction, so only one version of Equation 3.43 is required:

$$\varepsilon' = \mathscr{R}V = \frac{kVa_{out}b_{out}}{1 + k'c_{out}} \tag{3.59}$$

Equation 3.38 allows the various concentrations to be expressed in terms of the reaction coordinate:

$$Q_{out}\begin{pmatrix} a \\ b \\ c \\ d \end{pmatrix}_{out} - Q_{in}\begin{pmatrix} a \\ b \\ c \\ d \end{pmatrix}_{in} = \begin{pmatrix} -1 \\ -\tfrac{1}{2} \\ 1 \\ 0 \end{pmatrix}\varepsilon' \tag{3.60}$$

or

$$Q_{out}a_{out} - Q_{in}a_{in} = -\varepsilon'$$

$$Q_{out}b_{out} - Q_{in}b_{in} = \frac{-\varepsilon'}{2}$$

$$Q_{out}c_{out} - Q_{in}c_{in} = \varepsilon'$$ (3.61)

$$Q_{out}d_{out} - Q_{in}d_{in} = 0$$

where d denotes the concentration of inerts (nitrogen and argon) which must be considered in a variable-volume problem. Substituting Equations 3.61 into Equation 3.59 gives

$$\varepsilon' = \frac{kV[Q_{in}a_{in} - \varepsilon'][Q_{in}b_{in} - (\varepsilon'/2)]}{Q_{out}^2\left[1 + k'\left(\dfrac{Q_{in}c_{in} + \varepsilon'}{Q_{out}}\right)\right]}$$ (3.62)

Since $Q_{out} \neq Q_{in}$, an equation of state is required. The ideal gas law can be expressed as

$$\frac{Q_{out}P_{out}}{R_gT_{out}} = Q_{out}[\rho_{molar}]_{out} = Q_{out}[a_{out} + b_{out} + c_{out} + d_{out}]$$ (3.63)

Equations 3.61 are summed to give

$$Q_{out}[a_{out} + b_{out} + c_{out} + d_{out}] = -\frac{\varepsilon'}{2} + Q_{in}[a_{in} + b_{in} + c_{in} + d_{in}]$$ (3.64)

Thus

$$\frac{Q_{out}P_{out}}{R_gT_{out}} = -\frac{\varepsilon'}{2} + Q_{in}[a_{in} + b_{in} + c_{in} + d_{in}] = -\frac{\varepsilon'}{2} + \frac{Q_{in}P_{in}}{R_gT_{in}}$$ (3.65)

which gives

$$Q_{out} = \frac{-\varepsilon'R_gT_{out}}{2P_{out}} + Q_{in}\frac{T_{out}P_{in}}{T_{in}P_{out}}$$ (3.66)

as the additional equation that must be solved along with Equation 3.62. In a reactor design problem, we would presumably know T_{out}, P_{out}, and all inlet conditions. Equations 3.62 and 3.66 would then be solved for ε' and Q_{out} from which the outlet composition could be calculated. For the analysis of kinetic data, however, it is assumed that we know both the inlet and outlet compositions. The object is to estimate values for k and k'. A minimum of two runs is required.

Example 3.8 Suppose the following compositions (in mole percent) have been measured.

	Inlet		Outlet	
	Run 1	Run 2	Run 1	Run 2
SO_2	10	5	4.1	2.0
O_2	10	10	7.2	8.6
SO_3	0	5	6.2	8.1
Inerts	80	80	82.5	81.3

The operating conditions were $T_{in} = T_{out} = 300°C$, $P_{in} = P_{out} = 1.0$ atm, $Q_{in} = 1.0$ l/s, $V = 0.5$ l. Determine values for the rate constants k and k'.

Solution The reaction coordinate ε' can be calculated from the change in molar flow rates of any of the components A, B, or C. Which one to choose involves questions of analysis accuracy, data consistency, and experimental design. These questions are briefly addressed in Problems 3.7 and 3.8. We gloss over these questions in what follows, and choose component C, which is SO_3. We seek to calculate ε' using the C component of Equation 3.61, but Q_{out} has not been measured. Instead, Equation 3.66 must be used. This gives a pair of equations in ε' and Q_{out}. Solution for ε' gives

$$\varepsilon' = \frac{Q_{in}c_{out}(T_{out}P_{in}/T_{in}P_{out}) - Q_{in}c_{in}}{1 + (R_gT_{out}c_{out}/2P_{out})}$$

$$= \frac{Q_{in}[c_{out}(T_{out}P_{in}/T_{in}P_{out}) - c_{in}]}{1 + [(y_C)_{out}/2]}$$

(3.67)

where $c/\rho_{molar} = cR_gT/P = y_C$ is the mole fraction of C. Note that $\rho_{molar} = 0.0213$ mol/l at 1 atm and 573 K. Since $T_{in} = T_{out}$ and $P_{in} = P_{out}$, all quantities on the right-hand side of Equation 3.67 are known. For Run 1, $c_{in} = 0$, $c_{out} = 0.00132$ mol/l, and $\varepsilon' = .00128$. Equation 3.65 then gives $Q_{out} = 0.970$ l/s. Since $a_{out} = 0.00087$ mol/l and $b_{out} = 0.00153$ mol/l are known, Equation 3.59 can be used to give the first of the following equations:

$$1 + 0.00132k' = 0.00052k$$

$$1 + 0.00172k' = 0.00062k$$

(3.68)

The second of the above equations is based on the Run 2 results. Simultaneous solution gives $k = 5260$ mol l^{-1} s^{-1} and $k' = 1320$ l mol^{-1}.

In practice, many different runs should be made varying a_{in}, b_{in}, and c_{in} over broad limits and particularly over the range of conditions expected in the full-scale reactor. This would provide a check on the model and would generate best estimates for k and k' through regression analysis. See Section 4.3 and Appendix 4.2.

3.3   Combinations of Reactors

We have considered two types of ideal flow reactor: piston flow and perfect mixing. These two ideal types can be connected together in a variety of series and parallel arrangements to give new reactor types that are generally intermediate in performance compared with the ideal reactors. Often, the connections are only conceptual. They are used to create mathematical models of real reactors that have been experimentally determined to be intermediate in performance compared with the ideal reactors. Less often, real reactors that approximate the ideal types are physically connected.

3.3.1   Series and Parallel Connections

When reactors are connected in series, the output from one serves as the input for the other. Symbolically we write

$$(a_{out})_1 = (a_{in})_2 \tag{3.69}$$

Example 3.9   Find the yield for a first order reaction in a composite reactor that consists of a perfect mixer followed by a piston flow reactor. Assume a mean residence time of $\bar{t}_1$ in the perfect mixer and of $\bar{t}_2$ in the piston flow element.

Solution   The exit concentration from the perfect mixer is

$$(a_{out})_1 = \frac{a_{in}}{1 + k\bar{t}_1} \tag{3.70}$$

and that for the piston flow element is

$$a_{out} = (a_{in})_2 e^{-k\bar{t}_2} \tag{3.71}$$

Using Equation 3.69 gives

$$a_{out} = \frac{a_{in} e^{-k\bar{t}_2}}{1 + k\bar{t}_1} \tag{3.72}$$

This result may be compared to that of single, ideal reactors having the same input concentration, throughput, and total volume. Specifically, we compare Equation 3.72 with the outlet concentration for a single perfect mixer having mean residence time

$$\bar{t} = \frac{V}{Q} = \frac{V_1 + V_2}{Q} = \bar{t}_1 + \bar{t}_2 \tag{3.73}$$

and with that of a piston flow reactor having this same $\bar{t}$. Since, for all $k > 0$, $\bar{t}_1 > 0$, $\bar{t}_2 > 0$,

$$\frac{1}{1 + k(\bar{t}_1 + \bar{t}_2)} > \frac{e^{-k\bar{t}_2}}{1 + k\bar{t}_1} > e^{-k(\bar{t}_1 + \bar{t}_2)} \tag{3.74}$$

the combination reactor gives intermediate performance. The outlet concentration of reactant will be higher than that from a single piston flow reactor but lower than that from a single perfect mixer.

For two reactors in parallel, the output streams must be summed to find the average concentration:

$$\boxed{a_{\text{out}} = \frac{Q_1(a_{\text{out}})_1 + Q_2(a_{\text{out}})_2}{Q_1 + Q_2}} \tag{3.75}$$

Example 3.10 Find the conversion for a composite system that consists of a perfect mixer and a piston flow reactor in parallel.

Solution Using Equation 3.75,

$$a_{\text{out}} = \frac{a_{\text{in}}}{Q_1 + Q_2}\left(\frac{Q_1}{1 + k\bar{t}_1} + Q_2 e^{-k\bar{t}_2}\right) \tag{3.76}$$

The parallel reactor system has one degree of freedom more than the series system. Given fixed a_{in}, V, and Q, it is possible to vary both V_1/V and Q_1/Q for the parallel system. Even so, performance advantages are rare to nonexistent compared to a single reactor having the same V and Q, provided the single reactor remains isothermal. When significant amounts of heat must be transferred to or from the reactants, identical reactors in parallel are often used.

3.3.2 Tanks in Series

For the great majority of reaction schemes, piston flow is optimal. Thus the reactor designer normally wants to build a tubular reactor and to operate it at high Reynolds numbers so that piston flow is closely approximated. This may not be possible, however. There are a variety of situations where a tubular reactor is infeasible and where continuous-flow, stirred tank reactors must be used instead. Typical examples are reactions involving suspended solids and autorefrigerated reactors where the reaction mass is held at its boiling point. There will usually be a yield advantage from using several of these reactors in series rather than a single, large reactor.[2]

Example 3.11 Determine the fraction unreacted for two equal-volume perfect mixers in series with the second order reaction

$$2A \xrightarrow{k} B$$

Suppose $k\bar{t}_1 a_{in} = 0.5$ where $\bar{t}_1 = V_1/Q$ is the mean residence time in a single vessel. Compare this result with that obtainable in a single perfect mixer having the same total volume as the series combination, $V = 2V_1$. Assume constant mass density.

Solution Begin by considering the first perfect mixer. The rate of formation of A is

$$\mathscr{R}_A = -2ka^2 \tag{3.77}$$

For constant ρ, $Q_{in} = Q_{out} = Q$ and Equation 3.29 gives

$$a_{in} - 2k\bar{t}_1 a_{out}^2 = a_{out} \tag{3.78}$$

The solution is

$$\frac{a_{out}}{a_{in}} = \frac{-1 + \sqrt{1 + 8k\bar{t}_1 a_{in}}}{4k\bar{t}_1 a_{in}} \tag{3.79}$$

Applying this to the first perfect mixer having $k\bar{t}_1 a_{in} = 0.5$ gives

$$(a_{out})_1 = (a_{in})_2 = 0.618 a_{in} \tag{3.80}$$

[2]Unfortunately, the capital cost of the series combination will usually exceed that for the single large reactor. See Problem 3.10.

For the second perfect mixer, $\bar{t}_1 = \bar{t}_2$ but $(a_{in})_1 \neq (a_{in})_2$. Using $k\bar{t}_2(a_{in})_2 = (0.618)(0.5) = 0.309$ gives

$$a_{out} = (a_{out})_2 = 0.698(a_{in})_2 = 0.432 a_{in} \tag{3.81}$$

Thus $a_{out}/a_{in} = 0.432$ for the series combination. For the single reactor, $t = 2\bar{t}_1$ and $k\bar{t}a_{in} = 1.0$. Thus $a_{out}/a_{in} = 0.500$.

A series collection of N equal-volume perfect mixers will closely approximate a piston flow reactor when N is large. For a first order reaction

$$(a_{out})_n = \frac{(a_{in})_n}{1 + (k\bar{t}/N)} \tag{3.82}$$

for each mixer in series, $n = 1, 2, \ldots N$. Applying Equation 3.67 repeatedly gives

$$\frac{a_{out}}{a_{in}} = \left(1 + \frac{k\bar{t}}{N}\right)^{-N} \tag{3.83}$$

as the outlet concentration from the N-tank configuration. In the limit of many tanks in series,

$$\lim_{N \to \infty} \frac{a_{out}}{a_{in}} = e^{-k\bar{t}} \tag{3.84}$$

which is the same as a piston flow reactor with mean residence time $\bar{t}$. In practice, good improvements in yield are possible for fairly small N. The results shown in Table 3.1 give the fraction unreacted for a first order reaction.

A more dramatic comparison is the system volume required to obtain a given (high) conversion using N tanks in series compared to a piston flow reactor, as shown in Table 3.2. The consequence of these results is that high conversions, which are difficult to achieve in a single stirred tank, may suddenly become quite feasible using only two tanks in series.

Table 3.1   Fraction Unreacted for Tanks-in-Series Model

N	$k\bar{t} = 1$	$k\bar{t} = 3$	$k\bar{t} = 5$
1	0.500	0.333	0.200
2	0.444	0.160	0.082
3	0.422	0.125	0.053
5	0.402	0.095	0.031
10	0.386	0.073	0.017
∞	0.368	0.050	0.007

Table 3.2 Relative Total Reactor Volume Needed for
Tanks in Series
Compared to Piston Flow

	a_{out}/a_{in}		
N	0.1	0.01	0.001
1	3.9	21.5	144.6
2	1.9	3.9	8.9
3	1.5	2.4	3.9
5	1.3	1.6	2.2
10	1.1	1.3	1.4
∞	1.0	1.0	1.0

Example 3.12   The concentration of a toxic substance must be reduced by a factor of 1000. Assuming the substance decomposes with first order kinetics, compare the relative effectiveness of using one versus two stirred tanks in series.

Solution   A piston flow reactor requires $k\bar{t} = 6.9$ for this reaction. That is, $\exp(-6.9) = a_{out}/a_{in} = 0.001$. If a perfect mixer must be used, $k\bar{t} = 999$ for the same extent of reaction. Assuming operation with the same value of k and the same flow rate, a single perfect mixer requires 144.6 times the volume of the piston flow reactor. This drops to a factor of 8.9 when two tanks in series are used rather than one. Thus the savings in going to two tanks is a factor of $144.6/8.9 = 16$ reduction in volume. The required $k\bar{t}$ for the two tanks in series is 61.

3.3.3   Recycle Systems

Figure 3.2 shows a recycle system where some of the exiting reaction mass is returned to the inlet *without any separation or recovery steps*. This form of recycle exists within every stirred tank reactor and in this sense is quite common. An explicit, external recycle loop as shown in Figure 3.2 is rather uncommon. External recycle applied to a perfect mixer has no effect on performance. External recycle applied to a piston flow reactor usually degrades the performance by approaching that of a perfect mixer.

A material balance about the mixing point gives

$$a_{mix} = \frac{Q_{in}a_{in} + qa_{out}}{Q_{in} + q} \tag{3.85}$$

We next need a result for a_{out} from the reactor given an input concentration of

a_{mix} at a volumetric flow rate of $Q_{in} + q$. In the general case, this single-pass solution must be obtained numerically. Then the overall solution is iterative. One guesses an a_{mix}, solves numerically for a_{out}, and then uses Equation 3.85 to calculate a_{mix} for comparison with the original guess. The binary search technique of Section 3.2.2 may be used. The function to be zeroed is

$$a_{mix} - \frac{Q_{in}a_{in} + qa_{out}}{Q_{in} + q} = 0$$

(3.86)

where a_{out} denotes the solution of the single-pass problem. When a_{out} is known analytically, an analytical solution to the recycle reactor problem is usually possible.

Example 3.13   Determine a_{mix} and a_{out} for first order kinetics with a piston flow reactor in a recycle system.

Solution   For piston flow,

$$a_{out} = a_{mix} \exp\left(\frac{-kV}{Q_{in} + q}\right)$$

(3.87)

Substitution in Equation 3.85 and solution for a_{mix} gives

$$a_{mix} = \frac{Q_{in}a_{in}}{Q_{in} + q - q \exp\left[-kV/(Q_{in} + q)\right]}$$

(3.88)

and the solution for a_{out} is

$$a_{out} = \frac{Q_{in}a_{in} \exp\left[-kV/(Q_{in} + q)\right]}{Q_{in} + q - q \exp\left[-kV/(Q_{in} + q)\right]}$$

(3.89)

More common industrially is the situation of recycle **after** a product separation and recovery step. Any unreacted feedstocks may be separated and recycled to (ultimate) extinction.

Suggestions for Further Reading

Reactor models consisting of series and parallel combinations of ideal reactors are discussed at length in:

O. Levenspiel, *Chemical Reaction Engineering*, Wiley, New York, 1972.

Realistic examples of variable-property piston flow models, usually nonisothermal, are given in:

G. F. Froment, and K. B. Bischoff, *Chemical Reactor Analysis and Design*, Wiley, New York, 1979.

A mathematically rigorous treatment using the reaction coordinate methods of Chapters 2 and 3 is given in:

R. Aris, *Introduction to the Analysis of Chemical Reactors*, Prentice-Hall, Englewood Cliffs, NJ, 1965.

Problems

3.1 Repeat the ethylbenzene dehydration example of Section 3.1.2, assuming now an 8-to-1 molar ratio of steam to ethylbenzene at the reactor inlet and a void fraction of 0.40.

3.2 Suppose the following reaction network is occurring in a perfect mixer:

$$A \rightleftharpoons B \qquad \mathscr{R} = k_I a^{1/2} - k_I' b$$

$$B \rightarrow \tfrac{1}{2}C \qquad \mathscr{R} = k_{II} b^2$$

$$B + D \rightarrow E \qquad \mathscr{R} = k_{III} bd$$

Determine the composition of the outlet stream given $k_I = 3 \times 10^{-2}$ mol$^{1/2}$ m$^{-3/2}$ s^{-1}, $k_I' = 0.4$ s^{-1}, $k_{II} = 5 \times 10^{-4}$ mol^{-1} m^3 s^{-1}, $k_{III} = 3 \times 10^{-4}$ mol^{-1} m^3 s^{-1}, $a_0 = 3$ mol/m^3, $d_0 = 3$ mol/m^3, $b_0 = c_0 = e_0 = 0$, $\bar{t} = 1$ s.

3.3 Observed kinetics for the reaction

$$A + B \rightarrow 2C$$

are $\mathscr{R} = 0.43ab^{0.8}$, mol m^{-3} hr^{-1}. Suppose the reaction is run in a constant-density perfect mixer with $a_{in} = 15$ mol/m^3, $b_{in} = 20$ mol/m^3, $V = 3.5$ m^3, and $Q = 125$ m^3/hr. Determine the exit concentration of C.

3.4 Suppose a new catalyst is available for the oxidation of *o*-xylene to phthalic anhydride. You plan to evaluate the catalyst using the recycle reactor of Figure 3.2. Early runs show appreciable quantities of CO_2 and H_2O in addition to phthalic anhydride and unreacted xylene. You postulate the following reactions:

$$C_6H_4(CH_3)_2 + 3O_2 \rightarrow C_6H_4(CO)_2O + 3H_2O \qquad \text{(I)}$$

$$C_6H_4(CO)_2O + \tfrac{15}{2}O_2 \rightarrow 8CO_2 + 2H_2O \qquad \text{(II)}$$

(a) Formulate this problem in reaction coordinate form. Show any vectors or matrices explicitly.
(b) Show the two algebraic equations, containing only ε_I' and ε_{II}' as unknowns, that govern the exit composition from the reactor.

3.5 The low-temperature oxidation of hydrogen as in the cap of a lead–acid storage battery is an example of heterogeneous catalysis. It is proposed to model this reaction as if it were homogeneous:

$$H_2 + \tfrac{1}{2}O_2 \rightarrow H_2O \qquad \mathscr{R} = k[H_2][O_2] \qquad \text{(nonelementary)}$$

and to treat the cap as if it were a perfect mixer. The following data have been generated on a test rig:

$$T_{in} = 22°C \qquad T_{out} = 25°C$$
$$P_{in} = 2 \text{ atm} \qquad P_{out} = 1 \text{ atm}$$
$$H_2 \text{ in} = 2 \text{ gm/hr}$$
$$O_2 \text{ in} = 32 \text{ gm/hr} \quad (\tfrac{2}{1} \text{ excess})$$
$$N_2 \text{ in} = 160 \text{ gm/hr}$$
$$H_2O \text{ out} = 16 \text{ gm/hr}$$

(a) Determine k given $V = 25 \text{ cm}^3$.

(b) Calculate the adiabatic temperature rise for the observed extent of reaction. Is the measured rise reasonable? The test rig is exposed to natural convection. The room air is at 22°C.

3.6 Consider the prototypical autocatalytic reaction of Section 2.3.4 with $ka_0\bar{t} = kb_0\bar{t} = 1$. It has been proposed to run this reaction using a series combination of a perfect mixer followed by a piston flow reactor. How should $\bar{t}$ be divided between the reactors to maximize the yield of B?

3.7 Unless it springs a leak, the recycle reactor of Figure 3.2 should give a good overall mass balance and individual component balances that agree with the reaction stoichiometry.

(a) Is this true for the data of Run 1 in Example 3.8? What accuracy in the analytical results is suggested by the material balances? Is it better or worse than you would expect given an analytical precision of 0.1 mole-%?

(b) Determine k and k' as in the example, but now calculate ε' based on the consumption of component A (SO_2) rather than the formation of C (SO_3). Draw appropriate conclusions.

(c) Suppose a revised compositional analysis for Run I gave $(y_C)_{out} = 0.063$ rather than the original value of 0.062. Repeat the example calculation of k and k' using this new value but no other changes. Draw appropriate conclusions.

3.8 Suppose a repeat of Run 2 was made in the SO_2 oxidation example of Section 3.2.3. This time the gas analysis at the outlet was

SO_2	2.2%
O_2	8.7%
SO_3	7.9%
Inerts	81.2%

Find k and k'. Draw some conclusions.

3.9 (a) Annular flow reactors, such as that illustrated in Figure 3.3, are sometimes used for adiabatic, solid-catalyzed reactions where pressure drop must be minimized. Repeat the ethylbenzene example of Section 3.1.2 for an annular flow reactor of inner radius 0.1 m and an outer radius of 1.1 m.

(b) Do your calculations show an advantage for annular flow? Why or why not? Note that the actual dehydrogenation reaction is reversible.

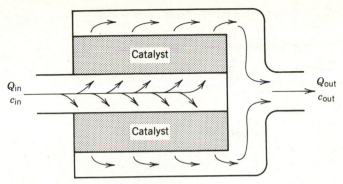

Figure 3.3 Annular packed-bed reactor.

3.10 Suppose the installed cost of stirred tank reactors varies as $V^{0.75}$. Determine the optimum number of tanks in series for a first order reaction going to 99% completion.

3.11 Find the limit of Equations 3.88 and 3.89 as $q \to \infty$ with Q_{in} fixed. Why would you expect this result?

3.12 A transpired wall reactor has porous walls through which a fluid can flow. Pumping a gas inward toward the reaction zone can cool the walls and protect them from corrosion in highly energetic reactions such as direct fluorinations. Outward flow is typical of membrane reactors. These may have microporous walls or walls permeable by diffusion. Permeability is usually selective so that separation occurs simultaneously with reaction. Modify Equation 3.17 to account for the inward or outward flow of one or more components. Assume the transwall flow rates to be independent of z, for example, that the pressure drop through the wall is large compared to that down the tube.

3.13 Current EPA guidelines require less than 20 ppm of BOD (biological oxygen demand) for process water to be discharged into a river. The untreated effluent from a certain plant contains 500 ppm of BOD. Pilot-scale results have shown it feasible to reduce the BOD to 80 ppm in an activated sludge aeration pond. (In an activated sludge process, biomass is separated in a clarifier and recycled to the pond. When aerobic fermentations are attempted without biomass recycle, greater retention times are needed for cell growth. See Section 10.4.)

(a) Suggest one or more strategies for meeting the EPA guidelines.

(b) Outline a program to confirm feasibility and to choose between alternative strategies.

3.14 The ethylbenzene-to-styrene dehydrogenation reaction of Example 3.1 is reversible with equilibrium constant

$$K_{equil} = \frac{P_{STY} P_{H_2}}{P_{EB}} = 0.61 \text{ atm}$$

at 700°C.

(a) Modify the marching-ahead scheme in Example 3.1 to account for this reversibility.

(b) Perform the calculations including extrapolation to $\Delta z \to 0$.

3.15 The oxidation of 2-butene over a vanadium phosphate catalyst is postulated[3] to involve the following reaction steps when conducted at 750 K and near-atmospheric pressure:

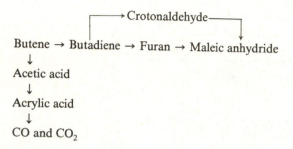

Experiments were conducted in a CSTR. By-product water was not measured directly but can be inferred from a hydrogen balance.

(a) Write stoichiometrically balanced equations for the above sequence. Use a variable stoichiometric coefficient to reflect the split between CO and CO_2.

(b) Write the algebraic equations governing steady state performance of the CSTR assuming constant physical properties.

Appendix 3.1

Solution of Simultaneous Algebraic Equations

Consider a set of N nonlinear equations of the form

$$F(a, b, \ldots) = 0$$

$$G(a, b, \ldots) = 0$$

(3.90)

$$\vdots$$

where $a, b, \ldots$ represent the N unknowns. We suppose none of these equations is explicitly solvable for any of the unknowns. If an original equation were solvable for an unknown, that unknown would be eliminated and the dimensionality of the set reduced by one (see Section 3.2.2).

---

[3] E. M. Breckner, S. Sundaresan, and J. B. Benziger, Paper presented at 1985 AIChE Annual Meeting, Chicago.

Consider some point $(a_0, b_0, \ldots)$ within the region of definition for the functions $F, G, \ldots$ and suppose the functions can be represented by an N-dimensional Taylor series about the point. Truncating this series after the first order derivatives gives

$$F(a, b, \ldots) = F(a_0, b_0, \ldots) + \frac{\partial F}{\partial a}\bigg|_0 (a - a_0) + \frac{\partial F}{\partial b}\bigg|_0 (b - b_0) + \cdots$$

$$(3.91)$$

$$G(a, b, \ldots) = G(a_0, b_0, \ldots) + \frac{\partial G}{\partial a}\bigg|_0 (a - a_0) + \frac{\partial G}{\partial b}\bigg|_0 (b - b_0) + \cdots$$

$$\vdots \qquad\qquad \vdots$$

This can be rewritten in matrix form as

$$\begin{bmatrix} \dfrac{\partial F}{\partial a}\bigg|_0 & \dfrac{\partial F}{\partial b}\bigg|_0 & \cdots \\ \dfrac{\partial G}{\partial a}\bigg|_0 & \dfrac{\partial G}{\partial b}\bigg|_0 & \cdots \\ \vdots & & \end{bmatrix} \begin{bmatrix} a - a_0 \\ b - b_0 \\ \vdots \end{bmatrix} = \begin{bmatrix} F - F_0 \\ G - G_0 \\ \vdots \end{bmatrix} \qquad (3.92)$$

We seek values for $a, b, \ldots$ which give $F = G = \cdots = 0$. Setting $F = G = \cdots = 0$ and solving for $a, b, \ldots$ gives

$$\begin{bmatrix} a \\ b \\ \vdots \end{bmatrix} = \begin{bmatrix} a_0 \\ b_0 \\ \vdots \end{bmatrix} - \begin{bmatrix} \dfrac{\partial F}{\partial a}\bigg|_0 & \dfrac{\partial F}{\partial b}\bigg|_0 & \cdots \\ \dfrac{\partial G}{\partial a}\bigg|_0 & \dfrac{\partial G}{\partial b}\bigg|_0 & \cdots \\ \vdots & & \end{bmatrix}^{-1} \begin{bmatrix} F_0 \\ G_0 \\ \vdots \end{bmatrix} \qquad (3.93)$$

For the special case of $N = 1$,

$$a = a_0 - \frac{F_0}{\dfrac{dF}{da}\bigg|_0} \qquad (3.94)$$

which is seen to be Newton's method for finding the roots of an equation. For

$N = 2$,

$$a = a_0 - \frac{F_0 \left.\frac{\partial G}{\partial \alpha}\right|_0 - G_0 \left.\frac{\partial F}{\partial a}\right|_0}{\left.\frac{\partial F}{\partial a}\right|_0 \left.\frac{\partial G}{\partial b}\right|_0 - \left.\frac{\partial F}{\partial b}\right|_0 \left.\frac{\partial G}{\partial a}\right|_0}$$

$$(3.95)$$

$$b = b_0 - \frac{-F_0 \left.\frac{\partial G}{\partial b}\right|_0 + G_0 \left.\frac{\partial F}{\partial b}\right|_0}{\left.\frac{\partial F}{\partial a}\right|_0 \left.\frac{\partial G}{\partial b}\right|_0 - \left.\frac{\partial F}{\partial b}\right|_0 \left.\frac{\partial G}{\partial a}\right|_0}$$

which is a two-dimensional generalization of Newton's method.

The above technique, or variants on it, is used to solve quite large sets of algebraic equations; but, like the ordinary one-dimensional form of Newton's method, the algorithm may diverge unless the initial guess $(a_0, b_0, \ldots)$ is quite close to the final solution. Thus it might be considered as a method for rapidly improving a good initial guess, with other techniques being necessary to obtain the initial guess.

For the one-dimensional case, dF/da can usually be estimated using values of F determined at previous guesses. Thus

$$a = a_0 - \frac{F_0}{(F_0 - F_{-1})/(a_0 - a_{-1})}$$

$$(3.96)$$

where $F_0 = F(a_0)$ is the value of F obtained one iteration ago when the guess was a_0 and $F_{-1} = F(a_{-1})$ is the value obtained two iterations ago when the guess was a_{-1}.

For two- and higher-dimensional solutions, it is probably best to estimate the first partial derivatives by a formula such as

$$\left.\frac{\partial F}{\partial a}\right|_0 \approx \frac{F(a_0, b_0, \ldots) - F(\gamma a_0, b_0, \ldots)}{a_0 - \gamma a_0}$$

$$(3.97)$$

where γ is a constant close to 1.0.

4

Thermal Effects and Energy Balances

4.1 Temperature Dependence of Reaction Rates

Most reaction rates are sensitive to temperature, and few if any laboratory studies fail to explore temperature as a means of improving reaction yields or selectivities. Our treatment has so far ignored this point. The reactors have been isothermal, and kinetic data were somehow available for exactly the temperature we chose to use. In practice, however, temperature effects should be considered even for isothermal reactor design since the operating temperature must be specified as part of the design. For nonisothermal reactors, where the temperature varies from point to point within the system, the temperature dependence of the reaction rate directly enters the design calculations.

4.1.1 Arrhenius and Non-Arrhenius Behavior

The rate constant for elementary reactions is almost always expressed as

$$k = k_0 T^m \exp\left(-\frac{E}{R_g T}\right) \tag{4.1}$$

where $m = 0, \frac{1}{2}, 1$, depending on the specific theoretical model being used. The quantity E is usually called the activation energy although the specific theories interpret this energy term in different ways. The case of $m = 0$ corresponds to classical Arrhenius theory; $m = \frac{1}{2}$ is derived from the collision theory of bimolec-

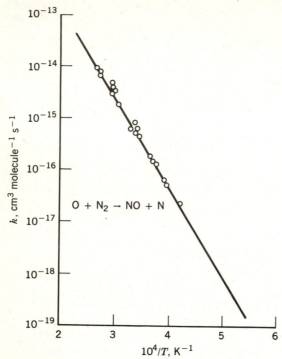

Figure 4.1 Example of normal Arrhenius behavior in a reaction of very large endothermicity. [J. P. Monat, R. K. Hanson, C. H. Kruger, *Seventeenth Symposium* (*International*) *on Combustion*, The Combustion Institute, Pittsburgh, 1979, p. 543.]

ular gas phase reactions; and $m = 1$ corresponds to activated complex or transition state theory. None of these theories is sufficiently well developed to predict reaction rates from first principles, and it is practically impossible to choose between them based on experimental measurements. The relatively small variation in rate constant due to the pre-exponential temperature dependence T^m is overwhelmed by the exponential dependence $\exp(-E/R_g T)$. For many reactions, a plot of $\ln(k)$ versus T^{-1} will be approximately linear, and the slope of this line can be used to calculate E. (More generally, plots of $\ln(k/T^m)$ will be approximately linear, which suggests the futility of determining m by this approach.) Figure 4.1 shows an Arrhenius plot for the reaction $O + N_2 \rightarrow NO + N$, which is linear over a temperature range of 2000 K. Note that the rate constant is expressed per molecule rather than per mole. This method for expressing k is favored by many chemical kineticists. It differs by a factor of Avagodro's number from the more usual k.

Not all reactions show the linear Arrhenius behavior of Figure 4.1. Complex reactions will have a different value of E for each elementary step, and the composite temperature behavior may be quite different than Arrhenius depen-

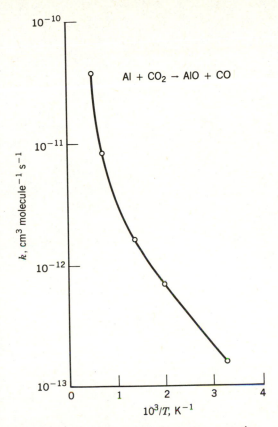

Figure 4.2 Apparently elementary reaction having pronounced curvature in Arrhenius plot. [A. Fontijn and W. Felder, *J. Chem. Phys.*, **67**, 1561 (1977).]

dence. A "normal" Arrhenius plot is neither necessary nor sufficient to prove that a reaction is elementary. It is not sufficient because complex reactions may have one dominant activation energy or several steps with similar activation energies which lead to an overall temperature dependence of the Arrhenius sort. It is not necessary since some low-pressure, gas phase, bimolecular reactions exhibit distinctly non-Arrhenius behavior even though the reactions are believed to be elementary. Figure 4.2 illustrates the gas phase reaction of aluminum vapor with carbon dioxide, where the Arrhenius plot has pronounced curvature; Figure 4.3 shows a similar reaction, aluminum vapor with oxygen, where the activation energy is apparently zero; and Figure 4.4 illustrates the reaction of fluorine atoms with hydrogen bromide, where the activation energy changes sign.

The behavior illustrated in Figures 4.2 through 4.4 shows that the activation energy must be regarded as a function of temperature, at least when large temperature ranges are considered. Transition state theory moves a step in the

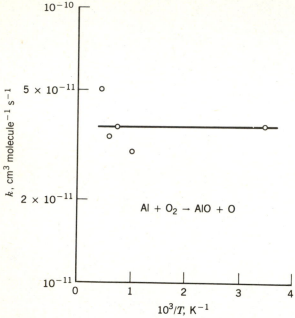

Figure 4.3 Apparently elementary reaction having an activation energy near zero. [A. Fontijn, W. Felder, and J. J. Houghton, *Sixteenth Symposium (International) on Combustion*, The Combustion Institute, Pittsburgh, 1977, p. 871.]

direction of allowing this temperature dependence in a logical way. Transition state theory gives $m = 1$ and interprets E as a free energy of activation:

$$E \approx \Delta G^{\#} = \Delta H^{\#} - T\Delta S^{\#}$$

(4.2)

Thus

$$k = k_0 T \exp\left(\frac{\Delta S^{\#}}{R_g}\right) \exp\left(\frac{-\Delta H^{\#}}{R_g T}\right)$$

(4.3)

Both $\Delta S^{\#}$ and $\Delta H^{\#}$ can be functions of temperature. Rate behavior such as that in Figures 4.2 through 4.4 can be explained by postulating different temperature dependences for $\Delta S^{\#}$ and $\Delta H^{\#}$. The theory also suggests that highly endothermic reactions—those with large, positive $\Delta H_{\mathscr{R}}$—will show normal Arrhenius behavior. This is borne out by the reaction of Figure 4.1, which has $\Delta H_{\mathscr{R}} = 315$ kJ/mol. In such endothermic reactions, $\Delta H^{\#} \approx \Delta H_{\mathscr{R}}$ is large compared with $T\Delta S^{\#}$. Further, $\Delta H_{\mathscr{R}}$ tends to be a weak function of temperature since the products and reactants will have similar heat capacities, $\Delta C_p \approx 0$. See Appendix 4.1.

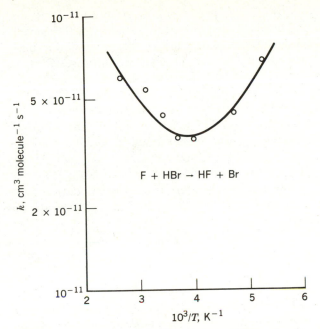

Figure 4.4 Apparently elementary reaction showing sign reversal in activation energy. [E. Wurzburg and P. L. Houston, *J. Chem. Phys.*, **72**, 1 (1980).]

Having made a partial case for transition state theory and thus for $m = 1$, we will revert back to simple Arrhenius theory with $m = 0$. This is the case usually used when fitting experimental rate data. It provides the same accuracy and has the merit of simplicity.

4.1.2 Optimal Temperatures for Isothermal Reactors

The behavior shown in Figures 4.3 and 4.4 is highly unusual. Reaction rates almost always increase with temperature. Thus the best temperature for an irreversible reaction, whether elementary or complex, is the highest possible temperature. Practical reactor designs must consider materials of construction limitations and economic trade-offs between heating costs and yield, but there is no optimal temperature from a strictly kinetic viewpoint.

Multiple reactions, and reversible reactions since these are a special form of multiple reaction, usually do exhibit an optimal temperature with respect to the yield of a desired product. The reaction energetics are not trivial even if the reactor is approximately isothermal. One must still find the best isotherm at which to operate. Consider the elementary, reversible reaction

$$A \underset{k_r}{\overset{k_f}{\rightleftharpoons}} P. \tag{4.4}$$

Suppose this reaction is occurring in a perfect mixer of fixed volume and throughput. It is desired to find the reaction temperature that maximizes the yield of product P. Before going through the necessary arithmetic, it is useful to make several observations. First, suppose $E_f > E_r$ as is normally the case when the forward reaction is endothermic. Then the forward reaction is favored by increasing temperature, and the best temperature is the highest possible one. For $E_f < E_r$, the equilibrium is shifted to the product by decreasing temperature, but the forward reaction is also retarded. Low temperatures improve the selectivity but lower the conversion. The lowest possible temperature is favored to maximize the moles of P formed per mole of A reacted, $p_{out}/(a_{in} - a_{out})$, but low temperatures will not maximize the total output of product as measured by p_{out}. For this maximization, an intermediate temperature is required, and an interior optimum exists. The outlet concentration from the stirred tank, assuming constant physical properties, is given by

$$p_{out} = \frac{k_f \bar{t} a_{in}}{1 + k_f \bar{t} + k_r \bar{t}} \tag{4.5}$$

where we have assumed $p_{in} = 0$. We assume $E_f < E_r$ and seek to maximize p_{out}. Both the forward and reverse reactions will be assumed to have the simple Arrhenius temperature dependence. Setting $dp_{out}/dT = 0$ gives

$$T_{optimal} = \frac{-E_r}{R_g \ln\left[\dfrac{E_f}{(E_r - E_f)(k_0)_r \bar{t}}\right]} \tag{4.6}$$

as the kinetically determined optimal temperature.

The same approach can be used to find the optimal temperature for a batch (or piston flow) reactor. The algebra becomes difficult, however, and one is left with a transcendental equation in $T_{optimal}$. This equation could be solved numerically, but it is easier to perform the entire optimization numerically. The binary search technique of Section 3.2.2 can be modified for use on optimization problems. We give it here as a brute force technique for optimizing a function of a single variable. The technique can obviously be improved, and the reader is free to make these improvements. The improvements will decrease the number of function evaluations needed to find the optimum, but at the expense of increased programming logic.

For concreteness, suppose we are attempting to maximize a function $f(x)$ and that a single maximum lies within the interval $a < x < b$. Evaluate $f(a)$, $f(b)$, and $f[(a + b)/2]$. Is one of the endpoints the highest of the three functional values? If so, discard the other endpoint and repeat the algorithm. If not, the midpoint is the maximum. Evaluate $f[(a + b)/4]$. Is this the highest value obtained so far? If so, the new range is $a < x < (a + b)/2$. Repeat the algorithm. If not, discard the point $x = a$. The new range is $(a + b)/4 < x < b$. Repeat the algorithm.

Example 4.1 Suppose $k_f = 10^8 e^{-5000/T}$ s^{-1} and $k_r = 10^{16} e^{-10,000/T}$ s^{-1}, where T is in degrees K. Find the value of T that maximizes the concentration of P for the reaction of Equation 4.4 conducted in a batch reactor with a 1s holding time. Assume $p_0 = 0$.

Solution For a batch reactor,

$$p_t = \frac{a_0 k_f [1 - e^{-(k_f + k_r)t}]}{k_f + k_r} \tag{4.7}$$

which is the function to be maximized. Since a_0 is a constant, it is sufficient to maximize $p_t/a_0 = f(T)$. The binary search algorithm described above will produce a result. However, trial and error calculations usually do as well. Following is a plausible trial-and-error sequence for finding the maximum:

T (K)	p_t/a_0
300	0.1475
400	0.0027
200	0.0014
350	0.0158
250	0.1825
275	0.4159
265	0.3961
270	0.4314
271	0.4325
272	0.4315

showing that the yield of product is maximized at T = 271 K.

Example 4.2 Suppose

$$A \xrightarrow{k_I} B \xrightarrow{k_{II}} C \tag{4.8}$$

with $k_I = 10^{15} \exp(-12,000/T)$ hr^{-1} and $k_{II} = 10^{16} \exp(-15,000/T)$ hr^{-1}. Find the temperature that maximizes b_{out} for a perfect mixer with $\bar{t} = 1$ hr.

Solution

$$b_{out} = \frac{a_{in} k_I \bar{t}}{(1 + k_I \bar{t})(1 + k_{II} \bar{t})} \tag{4.9}$$

so that b_{out}/a_{in} is to be maximized. A trial-and-error search gives $(b_{out})_{max}/a_{in} = 0.8907$ at $T = 377$ K.

This last problem introduces the concept of optimal temperatures for consecutive reactions. For a given $\bar{t}$, an intermediate optimal temperature always exists for the intermediate product. To see this, suppose T is very low. Then $a_{out} \approx a_{in}$ and $b_{out} \approx 0$. If T is very high, $a_{out} \approx 0$, $b_{out} \approx 0$, and $c_{out} \approx a_{in}$. The maximum for b_{out} must be at an intermediate temperature regardless of the relative magnitudes of E_I and E_{II}.

The competitive reactions

$$A \overset{k_I}{\rightarrow} P$$

$$A \overset{k_{II}}{\rightarrow} Q \tag{4.10}$$

will have an intermediate optimum for P only if $E_I < E_{II}$ and will have an intermediate optimum for Q only if $E_I > E_{II}$. Otherwise, the yield of the desired product is maximized at high temperatures: if $E_I > E_{II}$, high temperatures maximize the yield of P; if $E_I < E_{II}$, high temperatures maximize the yield of Q.

4.2 The Overall Heat Balance

Figure 4.5 illustrates the heat balance[1] for a steady state, flow reactor. In words,

Enthalpy in + Heat generated by reaction

$\qquad$ = Enthalpy out + Heat transferred out $\tag{4.11}$

and in mathematics,

$$Q_{in}\rho_{in}H_{in} - \Delta H_{\mathscr{R}}\hat{\mathscr{R}}V = Q_{out}\rho_{out}H_{out} + \hat{U}A_{ext}(\hat{T} - T_{ext}) \tag{4.12}$$

This balance is written for the system, including all components. The reaction rate $\hat{\mathscr{R}}$ is written "for the reaction" and is thus positive.

By thermodynamic convention, $\Delta H_{\mathscr{R}} < 0$ for exothermic reactions, so that a negative sign is attached to the heat generation term. Alternatively, one may write Equation 4.11 as

Enthalpy in = Enthalpy out + Heat absorbed by reaction

$\qquad$ + Heat transferred to the environment $\tag{4.13}$

so that the $\Delta H_{\mathscr{R}}\hat{\mathscr{R}}V$ term is shifted to the right side of Equation 4.12 and appears with a positive sign:

$$Q_{in}\rho_{in}H_{in} = Q_{out}\rho_{out}H_{out} + \Delta H_{\mathscr{R}}\hat{\mathscr{R}}V + \hat{U}A_{ext}(\hat{T} - T_{ext}) \tag{4.14}$$

---

[1] This form of energy balance is suitable for most problems in chemical reactor design. A more general case is considered in Problem 8.10.

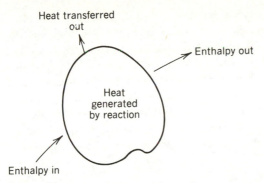

Figure 4.5   General heat balance for a reactor control volume.

Note that $\Delta H_{\mathscr{R}}\mathscr{R}$ represents an implicit summation over all reactions that may be occurring:

$$\Delta H_{\mathscr{R}}\mathscr{R} = \sum_{\text{Reactions}} (\Delta H_{\mathscr{R}})_{\text{I}}\mathscr{R}_{\text{I}} \qquad (4.15)$$

It is often useful to approximate the enthalpy terms as $H_{\text{in}} \approx \bar{C}_p T_{\text{in}}$ and $H_{\text{out}} \approx \bar{C}_p T_{\text{out}}$. Nominally, this chooses 0 K as the reference temperature, but $\bar{C}_p$ should really be the average specific heat over the range $(T_{\text{in}}, T_{\text{out}})$. As in the general mass balance, $\hat{\mathscr{R}}$ is the reaction rate averaged over the entire system. Similarly, $\hat{T}$ is the spatial average temperature:

$$\hat{T} = \frac{1}{V} \int\int\int_V T \, dV \qquad (4.16)$$

There may be a variety of ways in which heat is exchanged with the environment. For example, power input from an agitator is sometimes a significant source of heat. Symbolically, we have lumped all heat exchange into the term $\bar{U}A_{\text{ext}}(\hat{T} - T_{\text{ext}})$, where $\hat{U}A_{\text{ext}}$ is the product of the average heat transfer coefficient and the external surface area of the system. In practice, this formulation is directly useful only when the system is a stirred tank reactor. A more mechanistic approach to heat exchange with the environment is needed for other reactor configurations. When heat transfer *within* the system is important, the overall energy balance remains true but is not especially useful. Instead, a two- or three-dimensional, differential balance is needed. This will give rise to a partial differential equation governing the transfer of heat between points inside the system. Such equations are considered in Chapter 6. Here in Chapter 4, attention is restricted to the overall energy balance, which gives rise to an algebraic equation, or to a one-dimensional, differential balance, which gives rise to an ordinary differential equation. The overall balance is suitable for stirred tank reactors, but the differential balance is used for piston flow or batch reactors. This is exactly analogous to the mass balance equations for stirred tank and piston flow reactors as studied in Chapter 3. Steady state lumped systems are governed by algebraic equations. Steady state distributed systems, where concentrations and temperature vary spatially, are governed by ordinary or partial differential equations depending on whether one or more than one spatial dimension is involved. Adding time dependence to a lumped system converts its governing equations to ODEs. Adding time dependence to a distributed system ensures that the governing equations will be PDEs.

4.2.1 Nonisothermal Perfect Mixers

Perfect mixers are internally uniform in temperature as they are in composition. The same flow and mixing patterns that are assumed to eliminate concentration gradients must surely eliminate temperature gradients as well. True homogeneity on a molecular scale requires diffusion, and thermal diffusivities are typically higher than molecular diffusivities. Thus if one is willing to assume compositional uniformity, it is also reasonable to assume thermal uniformity.

For a perfect mixer, the energy balance is

$$Q_{in}\rho_{in}H_{in} - \Delta H_{\mathscr{R}}\mathscr{R}V = Q_{out}\rho_{out}H_{out} + UA_{ext}(T_{out} - T_{ext}) \tag{4.17}$$

and the component mass balance is

$$Q_{in}a_{in} + \mathscr{R}_A V = Q_{out}a_{out} \tag{4.18}$$

These algebraic equations are coupled by the temperature and compositional dependence of $\mathscr{R}$. They may also be weakly coupled through the temperature and compositional dependence of physical properties such as density and heat capacity, but the strong coupling is through the reaction rate.

In previous chapters, we assumed that $\mathscr{R}$ was somehow known at the appropriate temperature. If the perfect mixer were strictly isothermal, say, because $\Delta H_R \approx 0$, then $T_{out} = T_{in}$. By a *nonisothermal perfect mixer*, we mean only that $T_{out} \neq T_{in}$. The temperature within the system remains everywhere uniform, with $T = T_{out}$, except at the very inlet. For concreteness, suppose all inlet conditions are known. Then the outlet conditions can be found by the simultaneous solution of Equations 4.17 and 4.18.

Example 4.3 Find the steady states for styrene polymerization in a CSTR having a mean residence time of 2 hr. Assume cold monomer feed (300 K), adiabatic operation ($UA_{ext} = 0$), and a pseudo-first order reaction with simple

Arrhenius temperature dependence:

$$k = 10^{10} \exp\left(\frac{-10,000}{T}\right) \text{ hr}^{-1}$$

where T is in K. Assume constant density and heat capacity.

Solution   The material balance equation gives

$$Qa_{in} - a_{out}k_0V\exp\left(\frac{-E}{R_gT_{out}}\right) = Qa_{out} \tag{4.19}$$

and the energy balance becomes

$$Q\rho C_p T_{in} - a_{out}\Delta H_{\mathscr{R}}k_0V\exp\left(\frac{-E}{R_gT_{out}}\right) = Q\rho C_p T_{out} \tag{4.20}$$

or

$$T_{in} + \frac{-a_{in}\Delta H_{\mathscr{R}}}{\rho C_p}\frac{a_{out}}{a_{in}}k_0\bar{t}\exp\left(\frac{-E}{R_gT_{out}}\right) = T_{out} \tag{4.21}$$

The group $-a_{in}\Delta H_{\mathscr{R}}/\rho C_p$ has dimensions of temperature and is a measure of the reaction exotherm. It is the adiabatic temperature rise for complete conversion of the feed and is about 400 K for undiluted styrene monomer.

Substitution of the specified values gives the following pair of equations:

$$\frac{a_{out}}{a_{in}} = \frac{1}{1 + 2 \times 10^{10}\exp(-10,000/T_{out})}$$

$$T_{in} + 8 \times 10^{12}\left(\frac{a_{out}}{a_{in}}\right)\exp(-10,000/T_{out}) = T_{out} \tag{4.22}$$

Solution gives

T_{out}	a_{out}/a_{in}
300.03	≈ 1 (0.99993)
404	.738
699.97	≈ 0 (0.00008)

so that there are three steady state solutions. In the low-temperature solution, the CSTR acts as a styrene monomer storage vessel, and there is no significant reaction. The high-temperature solution represents an upper, runaway condition where the reaction goes to near completion. (In actuality, the styrene polymerization is reversible at very high temperatures, with a ceiling temperature of about

625 K.) The middle steady state is metastable. The reaction will tend toward either the upper or lower steady states, and a control system is needed to maintain operation around the metastable point. For the styrene polymerization, a common industrial practice is to operate at the metastable point, with temperature control through autorefrigeration (cooling by boiling). A combination of feed monomer preheating ($T_{in} > 300$ K) and jacket heating is used to ensure that the uncontrolled reaction would tend toward the upper, runaway condition. However, the reactor pressure is set so that the styrene boils when the desired operating temperature is exceeded. The latent heat of vaporization plus the return of subcooled condensate rapidly reduces the temperature of the reactor below the boiling point. Cooling then stops until the heat of polymerization returns the temperature to the boiling point.

The existence of three steady states, two stable and one metastable, is fairly common for highly exothermic reactions in stirred tanks. Perhaps more common is the existence of only one steady state. For the styrene polymerization example, all three steady states exist only for a limited range on T_{in}. When T_{in} is sufficiently low, no reaction occurs, and only the lower steady state is possible. If T_{in} is sufficiently high, only the upper, runaway condition can be realized. For intermediate values of T_{in}, all three steady states are possible.

The location and number of steady states can also be varied by changing the external heat transfer term $UA_{ext}(T_{out} - T_{ext})$ in Equation 4.17. At a steady state, the rate of heat generation due to the reaction must exactly equal the enthalpy change of the products plus any heat lost to the environment:

$$-\Delta H_{\mathscr{R}} \mathscr{R} V = Q_{out}\rho_{out}H_{out} - Q_{in}\rho_{in}H_{in} + UA_{ext}(T_{out} - T_{ext})$$

$$\begin{matrix} \text{Heat} \\ \text{generated} \end{matrix} = \begin{matrix} \text{Heat absorbed due to enthalpy} \\ \text{change and loss to ambient} \end{matrix} \qquad\qquad \textbf{(4.23)}$$

If this equality does not hold exactly, the reactor is not at steady state and T_{out} will change with time.

Consider Equation 4.23 as a function of T_{out}. The heat generation term will typically be the S-shaped curve shown in Figure 4.6. At low values of T_{out}, reaction rates are low, and little or no heat is generated. At high values of T_{out}, the reaction goes to completion, and the entire exotherm is released. For the case of constant physical properties, the right-hand side of Equation 4.23 becomes

Heat absorbed due to enthalpy change + Loss to ambient

$$= (-Q\rho C_p T_{in} - UA_{ext}T_{ext}) + (Q\rho C_p + UA_{ext})T_{out} \qquad\qquad \textbf{(4.24)}$$

which is linear in T_{out}. See Figure 4.7. Possible points of equality in heat generation and heat absorption are shown by superposition in Figure 4.8. Three steady states are possible for the case illustrated. The slope and intercept of the

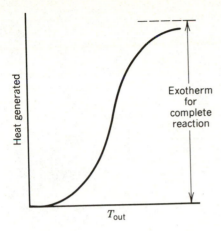

Figure 4.6 Heat generated by reaction.

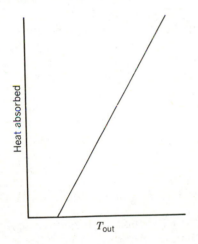

Figure 4.7 Heat absorbed by enthalpy change and loss to environment.

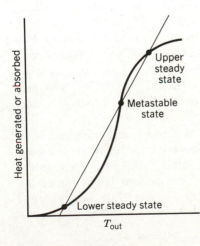

Figure 4.8 Multiple steady states for exothermic reaction in a CSTR.

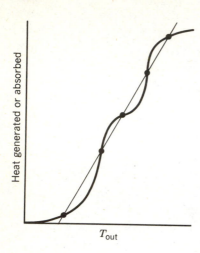

y-axis: Heat generated or absorbed

x-axis: T_{out}

Figure 4.9 Five steady states possible with complex kinetics.

heat absorption line, Equation 4.24, can be manipulated by changing operating variables such as Q, T_{in}, and T_{ext} or design variables such as $UA_{ext}/Q\rho C_p$. The magnitude of the reaction exotherm can be manipulated by changing the inlet reactant concentrations. Any of these manipulations can be used to vary the number and location of the possible steady states.

More than three steady states are sometimes possible. Consider the reaction sequence

$$A + B \rightarrow C \tag{I}$$

$$A \rightarrow D \tag{II}$$

where Reaction I occurs at a lower temperature than Reaction II. It is possible that Reaction I will go to near completion, consuming all the B, while still at temperatures below the point where Reaction II becomes significant. This situation can generate up to five steady states as illustrated in Figure 4.9. A practical example is styrene polymerization using component B as an initiator at low temperatures, $< 120°C$, and with thermal initiation at higher temperatures, $> 120°C$. The lower S-shaped portion of the heat generation curve consumes all the initiator B; but there remains unreacted styrene A. The higher S-shaped portion consumes the remaining styrene. In this particular case, products C and D are the same—polystyrene.

Suppose it has been decided to conduct a reaction in a continuous-flow, stirred tank. The methods of Section 4.1.2 can be used to determine the best operating temperature T_{opt}. Next, the design engineer must devise a physical scheme whereby the reactor actually operates at the desired temperature. An obvious requirement of this scheme is that $T_{out} = T_{opt}$ be one of the steady states. When T_{opt} is the only steady state, stable operation at the desired temperature is assured.

Example 4.4 Suppose that, to achieve a desired molecular weight, the styrene polymerization must be conducted at 413 K. Use external heat transfer to achieve this temperature as the single steady state in a stirred tank.

Solution Example 4.3 assumed adiabatic operation. With external heat transfer, the $UA_{ext}(T_{out} - T_{ext})$ term in Equation 4.17 is retained. Thus for the polystyrene example, Equations 4.22 become

$$\frac{a_{out}}{a_{in}} = \frac{1}{1 + 2 \times 10^{10}\exp\left(-10,000/T_{out}\right)}$$

$$T_{in} + 8 \times 10^{12}\left(\frac{a_{out}}{a_{in}}\right)\exp\left(-10,000/T_{out}\right) = T_{out} + \frac{UA_{ext}}{\rho QC_p}(T_{out} - T_{ext})$$

We consider T_{ext} to be an **operating variable** that will be manipulated to achieve $T_{out} = 413$ K. The dimensionless heat transfer group $UA_{ext}/\rho QC_p$ is considered a **design variable** that must be *large enough* so that $T_{out} = 413$ K becomes the only solution to the above equations. In small vessels, such as laboratory glassware, the heat transfer group is very large, and one simply sets $T_{ext} \approx T_{out}$ to achieve the desired steady state. In larger vessels, $UA_{ext}/\rho QC_p$ is finite, and one must set $T_{ext} < T_{out}$. Eventually, $UA_{ext}/\rho QC_p$ becomes so small and $T_{out} - T_{ext}$ becomes so large that stable operation is no longer possible:

$UA_{ext}/\rho QC_p$	Values of T_{ext} Needed to Give $T_{out} = T_{opt} = 413$ K
100	412.6
50	412.3
20	411.1
10	409.1
5	405.3
4	No solution

Thus the minimum value for $UA_{ext}/\rho QC_p$ is about 5. If the heat transfer group is any smaller than this, stable operation at $T_{out} = 413$ K by manipulation of T_{ext} is no longer possible.

Achievable values of $UA_{ext}/\rho QC_p$ tend to decrease as the size of the vessel increases. Note that ρ and C_p are physical properties and that U is an overall heat transfer coefficient which, to a first approximation, is independent of scale. Thus only A_{ext} and Q vary significantly with the size of the vessel. Suppose L is some characteristic dimension of the vessel, for example, the diameter. Then we expect A_{ext} to increase as L^2. However, Q will increase as L^3 since the volume increases as L^3 and since the mean residence time $\bar{t} = V/Q$ is usually held constant upon scaleup. Thus $UA_{ext}/\rho Q C_p$ will vary as L^{-1}, and external heat transfer becomes progressively more difficult as the vessel size is increased. For large vessels, the design engineer has three choices:

(i) Accept adiabatic operation.

(ii) Install additional heat transfer surface, using internal coils or external heat exchangers, so that A_{ext} increases as L^3.

(iii) Use a heat transfer mechanism such as autorefrigeration (cooling by boiling) that does not depend on external surface area.

Autorefrigeration represents a type of control that is truly automatic. Other operating variables, such as T_{in} or T_{ext} (with $UA_{ext} > 0$), can be manipulated to control a reaction around a metastable operating point. For example, the input stream could be cooled when the reactor contents become too hot and heated when the contents become too cold. For any such control scheme, operating conditions never stabilize exactly at the metastable point. Instead, temperatures and conversions follow a closed trajectory, known as a *limit cycle*, which has the metastable point at its center. The size of the limit cycle and assurances that the trajectory will indeed be closed depend on the dynamics of the control system, a subject that is discussed extensively in most courses on process control and briefly in Chapter 8.

To learn whether a particular steady state is stable, it is necessary to consider small deviations in operating conditions. Are they damped out or do they lead to larger deviations? Return to Figure 4.8 and suppose that the reactor has somehow achieved a value for T_{out} which is higher than the upper steady state. In this region, $T_{out} > (T_{out})_{upper}$, the heat absorption line is above the heat generation line so that the reactor will tend to cool, approaching $(T_{out})_{upper}$. Suppose on the other hand that the reactor becomes cooler than $(T_{out})_{upper}$ but remains in the region $(T_{out})_{intermediate} < T_{out} < (T_{out})_{upper}$. In this region the heat absorption line is below the heat generation line so that T_{out} will tend to increase, heading back to $(T_{out})_{upper}$. Thus the steady state at $(T_{out})_{upper}$ is stable. However, that at $(T_{out})_{intermediate}$ is unstable since, given a small deviation $T_{out} > (T_{out})_{intermediate}$, the reactor will continue heating so that the deviation grows.

Interestingly, systems exist that have but a single, metastable state and no stable steady states. Such systems come in two varieties. *Chemical oscillators* operate in a reproducible limit cycle about their metastable state. *Chaotic* systems have no discernible pattern but have temperature–composition trajectories that appear essentially random in nature. The dynamical behavior of nonisothermal perfect mixers is extremely complex and is the subject of considerable study. Occasionally, this dynamical behavior is of practical importance for industrial reactor design. A classic situation involved the emulsion polymerization of styrene–butadiene rubbers. This is a complex reaction involving both kinetic and mass transfer limitations, and a stable steady state conversion is difficult or impossible to achieve in a single perfect mixer. However, if enough perfect mixers were put in series, results would average out so that effectively stable, high conversions could be achieved. For the styrene–butadiene rubber

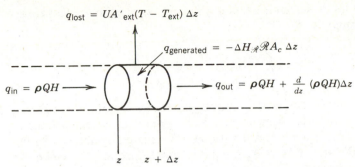

$q_{lost} = UA'_{ext}(T - T_{ext}) \Delta z$

$q_{generated} = -\Delta H_\mathscr{R} \mathscr{R} A_c \, \Delta z$

$q_{in} = \rho Q H \longrightarrow$

$q_{out} = \rho Q H + \frac{d}{dz} (\rho Q H) \Delta z$

$z \qquad z + \Delta z$

Figure 4.10 Nonisothermal piston flow differential reactor element.

example, enough stirred tanks turned out to be 25 to 40, and full-scale production units were actually built in this configuration!

4.2.2 Nonisothermal Piston Flow Reactors

Steady state temperatures along the length of a piston flow reactor are governed by an ordinary differential equation. Consider the differential reactor element shown in Figure 4.10. The words of Equation 4.11 remain correct, but now the mathematical expression of these words is

$$\rho Q H - \Delta H_\mathscr{R} \mathscr{R} A_c \Delta z = \rho Q H + \frac{d(\rho Q H)}{dz} \Delta z + UA'_{ext} \Delta z \, (T - T_{ext})$$

(4.25)

where A'_{ext} is the external surface area per unit length of reactor, m^2/m. Noting that $Q = \bar{u} A_c$,

$$\frac{1}{A_c} \frac{d(A_c \rho \bar{u} H)}{dz} = \frac{1}{A_c} \frac{d(A_c \Upsilon)}{dz} = -\Delta H_\mathscr{R} \mathscr{R} - \frac{UA'_{ext}}{A_c} (T - T_{ext})$$

(4.26)

where $\Upsilon = \rho \bar{u} H$ is the heat flux in $J\,m^{-2}\,s^{-1}$. It is this equation that is coupled to Equation 3.7 or 3.17 and is solved simultaneously to give temperature and composition as a function of axial position in the piston flow reactor. In the general case, ρ, $\bar{u}$, and H are all functions of $T, P, a, b, \ldots$. Using numerical methods, the nonisothermal problem is no more difficult to solve than the variable-density and variable-velocity problems treated in Section 3.2.3. One merely adds two additional equations to the coupled set considered in that section. One of the additions is Equation 4.26. The other is the thermodynamic relationship that gives enthalpy as a function of temperature, pressure, and composition.

With a constant, circular cross section and with constant physical properties, Equation 4.26 simplifies to

$$\frac{dT}{dz} = \frac{-\Delta H_{\mathscr{R}}\mathscr{R}}{\rho \bar{u} C_p} - \frac{2U}{R\rho \bar{u} C_p}(T - T_{ext})$$
(4.27)

which is the form usually encountered. However, the reader should recognize that Equation 4.26 is the appropriate version of the energy balance when the reactor cross section or physical properties are variable. The marching-ahead solution technique for either Equation 4.26 or 4.27 requires that T_{in}, a_{in}, $b_{in}, \ldots$ be known at $z = 0$. This allows reaction rates and physical properties to be calculated at $z = 0$ so that the right-hand side of Equation 4.26 or 4.27 can be evaluated. This gives ΔT, and thus $T(z + \Delta z)$, directly in the case of Equation 4.27. Equation 3.7 is evaluated similarly to give Δa, and thus $a(z + \Delta z)$. Equations similar to 3.7 are solved for any other compositional unknowns in the same way. Thus $T, a, b, \ldots$ are all determined at the next axial position along the tubular reactor. The axial position variable z can then be incremented and the entire procedure repeated to give temperatures and compositions at yet the next point. Thus we march down the tube.

Example 4.5 Hydrocarbon cracking reactions are endothermic, and many different techniques are used to supply heat to the system. The maximum inlet temperature is limited by materials of construction problems or by undesirable side reactions such as coking. Suppose the inlet temperature is fixed at T_{in}. Then $T(z) \leq T_{in}$ and the temperature will gradually decline as the reaction proceeds. This decrease, with the consequent reduction in reaction rate, can be minimized by using a high proportion of inerts in the feed stream.

Consider a cracking reaction with rate

$$\mathscr{R} = \left[10^{14} \exp\left(-\frac{24,000}{T} \right) \right] a \quad \text{gm m}^{-3}\,\text{s}^{-1}$$

where T is in degrees K and a is in gm/m^3. Suppose the reaction is conducted in an *adiabatic* tubular reactor having a mean residence time of 0.3 s. The crackable component and its products have a heat capacity of 0.4 cal gm^{-1} K^{-1}, and the inerts have a heat capacity of 0.5 cal gm^{-1} K^{-1}; the entering concentration of crackable component is 132 gm/m^3 and the concentration of inerts is 270 gm/m^3; $T_{in} = 525°C$. Calculate the exit concentration of A given $\Delta H_{\mathscr{R}} = 203$ cal/gm. Physical properties may be assumed constant.

Solution Aside from the temperature calculations, this example illustrates how the time elapsed since material entered the reactor can be used as a substitute for axial position:

$$\bar{u}\frac{da}{dz} = \frac{da}{d\alpha} = \mathscr{R}_A$$
(4.28)

It also illustrates the systematic use of mass rather than molar concentrations for reactor calculations. This is common practice for mixtures having ill-defined molecular weights.

In terms of α, the energy balance becomes

$$\frac{dT}{d\alpha} = \frac{-\Delta H_{\mathscr{R}}\mathscr{R}}{\rho C_p} \tag{4.29}$$

where we have set $UA'_{ext}(T - T_{ext}) = 0$ for adiabatic operation. Note that ρ and C_p are properties of the reaction mixture. Thus $\rho = 132 + 270 = 402$ gm/m^3 and $C_p = [0.4(132) + 0.5(270)]/402 = 0.467$ cal gm^{-1} K^{-1}. Substituting known values into Equations 4.28 and 4.29 and marching down the tube with a step size of 0.0375 s ($J = 8$) gives the following results

α (sec)	T (°C)	a (gm / m^3)
0	525	132
0.0375	478	89
0.0750	473	84
0.1125	470	81
0.1500	467	78
0.1875	464	76
0.2250	462	74
0.2625	460	72
0.3000	458	70

The large decrease in a and T for the first time increment suggests the use of a variable step size. As always, this will increase computational efficiency at the expense of greater programming effort. The alternative of simply decreasing $\Delta\alpha$ remains attractive:

J	$\Delta\alpha$ (sec)	a_{out} (gm / m^3)	Δ (gm / m^3)	T_{out} (°C)	Δ (°C)
1	0.3000	−214		151	
			171		187
2	0.1500	−41		338	
			85		92
4	0.0750	+44		430	
			26		28
8	0.0375	70		458	
			4		4
16	0.0188	74.0		462.3	
			1		1
32	0.0094	75.04		463.41	
			0.44		0.46
64	0.0047	75.48		463.87	
			0.21		0.23
128	0.00023	75.69		464.10	

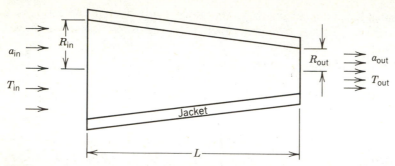

Figure 4.11 Reverse megaphone or converging tubular reactor.

Values of $a_{out} = 75.9$ gm/m^3 and $T_{out} = 464.3°C$ are correct to the indicated number of places. Note that the convergence $0(\Delta\alpha)$ is not achieved until fairly small mesh sizes are used. This is due to the highly nonlinear nature of the system.

When Equation 4.26 is used rather than Equation 4.27, the above procedure does not give ΔT and thus $T(z + \Delta z)$ directly. Instead, it gives $\Delta(\rho\bar{u}H)$ and thus the composite variable $\Upsilon = \rho\bar{u}H$ at the next axial point. However, this composite variable is a function of $T(z + \Delta z)$. Given the value for $\rho\bar{u}H$, solve for T and proceed as before. In those cases where $\rho\bar{u}H$ depends on composition as well as temperature, it is usually satisfactory to evaluate the compositional dependence at the "old" values rather than the "new." (A similar point was made in Section 3.1.2 where both $\bar{u}$ and a were found from a pair of ordinary differential equations.)

Example 4.6 Consider a jacketed "reverse megaphone" reactor as illustrated in Figure 4.11. Approximate this situation as piston flow with a constant heat transfer coefficient to the jacket. Determine the outlet concentration and temperature for an exothermic, first order, liquid phase reaction

$$A \overset{k}{\rightarrow} B$$

The rate constant is available from laboratory kinetic studies:

$$k = 1.3 \times 10^{11} \exp\left(-\frac{10{,}550}{T}\right) \quad s^{-1}$$

and a quick trip to the library allowed $\Delta H_{\mathscr{R}}$ to be estimated from heat of formation data:

$$\Delta H_{\mathscr{R}} = -12{,}500 \text{ J/mol} \qquad \text{at } 100°C$$

Measurements and previous operating experience give

$R_{in} = 0.05$ m

$R_{out} = 0.025$ m

$L = 1$ m

$U = 550$ $J\,m^{-2}\,s^{-1}\,K^{-1}$

It is desired to operate with $a_{in} = 2000$ mol/m³ at a mass flow rate of 0.06 kg/s. The inlet temperature will be 100°C, and the jacket will be held at 140°C.

   Solution   Since the reactor cross section varies and since physical properties may also vary over a 40°C temperature range, flux marching (rather than concentration and temperature marching) will be employed. The equations are

$$(\Phi_A)_{j+1} = (\Phi_A)_j + \Delta z \left[\mathcal{R}_A - \frac{\Phi_A}{A_c} \frac{dA_c}{dz} \right]_j \tag{4.30}$$

$$\Upsilon_{j+1} = \Upsilon_j + \Delta z \left[-\Delta H_{\mathcal{R}} \mathcal{R} - \frac{UA'_{ext}}{A_c}(T - T_{ext}) - \frac{\Upsilon}{A_c} \frac{dA_c}{dz} \right]_j \tag{4.31}$$

These correspond to the two ODEs, Equations 3.17 and 4.26, which must be solved simultaneously. The dependent variables are Φ_A and Υ, and the independent variable is z. However, Equations 4.30 and 4.31 contain a number of quantities that are not currently defined in terms of Φ_A, Υ, or z. This obviously has to be done before a solution can be obtained. The unknown quantities are

$$\mathcal{R}_A, \quad A_c, \quad \frac{dA_c}{dz}, \quad -\Delta H_{\mathcal{R}} \mathcal{R}, \quad \frac{UA'_{ext}}{A_c}, \quad T$$

Equations for them are

$$\mathcal{R}_A = -(1.3 \times 10^{11}) a \exp\left(-\frac{10,550}{T}\right) \quad mol\,m^{-3}\,s^{-1} \tag{4.32}$$

$$A_c = \pi \left[R_{in} + \left(\frac{R_{out} - R_{in}}{L} \right) z \right]^2$$

$$= \pi(0.5 - 0.025z)^2 \quad m^2 \tag{4.33}$$

$$\frac{dA_c}{dz} = 2\pi \left[R_{in} + \left(\frac{R_{out} - R_{in}}{L} \right) z \right] \left(\frac{R_{out} - R_{in}}{L} \right)$$

$$= -0.05\pi(0.05 - 0.025z) \quad m \tag{4.34}$$

$$-\Delta H_{\mathcal{R}} \mathcal{R} = (1.625 \times 10^{15}) a \exp\left(-\frac{10,550}{T}\right) \quad J\,m^{-3}\,s^{-1} \tag{4.35}$$

(But does $\Delta H_{\mathscr{R}}$ depend on T? Assume $\Delta C_p \approx 0$ so that $\Delta H_{\mathscr{R}}$ is approximately constant.)

$$\frac{UA'_{\text{ext}}}{A_c} = \frac{2U}{R} = \frac{44{,}000}{2-z} \quad \text{J m}^{-3}\,\text{s}^{-1}\,\text{K}^{-1} \tag{4.36}$$

$$H = \int_{T_{\text{in}}}^{T} C_p\,dT \tag{4.37}$$

which sets the enthalpy reference temperature equal to T_{in}.

The above auxiliary equations eliminate the unknowns that appear in the flux equations but introduce still new ones: a, H, C_p. Of these, C_p is a physical property. A quick trip to the library gives

$$C_p = 2100 + 2.0(T - T_{\text{in}}) \quad \text{J kg}^{-1}\,\text{K}^{-1} \tag{4.38}$$

For a and H we have

$$a = \frac{\Phi_A}{\bar{u}} \tag{4.39}$$

$$H = \frac{\Upsilon}{\rho\bar{u}} \tag{4.40}$$

which introduces ρ and $\bar{u}$ as unknowns. A final trip to the library suggests

$$\rho = 750 - 1.1(T - T_{\text{in}}) - 25\left(1 - \frac{a}{a_{\text{in}}}\right) \quad \text{kg/m}^3 \tag{4.41}$$

A total mass balance gives

$$\bar{u}A_c\rho = (\bar{u}A_c\rho)_{\text{in}} = 0.06 \quad \text{kg/s}$$

or

$$\bar{u} = \frac{0.06}{A_c\rho} = \frac{30.6}{(2-z)^2\rho} \quad \text{m/s} \tag{4.42}$$

This completes the formulation of the marching-ahead problem since every variable in the flux equations is now defined either directly or indirectly in terms of Φ_A, Υ, or z.

The various equations, 4.30 through 4.42, must now be sequenced so that the marching-ahead calculations can be performed given a set of initial values. A suitable sequence is

$$(\Phi_A)_{j+1} = (\Phi_A)_j + \Delta z \left[-(1.3 \times 10^{11}) a \exp\left(-\frac{10{,}550}{T} \right) + \frac{2\Phi_A}{2-z} \right]_j$$

(4.43)

$$\Upsilon_{j+1} = \Upsilon_j + \Delta z \left[(1.625 \times 10^{15}) a \exp\left(-\frac{10{,}550}{T} \right) \right.$$

$$\left. - \frac{44{,}000}{2-z}(T - 413.16) + \frac{2\Upsilon}{2-z} \right]_j$$

(4.44)

$$a_{j+1} = \frac{(\Phi_A)_{j+1}}{\bar{u}_j}$$

(4.45)

$$H_{j+1} = \frac{\Upsilon_{j+1}}{\rho_j \bar{u}_j}$$

(4.46)

$$T_{j+1} = T_{\text{in}} - 1050 + \sqrt{1050^2 + H_{j+1}}$$

(4.47)

(This results from integrating Equation 4.37 and solving the result for T.)

$$\rho_{j+1} = 750 - 1.1(T_{j+1} - T_{\text{in}}) - 25\left(1 - \frac{a_{j+1}}{a_{\text{in}}} \right)$$

(4.48)

$$\bar{u}_{j+1} = \frac{30.6}{(2 - z_{j+1})^2 \rho_{j+1}}$$

(4.49)

This set of equations, 4.43 through 4.49, is evaluated sequentially. First, Φ_A and Υ are marched ahead. Then a_{j+1}, H_{j+1}, and T_{j+1} are evaluated using the new values for Φ_A and Υ but previous values for ρ and $\bar{u}$. Finally, ρ and $\bar{u}$ are evaluated at the new point.

Initial values are

$\rho_0 = 750$ kg/m^3
$\bar{u}_0 = 0.0102$ m/s
$T_0 = 373.16$ K
$H_0 = 0$ J/kg
$a_0 = 2000$ mol/m^3
$\Upsilon_0 = 0$ J m^{-2} s^{-1}
$\Phi_A = 20.4$ mol m^{-2} s^{-1}

Following are a few steps taken with $\Delta z = 0.0625$ m:

Φ_A	Υ	a	H	T	ρ	$\bar{u}$	z
20.4	0	2000	0	373	750	0.0102	0
13.1	162,000	1290	21,200	383	730	0.0105	0.0625
2.44	359,000	232	46,900	395	704	0.0116	0.1250
−2.24	580,000	−193	71,200	407	686	0.0127	0.1875

Thus the chosen Δz was too large. The reaction goes to near completion. Results at $z = 1$ obtained with $\Delta z = \frac{1}{64}$ are: $\Phi_A = 10^{-10}$, $\Upsilon = 2 \times 10^6$, $a = 10^{-9}$, $H = 7 \times 10^4$, $T = 406$, $\rho = 689$, and $\bar{u} = 0.043$.

In the general case of a piston flow reactor, one must solve a fairly small set of simultaneous, ordinary differential equations. The minimum set (of one) arises for a single, isothermal reaction with constant physical properties. Then only Equation 3.17 need be solved. In principle, one extra equation must be added to the set for each additional reaction. In practice, numerical solutions are somewhat easier to implement if a separate equation is written for each reactive component. This ensures that stoichiometry is accounted for and keeps the physics and chemistry of the problem rather more transparent than when the reaction coordinate method is used to obtain the smallest possible set of differential equations. Unlike the case of stirred tank reactors, there is little computational advantage to eliminating unnecessary members from the set. If the piston flow reactor is nonisothermal, a single equation such as 4.26 must be added, and if the pressure drop down the tube is significant, an equation such as 3.12 must be added. Thus in the general case of variable temperatures and pressures and with N reactive components, the *maximum* set of ordinary differential equations to be solved is $N + 2$, one for each of the state variables $T, P, a, b, \ldots$. Such solutions are now quite feasible, and one will rarely run into computational limitations for steady state design problems of this sort.

Also needed as auxiliary, algebraic equations are the equations of state and the thermodynamic functions that allow physical properties to be calculated from known values of the state variables. In actual design problems, the work involved in assembling the data and determining the necessary functions will be significant. Understanding the problem and acquiring the data will almost always take much longer than the computation.

4.2.3 Nonisothermal Batch Reactors

The energy balance for a batch reactor at constant pressure takes the form

$$\frac{d(H\rho V)}{d\alpha} = -\Delta H_\mathscr{R} \mathscr{R} V - U A_{\text{ext}}(T - T_{\text{ext}})$$

$$(4.50)$$

Solution techniques are similar to those used for piston flow reactors. For constant volume and physical properties,

$$\frac{dT}{d\alpha} = \frac{-\Delta H_{\mathscr{R}}\mathscr{R}}{\rho C_p} - \frac{UA_{ext}(T - T_{ext})}{\rho C_p V} \tag{4.51}$$

where A_{ext}/V is the ratio of area to volume in the batch reactor. This equation is identical to that governing heat transfer in a piston flow reactor with constant physical properties. When the cross section of the piston flow reactor is circular, $2/R = A_{ext}/V$; and when physical properties are constant, $dz = \bar{u}\,d\alpha$. Thus Equation 4.27 can be written exactly like Equation 4.51.

4.3   Analysis of Rate Data

This section considers the sometimes formidable problem of finding $\mathscr{R}(a, b, \ldots, T)$ from experimental results on the extent of reaction. In essence, we seek to measure $\mathscr{R}$ and then to fit the data as a function of concentration and temperature. This is usually done in two steps. First, use isothermal data to fit $\mathscr{R}$ as a function of concentration. This gives a model containing one or more rate constants. Next, the temperature is changed and the concentration fitting is repeated. Hopefully, the same functional form will apply at the new temperature as at the old, but the constants will have different values. The variation in rate constant as a function of temperature is then fit to some functional form, usually the simple Arrhenius form.

The fitting of reaction rate data is an important topic within chemical reactor design. Many authors would consider it of such primary importance that they would introduce it very early in a text. We have postponed the introduction until the fundamentals of temperature dependence and nonisothermal reactors were explained. Do not be misled by this postponement. The gathering and analysis of kinetic data are indeed of major importance and are activities where reaction engineers can expect to spend much of their time. Also, the theory of nonisothermal reactors should be used to **avoid** temperature variations in kinetic experiments. Reaction engineers should seek to design experiments that collectively cover a broad range of temperatures but that are individually isothermal. These experiments should be conducted in a reactor that closely approximates one of the three ideal types. Of the three types, perfect mixers provide the most easily interpreted data.

4.3.1   Stirred Tank or Differential Reactor Data

The advantage of using a CSTR is that the reaction rate is measured directly:

$$\mathscr{R}_A = \frac{Q_{out}a_{out} - Q_{in}a_{in}}{V} \quad \text{moles per unit volume per unit time} \tag{4.52}$$

The same is true for a differential reactor, but the chemical analysis will be more difficult due to the small conversions. The numerical value obtained for $\mathscr{R}_A$ corresponds to reaction at concentrations of $a_{out}, b_{out}, \ldots$. These outlet concentrations can be changed by varying the feed conditions. This should be done over a wide range of compositions, at least as wide as will be encountered in the full-scale reactor. It is usually desirable to include runs with product species in the feed stream to check for reversibility or product inhibition.

When $\mathscr{R}_A$ has been measured at many combinations of $a_{out}, b_{out}, \ldots$, the data are fit to some functional form. If the reaction mechanism is known or suspected, it is entirely appropriate to use the form suggested by that mechanism. However, an arbitrary form is likely to provide as good a fit. A commonly used form is

$$\mathscr{R}_A = k a^\alpha b^\beta \ldots \tag{4.53}$$

where $\alpha, \beta, \ldots$ are constants, hopefully independent of temperature. Equation 4.53 includes all the irreversible, elementary reactions among its possibilities. For example

$$A + B \xrightarrow{k} \text{Products}$$

has $\alpha = \beta = 1$, with all other exponents being zero. The fractional exponents associated with the chain reactions of Section 2.3.2 are directly accommodated. Product inhibition and even reversibility when not too near equilibrium can also be accounted for using Equation 4.53.

Aside from its versatility, Equation 4.53 has an advantage of being easy to fit to experimental data. Taking logarithms,

$$\ln \mathscr{R}_A = \ln k + \alpha \ln a + \beta \ln b + \cdots \tag{4.54}$$

so that $\ln \mathscr{R}_A$ is a linear function of $\ln a, \ln b, \ldots$. This means that the exponents can be found using *multivariate, linear regression analysis*, which is a commonly used statistical technique available on most mainframe computer systems. A brief description is given in Appendix 4.2. Statistical purists will prefer *nonlinear regression*, which allows fitting the data without the logarithmic transformation. However, if there is a big difference between these approaches, it probably means the data are poor or the model is inappropriate for the reaction.

Example 4.7 Suppose the following data on the iodination of ethane have been obtained at 1 atm and 603 K in a recirculating gas phase reactor such as that shown in Figure 3.2:

$[I_2]_{in}$	$[C_2H_6]_{in}$	$\bar{t}$	$[I_2]_{out}$	$[C_2H_6]_{out}$	$[HI]_{out}$	$[C_2H_5I]_{out}$
0.1	0.9	240	0.0830	0.884	0.0176	0.0162
0.1	0.9	1300	0.0402	0.841	0.0615	0.0594
0.1	0.9	2300	0.0221	0.824	0.0797	0.0770

Note that the indicated concentrations are actually partial pressures in atm.

Solution   The primary reaction is

$$I_2 + C_2H_6 \rightarrow HI + C_2H_5I$$

We would like to fit a rate equation having the form

$$\mathcal{R} = k[I_2]^\alpha [C_2H_6]^\beta \tag{4.55}$$

However, the experimental range on $[C_2H_6]_{out}$ is very small, and it is unlikely that β can be obtained within reasonable limits. We can proceed by any of several plausible approaches. One approach would be to *ignore* the variation in $[C_2H_6]$ and to fit $\mathcal{R}$ as a function of $[I_2]$ alone. In effect, this treats the reaction as zero order in ethane. Another approach is to *assume* $\beta = 1$ so that the order agrees with the stoichiometry. In either case we can plot

$$\ln\left(\frac{\mathcal{R}}{[C_2H_6]_{out}^\beta}\right) = \ln k + \alpha \ln [I_2]_{out} \tag{4.56}$$

to obtain a straight line with slope equal to α. The data give

t (s)	$\mathcal{R}$ (atm / s)	$\ln \mathcal{R}$	$\ln(\mathcal{R}/[C_2H_6]_{out})$	$\ln[I_2]_{out}$
240	7.08×10^{-5}	-9.56	-9.43	-2.49
1300	4.60×10^{-5}	-9.90	-9.81	-3.21
2300	3.39×10^{-5}	-10.29	-10.10	-3.81

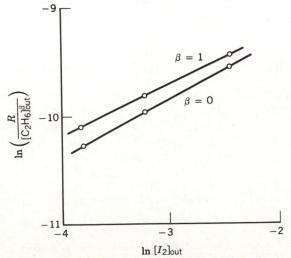

Figure 4.12   Rate data for ethane iodination.

where $\mathcal{R} = ([I_2]_{out} - [I_2]_{in})/\bar{t}$. Figure 4.12 gives plots for $\beta = 0$ and $\beta = 1$, and there is no clear choice between them. A straight line with slope $\alpha = 0.5$ fits both sets of data reasonably well. Thus, pending additional data, either of

$$\mathcal{R} = k[I_2]^{1/2} \quad \text{or} \quad \mathcal{R} = k[I_2]^{1/2}[C_2H_6]$$

is a reasonable choice for the rate expression.

The procedure for finding $k, \alpha, \beta, \ldots$ should be repeated at several different temperatures. For simple Arrhenius dependence, $\ln k$ will be a linear function of $1/T$:

$$\ln k = \ln k_0 - \frac{E}{R_g}\left(\frac{1}{T}\right) \tag{4.57}$$

so that linear regression can be used to evaluate the constants k_0 and E/R_g. A semilog plot of k versus $1/T$ is expected to give a straight line, but failure to do so is not unusual. Recall Section 4.1.1. It might also be hoped that the exponents $\alpha, \beta, \ldots$, are independent of temperature; but failure to be so is again not unusual since reaction mechanisms can shift with temperature. From the viewpoint of the reactor designer, any form for $\mathcal{R}_A(a, b, \ldots, T)$ is acceptable if it accurately portrays the data. However, there is certainly more confidence in the model if it agrees with a theoretical mechanism.

Extrapolation of a model outside its range of experimental validation is always dangerous. It is slightly less dangerous when the model is based on sound theory.

Example 4.8[2]   The following data have been obtained for the elementary reaction:

$$NO + NO_2Cl \xrightarrow{k} NO_2 + NOCl$$

T (K)	k (m^3 mol^{-1} s^{-1})
300	7.9
311	12.5
323	16.4
334	25.6
344	34.0

Fit these data to various forms ($m = 0, \frac{1}{2}, 1$) of Arrhenius temperature dependence.

---

[2] This problem was adapted from J. H. Espenson, *Chemical Kinetics and Reaction Mechanisms*, McGraw-Hill, New York, 1981, pp. 118–120. The experimental results are from E. C. Treiling, H. C. Johnson, and R. A. Ogg, Jr., *J. Chem. Phys.*, **20**, 327 (1952).

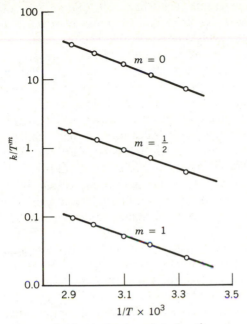

Figure 4.13 Arrhenius plots for the reaction $NO + NO_2Cl \xrightarrow{k} NO_2 + NOCl$.

Solution   Rearrange Equation 4.1 and take logarithms to give

$$\ln \frac{k}{T^m} = \ln k_0 - \frac{E}{R_g T} \tag{4.58}$$

Thus a plot of $\ln(kT^{-m})$ versus T^{-1} should give a straight line with slope $-E/R_g$. See Figure 4.13. It is seen that straight lines fit the data quite well for all values of m. Numerical results are

m	k_0 (m³ mol⁻¹ s⁻¹ K⁻¹)	E/R_g (K)
0	6.5×10^5	3390
$\frac{1}{2}$	2.1×10^4	3220
1	7.1×10^2	3060

Although the individual parameters k_0 and E/R_g are different, the product $K_0 T^m \exp(-E/R_g T)$ is quite similar for T within the experimental range.

Most of this book is concerned with homogeneous reactions that can occur within a single, fluid phase. The above analysis of rate data assumes that the various components are well mixed throughout the CSTR or differential reactor. This is usually true in practice, but exceptions are found in high-viscosity, low-diffusion systems which exhibit *segregation*. See Chapter 9. The analysis of rate data is even more complicated for heterogeneous reactions where reactive components must be transferred between phases or where the reaction occurs at an interface. Gas–solid heterogeneous catalysis is common and is discussed in Chapter 10. Gas–liquid reactions are also common and are discussed in Chapter 11. The alkylation of liquid phenol with gas phase butene is a typical example. The hydrodesulfurization of heavy hydrocarbons is frequently done in a trickle bed reactor which involves three phases: gas, liquid, and a catalytic solid. In any of these heterogeneous examples, measured reaction rates include mass transfer steps as well as true reaction steps. The design methods in this book are generally applicable to the case where mass transfer is fast compared to chemical reaction. When mass transfer is slow, the reaction is said to be **mass transfer controlled**. Reactor design and scaleup is then concerned more with interphase mixing and contacting than with chemical kinetics, and special techniques are required. See Chapter 11.

Laboratory rate data on heterogeneous reactions may well be in the regime of mass transfer control. Such reactions will usually appear first order in the component being transferred since driving forces for mass transfer are approximately linear in concentration. The activation energy for the pseudo-first order rate constant will reflect the temperature sensitivity of diffusion rather than that of the chemical reaction. Thus $E \approx 5000$ cal/mol = 20,934 J/mol for mass transfer controlled reactions as compared to $E \approx 10,000$ cal/mol = 41,868 J/mol or more for a kinetically controlled reaction.[3] The low activation energy for mass transfer means that systems that are kinetically controlled at low temperatures may become mass transfer controlled at high temperatures. Many industrial reactors operate in the regime of mass transfer control.

4.3.2 Integral Reactor Data

Batch reactors are easier to implement on a laboratory scale than are CSTRs. Thus they are widely used. Unfortunately, the rate data are harder to interpret since $\mathscr{R}_A$ is measured as an integral over time:

$$\int_0^t \mathscr{R}_A \, d\alpha = a_t - a_0 \tag{4.59}$$

which was obtained by integrating Equation 1.31. For the variable-volume case, Equation 1.28 integrates to give

$$\int_0^t \mathscr{R}_A V \, d\alpha = V_t a_t - V_0 a_0 \tag{4.60}$$

In either case, the rate can be found by numerical differentiation, but this is a notoriously inaccurate process. A better procedure is to assume a functional form for $\mathscr{R}_A$, to integrate this form, and then to fit the constants to the experimental

---

[3] Many reactions have activation energies in range 40 to 50 kJ/mol. Rates for such reactions roughly double for each 10 K rise in temperature when T is around 300 K.

data in the integral domain using nonlinear regression. This approach works fairly well for reactions that are nth order in a single component, but it becomes difficult when $\mathscr{R}_A$ depends on the concentrations of several components.

Example 4.9 Acetaldehyde decomposes according to the following stoichiometry:

$$CH_3CHO \rightarrow CH_4 + CO$$

This reaction was studied by passing vaporized acetaldehyde through a piston flow reactor at atmospheric pressure and a temperature of 600°C. The following data were obtained:

$\bar{t}$ (s)	$\dfrac{[CH_3CHO]_{out}}{[CH_3CHO]_{in}} = \dfrac{a_t}{a_0}$
10	0.917
20	0.842
40	0.733
80	0.577
160	0.401
320	0.254

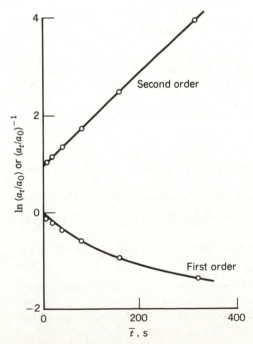

Figure 4.14 Determination of reaction order for acetaldehyde decomposition.

Determine the reaction order.

Solution First and second order reactions will be considered. For the first order reaction, a plot of $\ln(a_t/a_0)$ versus $\bar{t}$ should give a straight line with a slope of $-k$. For the second order reaction, a plot of $(a_t/a_0)^{-1}$ versus $\bar{t}$ should give a straight line with a slope of $a_0 k$.

Figure 4.14 shows the results. The first order plot shows distinct curvature whereas the second order plot is creditably linear.

Suggestions for Further Reading

An excellent account of reaction rate theory is given by:

K. J. Laidler, *Theories of Chemical Reaction Rates*, McGraw-Hill, New York, 1969.

Heat balances in CSTRs are a typical topic in chemical engineering control courses. A representative text is:

G. Stephanopoulos, *Chemical Process Control: An Introduction to Theory and Practice*, Prentice-Hall, Englewood Cliffs, NJ, 1984.

An article containing a comprehensive summary of steady state multiplicity and stability in stirred tank reactors is:

A. Varma and R. Aris, Chapter 2 in *Chemical Reactor Theory: A Review*, L. Lapidus and N. R. Amundson, Editors, Prentice-Hall, Englewood Cliffs, NJ, 1977.

The analysis of rate data and the use of rate data to determine true reaction mechanisms is an extremely important topic in chemical kinetics. An introductory account is provided in:

C. G. Hill, Jr., *An Introduction to Chemical Engineering Kinetics and Reactor Design*, Wiley, New York, 1977.

A very readable description of chemical kinetics and data fitting techniques for integral reactors is the following two volume set. It contains a wealth of chemical insights:

K. J. Laidler, *Reactor Kinetics* (in two volumes), Pergamon, London, 1963.

Regression analysis is briefly outlined in Appendix 4.2. A comprehensive treatment is provided in

N. R. Draper and H. Smith, *Applied Regression Analysis*, 2nd ed., Wiley, New York, 1981.

The treatment of variable cross section reactors was included to illustrate general principles. However, they have been seriously proposed. See:

L. M. Akella, J. C. Hong, and H. H. Lee, "Variable Cross-Section Reactors for Highly Exothermic Reactions," *Chem. Eng. Sci.*, **40**, 1011 (1985).

Problems

4.1   Consider the following pair of reactions occurring in a batch system:

$$A \rightarrow P \tag{I}$$

$$A + B \rightarrow Q \tag{II}$$

Suppose both reactions are pseudo-first order in A with $k_I = 10^{18} \exp(-20\,000/T)$ and $k_{II} = 10^{20} \exp(-22\,000/T)$, both in hr^{-1}.
(a) Find the reaction temperature that maximizes p_t/a_0 for a reaction time of 1 hr.
(b) Repeat for a reaction time of 10 hr.
(c) How might a pseudo-first order rate be explained for Reaction II?

4.2   For the styrene polymerization example of Section 4.2.1, determine that value of T_{in} below which only the lower steady state is possible. Also determine that value of T_{in} above which only the upper steady state is possible.

4.3   For the styrene polymerization example of Section 4.2.1, determine those values of the mean residence time $\bar{t}$ which give one, two, or three steady states.

4.4   Repeat the analysis of hydrocarbon cracking in Example 4.5 for the case where there are no inerts. Use $a_{in} = 132 \text{ gm/m}^3$ as before.

4.5   Repeat the analysis of hydrocarbon cracking in Example 4.5 for the case where the inerts are present and there is also external heat exchange. Use the following values: $R = 0.5$ in., $U = 7.0$ Btu ft^2 hr^{-1} $°F^{-1}$, $L = 10$ ft, $T_{ext} = 525°C$. (This mess of inconsistent units represents the way many engineers think. You don't have to join them, but you should at least be able to understand them.)

4.6   Your company is developing a highly proprietary new product. The chemistry is complicated, but the last reaction step is a dimerization:

$$2A \xrightarrow{k} B$$

Laboratory kinetic studies gave $a_0 k = 1.7 \times 10^{13} \exp(-14\,000/T)$ s^{-1}. The reaction was then translated to the pilot plant and reacted in a 10-liter batch reactor according to the following schedule:

Time from Start of Batch (Minutes)	Action
0	Begin charging raw materials.
15	Seal vessel; turn on jacket heat (140°C steam).
90	Vessel reaches 100°C and reflux starts.
180	Reaction terminated; vessel discharge begins.
195	Vessel empty; Washdown begins.
210	Reactor clean, empty, and cool.

Management likes the product and has begun to sell it enthusiastically. The pilot plant vessel is being operated around the clock and produces two batches per shift for a total of 42 batches per week.

It is desired to increase production by a factor of 1000, and the engineer[4] assigned to the job ordered a geometrically similar vessel which has a working capacity of 10,000 liters.

(a) What production rate will actually be realized in the larger unit? (Assume $\Delta H_{\mathscr{R}}$ is negligible.)

(b) You have replaced the original engineer[5] and been told to achieve the forecast production rate of 1000 times the pilot rate. What might you do to achieve this?

(c) How would you work this problem if $\Delta H_{\mathscr{R}}$ were significant?

4.7 Draw the curve of heat absorption versus T_{out} for a CSTR with autorefrigeration. Superimpose this on the S-shaped heat generation curve and argue how autorefrigeration can give rise to a stable steady state.

4.8 The reaction

$$A \xrightarrow{k_{\mathrm{I}}} B \xrightarrow{k_{\mathrm{II}}} C$$

is occurring in an isothermal, piston flow reactor having a mean residence time of 2 min. Assume constant cross section and physical properties and

$$k_{\mathrm{I}} = 1.2 \times 10^{15} e^{-12,000/T} \quad \min^{-1}$$

$$k_{\mathrm{II}} = 9.4 \times 10^{15} e^{-14,700/T} \quad \min^{-1}$$

(a) Find the operating temperature that maximizes b_{out} given $b_{\text{in}} = 0$. Your answer should be precise to the nearest 1 K.

(b) The laboratory data were confused: k_{I} was interchanged with k_{II}. Revise your answer accordingly.

4.9 Your kinetic studies on the partial oxidation of *o*-xylene (see Problem 3.4) are successfully complete. The new catalyst provides good yields of phthalic anhydride with minimum CO_2 formation. You are now ready to design a single tube of what will ultimately be a multitubular reactor having 10,000 tubes. The xylene concentration must be low to remain outside the explosive limits. Thus oxygen is in great excess, and you found both reactions to be pseudo-first order with respect to the organics. They have normal Arrhenius temperature dependencies. Air is used as the oxygen source. The tube has a constant radius and will be jacketed with molten salts to maintain a constant value for T_{ext}. All thermal properties are constant, as is the gas viscosity. Pressure drop down the tube is significant and must be considered. Ideal gas behavior may be assumed. Set up the marching-ahead equations needed to find the yield of phthalic anhydride, the temperature, and the pressure at

[4] Presumably not from your school.
[5] You might think the engineer was fired. More likely, he was promoted based on the commercial success of the pilot plant work, is now your boss, and will expect you to deliver planned capacity from the new reactor.

the reactor outlet. Show all necessary equations and the sequence in which they would be evaluated.

4.10 Consider the reversible reaction

$$A \underset{k_r}{\overset{k_f}{\rightleftharpoons}} B$$

Suppose it is conducted in an isothermal piston flow reactor having $\bar{t} = 3$ s. Assume $b_{in} = 0$.

(a) Find the temperature T_{opt} that will maximize b_{out} given $k_f = 10^{10} \exp(-10,000/T)$ s^{-1} and $k_r = 10^{12} \exp(-12,000/T)$ s^{-1} where T is in degrees Kelvin.

(b) You were working late and weren't minding your r's and f's. The real data are $k_f = 10^{12} \exp(-12,000/T)$ s^{-1} and $k_r = 10^{10} \exp(-10,000/T)$ s^{-1}. What does this do to your optimization?

4.11 Dilute acetic acid is to be made by the hydrolysis of acetic anhydride at 25°C. Pseudo-first order rate constants are available at 10°C and 40°C. They are $k = 3.40$ hr^{-1} and 22.8 hr^{-1}, respectively. Estimate k at 25°C.

4.12 The disproportionation of p-toluenesulfonic acid has the following stoichiometry:

$$3(CH_3C_6H_4SO_2H) \rightarrow CH_3C_6H_4SO_2SC_6H_4CH_3 + CH_3C_6H_4SO_3H + H_2O$$

The following batch data[6] were obtained at 70°C in a reaction medium consisting of acetic acid plus 0.56-molar H_2O plus 1.0-molar H_2SO_4:

t (hr)	$[CH_3C_6H_4SO_2H]^{-1}$ (molarity^{-1})
0	5
0.5	8
1.0	12
1.5	16
4.0	36
5.0	44
6.0	53

Note that reciprocal concentrations are often cited in the chemical kinetics literature for second order reactions. Confirm that second order kinetics do provide a good fit and determine the rate constant.

4.13 Hinshelwood and Green[7] studied the homogeneous, gas phase reaction

$$2NO + 2H_2 \rightarrow N_2 + 2H_2O$$

at 1099 K in a constant volume batch reactor. The reactor was charged with known partial pressures of NO and H_2, and the course of the reaction was monitored by

[6]J. L. Kice and K. W. Bowers, *J. Am. Chem. Soc.*, **84**, 605 (1962).
[7]C. N. Hinshelwood and T. W. Green, *J. Chem. Soc.*, **1926**, 730 (1926).

the total pressure. Following are the data for one of their runs:

$$P_{NO}^0 = 406 \text{ mm} \qquad P_{H_2}^0 = 289 \text{ mm} \qquad P_0 = 695 \text{ mm}$$

t (s)	$\Delta \text{mm} = P_0 - P_t$
8	10
13	20
19	30
26	40
33	50
43	60
54	70
69	80
87	90
110	100
140	110
204	120
310	127
∞	144.5

Suppose $\mathcal{R} = k[NO]^\alpha[H_2]^\beta$. Use these data to estimate the overall reaction order $\alpha + \beta$.

4.14 The kinetic study by Hinshelwood and Green cited in Problem 4.13 also included initial rate measurements:

P_{NO}^0 (mm)	$P_{H_2}^0$ (mm)	$\mathcal{R}^0$ (mm/s)
359	400	1.50
300	400	1.03
152	400	0.25
400	300	1.74
310	300	0.92
232	300	0.45
400	289	1.60
400	205	1.10
400	147	0.79

(a) Use these to estimate α and β assuming $\mathcal{R} = k[NO]^\alpha[H_2]^\beta$.

(b) Are these results consistant with those of Problem 4.13?

4.15 At extreme pressures, liquid phase reactions exhibit pressure effects. A suggested means for correlation is the **activation volume**, $\Delta V^{\#}$. Thus

$$k = k_0 \exp\left(\frac{-E}{R_g T}\right) \exp\left(\frac{-\Delta V^{\#} P}{R_g T}\right)$$

Di-t-butyl peroxide is a commonly used free-radical initiator that decomposes according to first order kinetics. Use the following data[8] to estimate $\Delta V^{\#}$ for the

---

[8] C. Walling and G. Metzger, *J. Am. Chem. Soc.*, **81**, 5365 (1959).

decomposition in toluene at 120°C.

P (kg/cm²)	k (s^{-1})
1	13.4×10^{-6}
2040	9.5×10^{-6}
2900	8.0×10^{-6}
4480	6.6×10^{-6}
5270	5.7×10^{-6}

4.16 The ethane iodination reactor in Example 4.7 has been rerun at two atmospheres total pressure to produce the following data:

$[I_2]_{in}$	$[C_2H_6]_{in}$	$\bar{t}$	$[I_2]_{out}$	$[C_2H_6]_{out}$	$[HI]_{out}$	$[C_2H_6I]_{out}$
0.1	1.9	150	0.0783	1.878	0.0222	0.0220
0.1	1.9	650	0.0358	1.839	0.0641	0.0609
0.1	1.9	1150	0.0200	1.821	0.082	0.0803

(a) Combine these data with those in Example 4.7 to estimate α and β.

(b) Suppose the reaction mechanism is

$$I_2 + M \underset{k_{-1}}{\overset{k_1}{\rightleftharpoons}} 2I\cdot + M \tag{I}$$

$$I\cdot + C_2H_6 \overset{k_2}{\rightarrow} C_2H_5\cdot + HI \tag{II}$$

$$C_2H_5\cdot + I_2 \overset{k_3}{\rightarrow} C_2H_5I + I\cdot \tag{III}$$

Apply the steady state hypothesis to determine a functional form for $\mathcal{R}$.

4.17 The decomposition reaction

$$C_2H_5I \rightarrow C_2H_4 + HI$$

may be significant at 603 K.

(a) Would it affect the kinetic analysis in Example 4.7 and Problem 4.16?

(b) Do the data in Example 4.7 and Problem 4.16 provide any evidence for the decomposition reaction?

(c) Hydrogen iodide itself decomposes according to

$$2HI \underset{k_r}{\overset{k_f}{\rightleftharpoons}} H_2 + I_2$$

The forward rate constant is $k_f = 10^{14} \exp(-44,000/R_gT)$. Is this decomposition significant in the ethane iodination study?

4.18 The ordinary burning of sulfur produces SO_2. As a first step in the manufacture of sulfuric acid, SO_2 is oxidized to SO_3 in a gas–solid catalytic reaction. The catalyst increases the reaction rate but does not shift the gas phase equilibrium.

(a) Determine the heat of reaction for SO_2 oxidation at 600 K and 1 atm.

(b) Determine the mole fractions at equilibrium of N_2, O_2, SO_2, and SO_3 at 600 K and 1 atm given an initial composition of 79 mole-% N_2, 15 mole-% O_2 and 6 mole-% SO_2. Assume that the nitrogen is inert.

4.19 (a) Determine the equilibrium distribution of the three pentane isomers given the following data on free energies of formation at 600 K:

$$\Delta G_{\mathscr{F}}^0 = 40{,}000 \text{ J/mol of } n\text{-pentane}$$

$$\Delta G_{\mathscr{F}}^0 = 34{,}000 \text{ J/mol of isopentane}$$

$$\Delta G_{\mathscr{F}}^0 = 37{,}000 \text{ J/mol of neopentane}$$

The ideal gas law may be assumed.

(b) A mixture containing 76 mole-% n-pentane, 22 mole-% isopentane, and 2 mole-% neopentane is converted to a near equilibrium mixture. How many phase rule variables can be independently specified for the reactor effluent?

4.20 Example 4.15 in Appendix 4.1 treats the high-temperature equilibrium of four chemical species: N_2, O_2, NO, and NO_2. There are two other oxides of nitrogen that could have been included: N_2O and N_2O_4. (One might also add N, O, N_2O_2, N_2O_3, N_2O_5, NO_3, and O_3 but we will leave that for another day.)

(a) Find a set of independent chemical reactions that can be used to determine equilibria for the six species N_2, O_2, NO, NO_2, N_2O, and N_2O_4.

(b) Using the data of Tables 4.1 and 4.2, determine the thermodynamic equilibrium constants at 1500 K for each of the reactions in (a).

(c) Find the equilibrium mole fractions at 1500 K and 1 atm of the six chemical species given an initial mixture of 79 mole-% nitrogen and 21 mole-% oxygen.

4.21 Equal volumes of carbon dioxide and water are reacted at 2000 K and 10^5 Pa. Use the data of Figure 4.15 to construct an equilibrium model for this reaction. Exclude any species with equilibrium concentrations less than 1 ppm by volume.

4.22 The following reaction has been used to eliminate NO_x from the stack gases of stationary power plants:

$$NO_x + NH_3 + \tfrac{1}{2}\left(\tfrac{3}{2} - x\right)O_2 \rightleftharpoons N_2 + \tfrac{3}{2}H_2O$$

A zeolite catalyst operated at 1 atm and 325 to 500 K is sufficiently active that the above reaction approaches equilibrium. Use the results of Example 4.15 and other reasonable assumptions to estimate the amount of ammonia needed, per KWH of power generated, to reduce NO_x emissions by a factor of 10. Do an order of magnitude calculation. A precise answer is not required.

4.23 Suppose that the bimolecular, solid catalyzed, gas phase reaction

$$A + B \rightarrow C + D$$

has a rate expression of the form:

$$\mathcal{R} = \frac{kab}{\left(1 + k_A a + k_B b + k_C c + k_D d\right)^2}$$

Show how *linear* regression analysis can be used to determine k, k_A, k_B, k_C, and k_D from experimental rate data.

Appendix 4.1

Thermodynamics of Chemical Reactions

Thermodynamics is a fundamental engineering science that has many applications to chemical reactor design. Here we give a summary of two important topics: determination of heats of reaction for inclusion in energy balances and determination of free energies of reaction to calculate equilibrium compositions. The treatment is necessarily brief. The interested reader is referred to any standard textbook on thermodynamics.[9] Tables 4.1 and 4.2 provide selected thermodynamic data for use in the examples and for general utility in reaction engineering.

Heats of Reaction

Chemical reactions absorb or liberate energy, usually in the form of heat. The *heat of reaction*, $\Delta H_\mathcal{R}$, is defined as the amount of energy absorbed or liberated if the reaction goes to completion at a fixed temperature and pressure. When $\Delta H_\mathcal{R} > 0$, energy is absorbed and the reaction is said to be *endothermic*. When $\Delta H_\mathcal{R} < 0$, energy is liberated and the reaction is said to be *exothermic*. The magnitude of $\Delta H_\mathcal{R}$ depends on the temperature and pressure of the system and on the phases (e.g., gas, liquid, solid) of the various components. It also depends on an arbitrary constant multiplier in the stoichiometric equation.

Example 4.10   The reaction of hydrogen and oxygen is highly exothermic. At 298 K and 1 atm,

$$H_2(g) + \tfrac{1}{2}O_2(g) \rightarrow H_2O(g) \qquad \Delta H_\mathcal{R} = -197{,}806 \text{ J} \qquad \text{(I)}$$

Alternatively, we can write

$$2H_2(g) + O_2(g) \rightarrow 2H_2O(g) \qquad \Delta H_\mathcal{R} = -395{,}612 \text{ J} \qquad \text{(II)}$$

[9]See, for example, J. M. Smith and H. C. Van Ness, *Introduction to Chemical Engineering Thermodynamics*, 4th ed., McGraw-Hill, New York, 1986.

Table 4.1 **Standard Heats of Formation and Gibbs Free Energies of Formation at 298 K[a]**

Chemical species		State	$\Delta H_{\mathscr{F}}^0$	$\Delta G_{\mathscr{F}}^0$
Paraffins:				
Methane	CH_4	g	$-74,520$	$-50,460$
Ethane	C_2H_6	g	$-83,820$	$-31,855$
Propane	C_3H_8	g	$-104,680$	$-24,290$
n-Butane	C_4H_{10}	g	$-125,790$	$-16,570$
n-Pentane	C_5H_{12}	g	$-146,760$	$-8,650$
n-Hexane	C_6H_{14}	g	$-166,920$	150
n-Heptane	C_7H_{16}	g	$-187,780$	8,260
n-Octane	C_8H_{18}	g	$-208,750$	16,260
1-Alkenes:				
Ethylene	C_2H_4	g	52,510	68,460
Propylene	C_3H_6	g	19,710	62,205
1-Butene	C_4H_8	g	-540	70,340
1-Pentene	C_5H_{10}	g	$-21,280$	78,410
1-Hexene	C_6H_{12}	g	$-41,950$	86,830
Miscellaneous Organics:				
Acetaldehyde	C_2H_4O	g	$-166,190$	$-128,860$
Acetic acid	$C_2H_4O_2$	l	$-484,500$	$-389,900$
Acetylene	C_2H_2	g	227,480	209,970
Benzene	C_6H_6	g	82,930	129,665
Benzene	C_6H_6	l	49,080	124,520
1,3-Butadiene	C_4H_6	g	109,240	149,795
Cyclohexane	C_6H_{12}	g	$-123,140$	31,920
Cyclohexane	C_6H_{12}	l	$-156,230$	26,850
1,2-Ethanediol	$C_2H_6O_2$	l	$-454,800$	$-323,080$
Ethanol	C_2H_6O	g	$-235,100$	$-168,490$
Ethanol	C_2H_6O	l	$-277,690$	$-174,780$
Ethylbenzene	C_8H_{10}	g	29,920	130,890
Ethylene oxide	C_2H_4O	g	$-52,630$	$-13,010$
Formaldehyde	CH_2O	g	$-108,570$	$-102,530$
Methanol	CH_4O	g	$-200,660$	$-161,960$
Methanol	CH_4O	l	$-238,660$	$-166,270$
Styrene	C_8H_8	g	147,360	213,900
Toluene	C_7H_8	g	50,170	122,050
Toluene	C_7H_8	l	12,180	113,630
Miscellaneous Inorganics:				
Ammonia	NH_3	g	$-46,110$	$-16,450$
Calcium carbide	CaC_2	s	$-59,800$	$-64,900$
Calcium carbonate	$CaCO_3$	s	$-1,206,920$	$-1,128,790$
Calcium chloride	$CaCl_2$	s	$-795,800$	$-748,100$
Calcium hydroxide	$Ca(OH)_2$	s	$-986,090$	$-898,490$

Table 4.1 Continued

Chemical species		State	$\Delta H_{\mathscr{F}}^0$	$\Delta G_{\mathscr{F}}^0$
Miscellaneous Inorganics:				
Calcium oxide	CaO	s	−635,090	−604,030
Carbon dioxide	CO_2	g	−393,509	−394,359
Carbon monoxide	CO	g	−110,525	−137,169
Hydrogen chloride	HCl	g	−92,307	−95,299
Hydrogen cyanide	HCN	g	135,100	124,700
Hydrogen sulfide	H_2S	g	−20,630	−33,560
Iron oxide (hematite)	Fe_2O_3	s	−824,200	−742,200
Iron oxide (magnetite)	Fe_3O_4	s	−1,118,400	−1,015,400
Nitric acid	HNO_3	l	−174,100	−80,710
Nitric oxide	NO	g	90,250	86,550
Nitrogen dioxide	NO_2	g	33,180	51,310
Dinitrogen oxide	N_2O	g	82,050	104,200
Dinitrogen tetroxide	N_2O_4	g	9,160	97,540
Sodium carbonate	Na_2CO_3	s	−1,130,680	−1,044,440
Sodium chloride	NaCl	s	−411,153	−384,138
Sodium hydroxide	NaOH	s	−425,609	−379,494
Sulfur dioxide	SO_2	g	−296,830	−300,194
Sulfur trioxide	SO_3	g	−395,720	−371,060
Sulfuric acid	H_2SO_4	l	−813,989	−690,003
Water	H_2O	g	−241,818	−228,572
Water	H_2O	l	−285,830	−237,129

[a]Selected from J. M. Smith and H. C. Van Ness, Introduction to Chemical Engineering Thermodynamics, 4th ed., McGraw-Hill, New York, 1986.

which differs from the first reaction by a factor of $+2$. The decomposition of water would be highly endothermic:

$$H_2O(g) \rightarrow H_2(g) + \tfrac{1}{2}O_2(g) \qquad \Delta H_{\mathscr{R}} = +197,806 \text{ J} \tag{III}$$

Reaction III was obtained from Reaction I using a constant multiplier of -1.

Suppose $\Delta H_{\mathscr{R}}$ for Reaction I was measured in a calorimeter. Hydrogen and oxygen were charged at 298 K and 1 atm. The reaction occurred, the system was restored to 298 K and 1 atm but the product water was not condensed. Had it been condensed, the measured exothermicity would have been larger:

$$H_2(g) + \tfrac{1}{2}O_2(g) \rightarrow H_2O(l) \qquad \Delta H_{\mathscr{R}} = -241,818 \text{ J} \tag{IV}$$

Reactions I and IV differ by the heat of vaporization:

$$H_2O(l) \rightarrow H_2O(g) \qquad \Delta H_{\mathscr{R}} = +44,012 \text{ J} \tag{V}$$

Reactions IV and V can obviously be summed to give Reaction I.

Table 4.2 **Heat Capacity Constants for Various Substances at Low Pressures**[a]
$C_p = R_g(A + BT + CT^2 + D/T^2)$, for T from 298 K to T_{max}

Chemical Species		State	T_{max}	A	$B \times 10^2$	$C \times 10^6$	$D \times 10^5$
Paraffins							
Methane	CH_4	g	1500	1.702	9.081	-2.164	0
Ethane	C_2H_6	g	1500	1.131	19.225	-5.561	0
Propane	C_3H_8	g	1500	1.213	28.785	-8.824	0
n-Butane	C_4H_{10}	g	1500	1.935	36.915	-11.402	0
n-Pentane	C_5H_{12}	g	1500	2.464	45.351	-14.111	0
n-Hexane	C_6H_{14}	g	1500	3.025	53.722	-16.791	0
n-Heptane	C_7H_{16}	g	1500	3.570	62.127	-19.486	0
n-Octane	C_8H_{18}	g	1500	8.163	70.567	-22.208	0
1-Alkenes:							
Ethylene	C_2H_4	g	1500	1.424	14.394	-4.392	0
Propylene	C_3H_6	g	1500	1.637	22.706	-6.915	0
1-Butene	C_4H_8	g	1500	1.967	31.630	-9.873	0
1-Pentene	C_5H_{10}	g	1500	2.691	29.753	-12.447	0
1-Hexene	C_6H_{12}	g	1500	3.220	48.189	-15.157	0
Miscellaneous Organics							
Actaldehyde	C_2H_4O	g	1000	1.693	17.978	-6.158	0
Acetylene	C_2H_2	g	1500	6.132	1.952	0	-1.299
Benzene	C_6H_6	g	1500	-0.206	39.064	-13.301	
Benzene	C_6H_6	l	373	-0.747	67.96	-37.78	0
1,3-Butadiene	C_6H_6	g	1500	2.734	26.786	-8.882	0
Cyclohexane	C_6H_{12}	g	1500	-3.876	63.249	-20.928	0
Cyclohexane	C_6H_{12}	l	373	-9.048	141.38	-161.62	0
Ethanol	C_2H_6O	g	1500	3.518	20.001	-6.002	0
Ethanol	C_2H_6O	l	373	33.866	-172.60	349.17	0
Ethylbenzene	C_8H_{10}	g	1500	1.124	55.380	-18.476	0
Ethylene oxide	C_2H_4O	g	1000	-0.385	23.463	-9.296	0
Formaldehyde	CH_2O	g	1500	2.264	7.022	-1.877	0
Methanol	CH_4O	g	1500	2.211	12.216	-3.450	0
Methanol	CH_4O	l	373	13.431	-51.28	131.13	0
Styrene	C_8H_8	g	1500	2.050	50.192	-16.662	0
Toluene	C_7H_8	g	1500	0.290	47.052	-15.716	0
Toluene	C_7H_8	l	373	15.133	6.79	16.35	0
Miscellaneous Inorganics							
Ammonia	NH_3	g	1800	3.578	3.020	0	-0.186
Calcium carbide	CaC_2	s	720	8.254	1.429	0	-1.042
Calcium carbonate	$CaCO_3$	s	1200	12.572	2.637	0	-3.120
Calcium chloride	$CaCl_2$	s	1055	8.646	1.530	0	-0.302
Calcium hydroxide	$Ca(OH)_2$	s	700	9.597	5.435	0	0
Calcium oxide	CaO	s	2000	6.104	0.443	0	-1.047
Carbon dioxide	CO_2	g	2000	5.457	1.045	0	-1.157

Table 4.2 Continued

Chemical Species		State	T_{max}	A	$B \times 10^2$	$C \times 10^6$	$D \times 10^5$
Miscellaneous Inorganics							
Carbon monoxide	CO	g	2500	3.376	0.557	0	−0.031
Hydrogen chloride	HCl	g	2000	3.156	0.623	0	0.151
Hydrogen cyanide	HCN	g	2500	4.736	1.359	0	−0.725
Hydrogen sulfide	H_2S	g	2300	3.931	1.490	0	−0.232
Iron oxide (hematite)	Fe_2O_3	s	960	11.812	9.697	0	−1.976
Iron oxide (magnetite)	Fe_3O_4	s	850	9.594	27.112	0	0.409
Nitric oxide	NO	g	2000	3.387	0.629	0	0.014
Nitrogen dioxide	NO_2	g	2000	4.982	1.195	0	−0.792
Dinitrogen oxide	N_2O	g	2000	5.328	1.214	0	−0.928
Dinitrogen tetroxide	N_2O_4	g	2000	11.660	2.257	0	−2.787
Sodium chloride	NaCl	s	1073	5.526	1.963	0	0
Sodium hydroxide	NaOH	s	566	0.121	16.316	0	1.948
Sulfur dioxide	SO_2	g	2000	5.699	0.801	0	−1.015
Sulfur trioxide	SO_3	g	2000	8.060	1.056	0	−2.028
Water	H_2O	g	2000	3.470	1.450	0	0.121
Water	H_2O	l	373	8.712	1.25	−0.18	0
Elements							
Carbon (graphite)	C	s	2000	1.771	0.771	0	−0.867
Chlorine	Cl_2	g	3000	4.442	0.089	0	−0.344
Hydrogen	H_2	g	3000	3.249	0.422	0	0.083
Iron (alpha)	Fe	s	1043	−0.111	6.111	0	1.150
Nitrogen	N_2	g	2000	3.280	0.593	0	0.040
Oxygen	O_2	g	2000	3.639	0.506	0	−0.227
Sodium	Na	s	371	1.988	4.688	0	0
Sulfur (rhombic)	S	s	368	4.114	−1.728	0	−0.783

[a]Selected from J. M. Smith and H. C. Van Ness, Introduction to Chemical Engineering Thermodynamics, 4th ed., McGraw-Hill, New York, 1986.

The heats of reaction associated with stoichiometric equations are additive just as the equations themselves are additive.[10] This fact lends itself to tabulating $\Delta H_{\mathscr{R}}$ for a relatively few canonical reactions that can be algebraically summed to

---

[10] Some authors illustrate this fact by treating the heat of reaction as a pseudoreactant. Thus they write

$$H_2(g) + \tfrac{1}{2}O_2(g) \rightarrow H_2O(l) + 241,818 \text{ J}$$

This approach is marred by the thermodynamic convention that regards heat as positive when absorbed by the system. The convention is logical for mechanical engineers concerned with heat engines. Chemical reaction engineers would have chosen the opposite convention.

give $\Delta H_{\mathscr{R}}$ for the reaction of interest. The canonical reactions are the **heats of formation** from the elements. The participating species in these reactions are the elements as reactants and a single chemical compound as the product. Table 4.1 gives **standard heats of formation** $\Delta H_{\mathscr{F}}^0$ at 298 K for a variety of compounds. The reacting elements and the product compound are all assumed to be in **standard states**. The standard states are usually the pure compound at 1 atm (or perhaps 10^5 Pa).[11] The definition of standard state includes a specified temperature. It is usually 298 K for tabulated values of $\Delta H_{\mathscr{F}}^0$.

Example 4.11   Determine $\Delta H_{\mathscr{R}}$ for the dehydrogenation of ethylbenzene to styrene at 298 K and 1 atm.

    Solution   Table 4.1 gives $\Delta H_{\mathscr{F}}^0$ for styrene at 298 K. The formation reaction may be written as

$$8C(\text{graphite}) + 4H_2(g) \rightarrow \text{Styrene}(g) \qquad \Delta H_{\mathscr{R}} = 147{,}360 \text{ J} \tag{I}$$

For ethylbenzene, $\Delta H_{\mathscr{F}}^0 = 29{,}920$ J, but we write the stoichiometric equation using a multiplier of -1. Thus

$$-8C(\text{graphite}) - 5H_2(g) \rightarrow -\text{Ethylbenzene}(g) \qquad \Delta H_{\mathscr{R}} = -29{,}920 \text{ J} \tag{II}$$

Reactions I and II are algebraically summed to give

$$\text{Ethylbenzene}(g) \rightarrow \text{Styrene}(g) + H_2(g) \qquad \Delta H_{\mathscr{R}} = 117{,}440 \text{ J} \tag{III}$$

so that $\Delta H_{\mathscr{R}} = 117{,}440$ J per mole of ethylbenzene reacted. Note that the species participating in Reaction III are in their standard states since we used standard heats of formation for Reactions I and II. Thus we have actually obtained the **standard heat of reaction** $\Delta H_{\mathscr{R}}^0$.

    It does not matter that there is no known catalyst that can accomplish Reaction I directly. Heats of reaction, including heats of formation, depend on conditions before and after the reaction but not on the specific reaction path. Thus one might imagine a very complicated chemistry that has Reaction I as its overall result. Then $\Delta H_{\mathscr{R}} = +147{,}360$ J per mole of styrene formed is the net heat effect associated with this overall reaction.

    Reaction III is feasible as written but certainly not at temperatures as low as 298 K. It is adjusted for more realistic conditions in Example 4.12.

    If $\Delta H_{\mathscr{R}}$ is known at one temperature and pressure, it can be adjusted for another temperature and pressure using enthalpy data for the pure components.

---

[11] The standard state for gases is actually that for a hypothetical, ideal gas. Real gases are not perfectly ideal at 1 atm (or 10^5 Pa). Thus $\Delta H_{\mathscr{F}}^0$ at, say, 298 K will be slightly different than the actual $\Delta H_{\mathscr{F}}$ that would be measured at 1 atm (or 10^5 Pa) and 298 K. The difference is usually negligible.

This adjustment is usually done in separate steps: one in which temperature is varied at constant pressure and one in which pressure is varied at constant temperature. At constant pressure,

$$\frac{d \Delta H_{\mathscr{R}}}{dT} = \sum_{\text{Species}} \nu_{\text{A}}(C_p)_{\text{A}} = \Delta C_p \tag{4.61}$$

This equation governs the temperature dependence of $\Delta H_{\mathscr{R}}$. Selected specific heat data are provided in Table 4.2. Equation 4.61 can also be used to account for phase changes by adding the appropriate heats of vaporization, heats of crystallization, and so on into the integral of ΔC_p. If the reaction pressure is substantially different than the pressure for which $\Delta H_{\mathscr{R}}$ is known, the following equation can be used at constant temperature:

$$\frac{d \Delta H_{\mathscr{R}}}{dP} = \sum_{\text{Species}} \nu_{\text{A}}\left[\left(\frac{\partial H}{\partial P}\right)_T\right]_{\text{A}} = \Delta\left(\frac{\partial H}{\partial P}\right)_T \tag{4.62}$$

Example 4.12 Determine $\Delta H_{\mathscr{R}}$ for the ethylbenzene dehydrogenation reaction at 973 K and 0.5 atm.

Solution From Example 4.10, $\Delta H_{\mathscr{R}}^0 = 117{,}440$ J at 298 K. We need to calculate ΔC_p.

$$\Delta C_p = \sum_{\text{Species}} \nu_{\text{A}}(C_p)_{\text{A}} = (C_p)_{\text{styrene}} + (C_p)_{\text{hydrogen}} - (C_p)_{\text{ethylbenzene}}$$

Using the data of Table 4.2 gives

$$\frac{\Delta C_p}{R_g} = 4.175 - \frac{(4.766)T}{10^3} + \frac{(1.814)T^2}{10^6} + \frac{8300}{T^2}$$

Integrating Equation 4.61 gives

$$\Delta H_{\mathscr{R}} = \Delta H_{\mathscr{R}}^0 + R_g \int_{298}^{T} \Delta C_p \, dT'$$

$$= 117{,}400 + 8.314\left[4.175(T - 298) - \frac{4.776}{2 \times 10^3}(T^2 - 298^2)\right.$$

$$\left. + \frac{1.814}{3 \times 10^6}(T^3 - 298^3) + 8300\left(\frac{1}{T} - \frac{1}{298}\right)\right]$$

This result gives $\Delta H_{\mathscr{R}}$ as a function of T. Setting $T = 973$ K gives $\Delta H_{\mathscr{R}} = 128{,}130$ J.

Regarding the pressure change from 1 atm to 0.5 atm, the temperature is high and the pressure is low relative to critical conditions for all three components. Thus an ideal gas assumption is reasonable, and $(\partial H / \partial P)_T \approx 0$ so that there is no pressure correction. Thus $\Delta H_{\mathscr{R}} \approx \Delta H_{\mathscr{R}}^0$ at 973 K and 0.5 atm.

Reaction Equilibria

We consider homogeneous reactions only. A general condition of equilibrium is that the Gibbs free energy assume a minimum value. For a multicomponent reaction mixture, this condition can be restated as

$$\sum_{\text{Species}} \nu_A \bar{G}_A = 0 \tag{4.63}$$

where $\bar{G}_A$ is the partial molar free energy of component A. The fugacity of A in the reaction mixture, $\hat{f}_A$, is related to G_A through the defining equation

$$d\bar{G}_A = R_g T \, d\left(\ln \hat{f}_A\right) \tag{4.64}$$

This equation can be integrated subject to the boundary condition that

$$\bar{G}_A = G_A^0 \qquad \text{when } \hat{f}_A = f_A^0 \tag{4.65}$$

where G_A^0 and f_A^0 denote the free energy and fugacity of A in the standard state. Solution of Equations 4.64 and 4.65 gives

$$\bar{G}_A = G^0 + R_g T \ln \frac{\hat{f}_A}{f_A^0} \tag{4.66}$$

Substitution into Equation 4.63 gives

$$\sum_{\text{Species}} \nu_A G_A^0 = -R_g T \sum_{\text{Species}} \nu_A \ln \left(\frac{\hat{f}_A}{f_A^0}\right) = -R_g T \ln \prod_{\text{Species}} \left(\frac{\hat{f}_A}{f_A^0}\right)^{\nu_A} \tag{4.67}$$

The left-hand side of the result is the *standard Gibbs free energy of reaction*:

$$\Delta G_{\mathscr{R}}^0 = \sum_{\text{Species}} \nu_A G_A^0 \tag{4.68}$$

It can be determined from standard free energies of formation $\Delta G_{\mathscr{F}}^0$ in the same way as $\Delta H_{\mathscr{R}}^0$ is determined from heats of formation. Table 4.1 provides selected values for $\Delta G_{\mathscr{F}}^0$. Like $\Delta H_{\mathscr{R}}^0$, $\Delta G_{\mathscr{R}}^0$ is a function of temperature but not of pressure.

The right-hand side of Equation 4.67 is used to define the *thermodynamic equilibrium constant*:

$$K = \prod_{\text{Species}} \left(\frac{\hat{f}_A}{f_A^0} \right)^{\nu_A} \tag{4.69}$$

so that the equilibrium condition becomes

$$\Delta G_{\mathscr{R}}^0 = -R_g T \ln K \tag{4.70}$$

Since $\Delta G_{\mathscr{R}}^0$ is a function only of temperature, so must be K. This does not mean that equilibrium compositions are independent of pressure but only that the product of fugacities in Equation 4.69 is independent of pressure. As will be seen below, a pressure dependence may arise when the fugacities are replaced by more normal concentration variables such as mole fractions or concentrations. The reader may also note that, while numerical values for $\Delta G_{\mathscr{F}}^0$ and K will depend on the choice of standard state, the equilibrium composition is independent of this choice. Figure 4.15 shows the equilibrium constant for some common gas phase reactions.

We have chosen to call K the *thermodynamic* equilibrium constant to distinguish it from the *chemical* equilibrium constant:

$$K_{\text{equil}} = \prod_{\text{Species}} a^{\nu_A} \tag{4.71}$$

Neither is truly constant, but K depends only on temperature and is dimensionless whereas K_{equil} may depend on pressure and will have dimensions except when $\Sigma \nu_A = 0$. Furthermore, K_{equil} is equal to the ratio of forward and reverse rate constants for an elementary reaction whereas K need not support this interpretation. The thermodynamic equilibrium constant represents a theoretically sound approach to predicting reaction equilibrium. Unfortunately, it may not be useful in nonideal systems since the necessary thermodynamic data are seldom available. The chemical equilibrium constant represents an empirical approach to correlating equilibrium data. Unfortunately, it may not be accurate even at a fixed temperature and pressure if the initial reactant concentrations change too much.

The fugacity ratios appearing in Equation 4.69 can, of course, be related to direct measures of composition such as mole fractions or concentrations. We start with an approach suitable for gas phase reactions:

$$\hat{f}_A = y_A P \hat{\phi}_A \tag{4.72}$$

where y_A is the mole fraction and $\hat{\phi}_A$ the fugacity coefficient for component A.

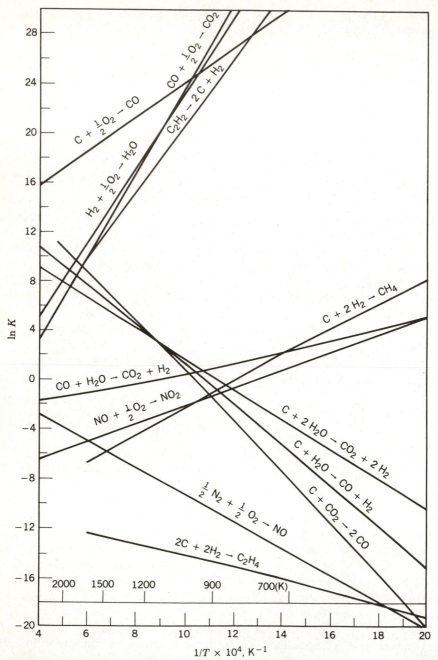

Figure 4.15 Thermodynamic equilibrium constants for some gas phase reactions. [J. M. Smith and H. C. Van Ness, *Introduction to Chemical Thermodynamics*, 4th ed., McGraw-Hill, New York, 1986, reproduced with permission.]

Note that the standard state for gaseous reactants is the pure gas in a hypothetical ideal gas state at a pressure of P^0. Thus $f_A^0 = P^0$, and Equation 4.69 becomes

$$K = \prod_{\text{Species}} \left[\frac{y_A P \hat{\phi}_A}{P^0} \right]^{\nu_A} = \left[\frac{P}{P^0} \right]^{\nu} \prod_{\text{Species}} [y_A \hat{\phi}_A]^{\nu_A} \qquad (4.73)$$

where $\nu = \Sigma \nu_A$.[12] For an ideal gas, $\hat{\phi}_A = 1$ and

$$K = \left[\frac{A}{P^0} \right]^{\nu} \prod_{\text{Species}} (y_A)^{\nu_A} \qquad (4.74)$$

or, converting from mole fractions to concentrations,

$$K = \left[\frac{R_g T}{P^0} \right]^{\nu} \prod_{\text{Species}} a^{\nu_A} = \left[\frac{R_g T}{P^0} \right]^{\nu} K_{\text{equil}} \qquad (4.75)$$

For incompressible liquid systems, the analog of Equation 4.73 is

$$K = \exp \left[\frac{P - P^0}{R_g T} \sum_{\text{Species}} \frac{\nu_A}{\rho_A} \right] \prod_{\text{Species}} (x_A \gamma_A)^{\nu_A} \qquad (4.76)$$

where x_A is the mole fraction and γ_A is the activity coefficient of component A in the equilibrium mixture. Should the mixture be ideal, $\gamma_A = 1$ and

$$K = \exp \left[\frac{P - P_0}{R_g T} \sum_{\text{Species}} \frac{\nu_A}{\rho_A} \right] \prod_{\text{Species}} x_A^{\nu_A}$$

$$= \rho_{\text{molar}}^{-\nu} \exp \left[\frac{P - P_0}{R_g T} \sum_{\text{Species}} \frac{\nu_A}{\rho_A} \right] K_{\text{equil}} \qquad (4.77)$$

Example 4.13 Estimate the equilibrium composition of the ethylbenzene dehydrogenation reaction at 298 K and 0.5 atm. Consider two cases:

(a) The initial composition was pure ethylbenzene.
(b) The initial composition was 1 mole each of ethylbenzene and styrene and 0.5 moles of hydrogen.

[12] Some authors set $P^0 = 1$ in an apparent simplification of Equation 4.73. This is formally incorrect, since the resulting equation is dimensionally inconsistent. It gives the correct numerical result only when $P^0 = 1$ in a given set of pressure units. It is inappropriate with SI units, since $P^0 = 101,325$ Pa (or perhaps 10^5 Pa).

Solution We first evaluate $\Delta G_{\mathcal{R}}^0$ using values of $\Delta G_{\mathcal{F}}^0$ from Table 4.1. Using the same reaction combination as in Example 4.11,

$$\Delta G_{\mathcal{R}}^0 = 213,900 - 130,890 = 83,010 \text{ J at 298 K}$$

Equation 4.70 gives $K = 2.8 \times 10^{-15}$ so that equilibrium at 298 K overwhelmingly favors ethylbenzene. Suppose the ideal gas assumption is not too bad even at this low temperature. Then equation 4.74 gives

$$2.8 \times 10^{-15} = \left[\frac{0.5}{1.0}\right]^{1.0} \frac{y_{H_2}^{1.0} y_{styrene}^{1.0}}{y_{ethylbenzene}^{1.0}}$$

The reaction coordinate method of Section 2.5 is a simple means for determining the various mole fractions when subject to the above equation and the constraints of stoichiometry. For part (a), assume 1 mole of ethylbenzene initially. At equilibrium there will be $1 - \varepsilon$ moles ethylbenzene and ε moles of each of styrene and hydrogen. The various mole fractions are

$$y_{ethylbenzene} = \frac{1 - \varepsilon}{1 + \varepsilon} \qquad y_{styrene} = y_{hydrogen} = \frac{\varepsilon}{1 + \varepsilon}$$

Thus

$$\frac{2.8 \times 10^{-15}}{0.5} = \frac{\varepsilon^2}{(1 - \varepsilon)(1 + \varepsilon)}$$

Solution gives $\varepsilon = 7.5 \times 10^{-8}$. Thus

$$y_{ethylbenzene} \approx 1.0 \qquad y_{styrene} = y_{hydrogen} = 7.8 \times 10^{-8}$$

For part (b), the equilibrium mixture will contain $1 - \varepsilon$ moles of ethylbenzene, $1 + \varepsilon$ moles of styrene, and $0.5 + \varepsilon$ moles of hydrogen. The various mole fractions are

$$y_{ethylbenzene} = \frac{1 - \varepsilon}{2.5 + \varepsilon} \qquad y_{styrene} = \frac{1 + \varepsilon}{2.5 + \varepsilon} \qquad y_{H_2} = \frac{0.5 + \varepsilon}{2.5 + \varepsilon}$$

Thus

$$5.6 \times 10^{-15} = \frac{(1 + \varepsilon)(0.5 + \varepsilon)}{(2.5 + \varepsilon)(1 - \varepsilon)}$$

Solution gives $\varepsilon \approx -0.5$. Thus

$$y_{ethylbenzene} \approx 0.75 \qquad y_{styrene} \approx 0.25 \qquad y_{H_2} \approx 0$$

We leave the problem of finding a more exact estimate for y_{H_2} as an exercise to the reader.

We now consider the problem of adjusting K for temperature. The governing equation is

$$\frac{d \ln K}{dT} = \frac{\Delta H_{\mathscr{R}}^0}{R_g T^2} \qquad\qquad (4.78)$$

which requires that $\Delta H_{\mathscr{R}}^0$ be determined as a function of temperature using Equation 4.61. To obtain a boundary condition for Equation 4.78, we must know a K' at some temperature T'. Table 4.2 allows $\Delta G_{\mathscr{R}}^0$ and thus K to be calculated at 298 K so that this is the usual boundary condition. When $\Delta C_p \approx 0$, $\Delta H_{\mathscr{R}}^0$ is constant, and Equation 4.78 can be integrated analytically:

$$\ln K = \left[\ln K' + \frac{\Delta H_{\mathscr{R}}^0}{R_g T'} \right] - \left(\frac{\Delta H_{\mathscr{R}}^0}{R_g} \right) \frac{1}{T} \qquad (4.79)$$

Thus a plot of $\ln K$ versus $1/T$ should yield a straight line. This is in fact true for many reactions (see Figure 4.15).

For exothermic reactions, $\Delta H_{\mathscr{R}}^0 < 0$, and Equation 4.78 shows that K will decrease with temperature, thus shifting the equilibrium composition toward the left. For endothermic reactions, $\Delta H_{\mathscr{R}}^0 > 0$, and K will increase with temperature, thus shifting the equilibrium toward the right.

Example 4.14   Estimate the equilibrium composition from the ethylbenzene dehydrogenation reaction at 973 K and 0.5 atm. The starting composition was pure ethylbenzene.

Solution   We first need $\Delta H_{\mathscr{R}}^0$ as a function of temperature. From Example 4.12,

$$\Delta H_{\mathscr{R}}^0 = 108,450 + 8.314 \left[4.175T - \frac{4.776}{2 \times 10^3} + \frac{1.814T^3}{3 \times 10^6} + \frac{8300}{T} \right]$$

This is substituted into Equation 4.78. The boundary condition follows from

$$\ln K' = -33.505 \qquad \text{at} \qquad T' = 298$$

Integrating Equation 4.76 gives

$$\ln K = -33.505 + \left[\frac{-108450}{8.314T} + 4.175 \ln T - \frac{4.776}{2 \times 10^3} T \right.$$
$$\left. + \frac{1.814T^2}{6 \times 10^6} - \frac{8300}{2T^2} \right]_{298}^{973}$$
$$= -0.492$$

so that $K = 0.61$ and, as in Example 4.13, part (a),

$$\frac{0.61}{0.5} = \frac{\varepsilon^2}{(1 - \varepsilon)(1 + \varepsilon)}$$

Solution gives $\varepsilon = 0.74$. The various mole fractions are

$$y_{\text{ethylbenzene}} = .15 \qquad y_{\text{styrene}} = y_{\text{hydrogen}} = .425$$

Multireaction Equilibria

When there are two or more simultaneous reactions, Equation 4.70 is written for each reaction:

$$\left(\Delta G_{\mathscr{R}}^0 \right)_{\text{I}} = -R_g T \ln K_{\text{I}}$$

$$\left(\Delta G_{\mathscr{R}}^0 \right)_{\text{II}} = -R_g T \ln K_{\text{II}} \qquad \text{(4.80)}$$

$$\vdots \qquad \qquad \vdots$$

so that there are M thermodynamic equilibrium constants associated with M reactions involving N chemical components. The various equilibrium constants can be expressed in terms of the component mole fractions using Equation 4.73 or Equation 4.76. There will be N such mole fractions, but these can be expressed in terms of M reaction coordinates using the reaction coordinate method of Section 2.5. Thus Equations 4.80 become a set of M equations in M compositional unknowns: $\varepsilon_{\text{I}}, \varepsilon_{\text{II}}, \ldots$.

Example 4.15 At high temperature, atmospheric nitrogen can be converted to various oxides. Consider then only the two reactions

$$\tfrac{1}{2}N_2 + \tfrac{1}{2}O_2 \rightleftharpoons NO \qquad \text{(I)}$$

$$NO + \tfrac{1}{2}O_2 \rightleftharpoons NO_2 \qquad \text{(II)}$$

and suppose that at 1500 K, $K_{\text{I}} = 0.0033$ and $K_{\text{II}} = 0.011$. What is the equilibrium composition of air at 1500 K and 1 atm?

Solution   We approximate air as 21 mole percent O_2 and 79 mole percent N_2. Applying Equation 2.78 to an initial mixture consisting of 1 mole of air and no nitrogen oxides gives

$$
\begin{bmatrix} N_{N_2} \\ N_{O_2} \\ N_{NO} \\ N_{NO_2} \end{bmatrix} = \begin{bmatrix} .79 \\ .21 \\ 0 \\ 0 \end{bmatrix} + \begin{bmatrix} -.5 & 0 \\ -.5 & -.5 \\ +1 & -1 \\ 0 & +1 \end{bmatrix} \begin{pmatrix} \varepsilon_I \\ \varepsilon_{II} \end{pmatrix}
$$

or

$$N_{N_2} = 0.79 - 0.5\varepsilon_I$$

$$N_{O_2} = 0.21 - 0.5\varepsilon_I - 0.5\varepsilon_{II}$$

$$N_{NO} = \qquad + 1.0\varepsilon_I - 1.0\varepsilon_{II}$$

$$N_{NO_2} = \qquad\qquad\quad + 1.0\varepsilon_{II}$$

$$N_{total} = 1.00 \qquad\quad - 0.5\varepsilon_{II}$$

The various mole fractions are

$$y_{N_2} = \frac{0.79 - 0.5\varepsilon_I}{1 - \varepsilon_{II}}$$

$$y_{O_2} = \frac{0.21 - 0.5\varepsilon_I - 0.5\varepsilon_{II}}{1 - \varepsilon_{II}}$$

$$y_{NO} = \frac{\varepsilon_I - \varepsilon_{II}}{1 - \varepsilon_{II}}$$

$$y_{NO_2} = \frac{\varepsilon_{II}}{1 - \varepsilon_{II}}$$

The ideal gas law is an excellent approximation at 1500 K and 1 atm. Thus we use Equation 4.74 to give

$$0.0033 = \frac{\varepsilon_I - \varepsilon_{II}}{(0.79 - 0.5\varepsilon_I)^{1/2}(0.21 - 0.5\varepsilon_I - 0.5\varepsilon_{II})^{1/2}}$$

$$0.011 = \frac{\varepsilon_{II}(1 - \varepsilon_{II})^{1/2}}{(\varepsilon_I - \varepsilon_{II})(0.21 - 0.5\varepsilon_I - 0.5\varepsilon_{II})^{1/2}}$$

Solution gives $\varepsilon_I = 0.00135$, $\varepsilon_{II} = 6.7 \times 10^{-6}$. Thus the equilibrium composition is

$$y_{N_2} = 0.7893 \qquad y_{O_2} = 0.2093$$

$$y_{NO} = 0.0014 \qquad y_{NO_2} = 7 \times 10^{-6}$$

The above example illustrates the utility of the reaction coordinate method for solving equilibrium problems. There are no more equations than there are independent chemical reactions. However, in practical problems such as atmospheric chemistry and combustion, the number of reactions is very large; the various reaction coordinates will differ by many orders of magnitude; and the numerical solution can be fairly difficult. Problem 4.20 remains fairly simple but hints at the more general case.

Independent Reactions and the Phase Rule

In this section we consider the number of independent reactions that are necessary to develop equilibrium relationships between N chemical species. A systematic approach is the following:

1. List all chemical species, both elements and compounds, that are believed to exist at equilibrium. By "element" we mean the predominant species in the standard state, for example, O_2 for oxygen at 1 atm and 298 K.
2. Write the formation reactions from the elements for each compound. The term "compound" includes elemental forms other than the standard one, for example, we would consider monatomic oxygen a compound and write $\frac{1}{2}O_2 \rightarrow O$ as one of the reactions.
3. The stoichiometric equations are combined to eliminate any elements in their standard forms that are not believed to be present in significant amounts at equilibrium.

The result of the above procedures is M equations where $M \leq N$.

Example 4.16 Find a set of independent reactions to represent the equilibrium of CO, CO_2, H_2, and H_2O.

Solution Assume that only the stated species are present at equilibrium. Then there are three formation reactions:

$$H_2 + \tfrac{1}{2}O_2 \rightarrow H_2O$$

$$C + \tfrac{1}{2}O_2 \rightarrow CO$$

$$C + O_2 \rightarrow CO_2$$

The third reaction is subtracted from the second to eliminate carbon, giving the following set:

$$H_2 + \tfrac{1}{2}O_2 \to H_2O$$

$$-\tfrac{1}{2}O_2 \to CO - CO_2$$

$$C + O_2 \to CO_2$$

The second reaction is now added to the first to eliminate oxygen, giving the following set:

$$H_2 \to H_2O + CO - CO_2$$

$$-\tfrac{1}{2}O_2 \to CO - CO_2$$

$$C + O_2 \to CO_2$$

We discard those reactions used for the eliminations. This leaves only the first reaction, which can be rewritten as

$$CO + H_2O \to CO_2 + H_2$$

Thus $N = 4$ and $M = 1$ for this system. The final reaction can be recognized as the water–gas shift reaction.

Example 4.17 Find a set of independent reactions to represent the equilibrium products for a reaction between 1 mole of methane and 0.5 mole of oxygen.

Solution The most difficult part of this and many other reaction equilibrium problems is to decide what species will be present in significant concentrations. Experimental observations are the best guide to constructing an equilibrium model. Lacking this, exhaustive calculations or chemical insight must be used.[13] We shall assume that oxygen and hydrogen will not be present as elements but that carbon may be. Compounds to be considered are CH_4, CO_2,

[13]Except at very high temperatures, free-radical concentrations will be quite low, but free radicals could provide the reaction mechanisms by which equilibrium is approached. Reactions such as $2CH_3 \cdot \to C_2H_6$ will yield higher hydrocarbons so that the number of theoretically possible species is unbounded. In a low-temperature catalytic oxidation, some reactions might be impossible. However, this impossibility is based on kinetic considerations, not thermodynamics.

CO, H_2O, CH_3OH, and CH_2O. There are six formation reactions:

$$C + 2H_2 \rightarrow CH_4$$

$$C + O_2 \rightarrow CO_2$$

$$C + \tfrac{1}{2}O_2 \rightarrow CO$$

$$H_2 + \tfrac{1}{2}O_2 \rightarrow H_2O$$

$$C + 2H_2 + \tfrac{1}{2}O_2 \rightarrow CH_3OH$$

$$C + H_2 + \tfrac{1}{2}O_2 \rightarrow CH_2O$$

If carbon, hydrogen, and oxygen were all present as elements, none of the formation reactions could be eliminated. We would then have $N = 9$ and $M = 6$. With elemental hydrogen and oxygen assumed absent, two species and two equations can be eliminated, giving $N = 7$ and $M = 4$. Use the third and fourth equations to eliminate O_2 and H_2:

$$3C \rightarrow CH_4 + 2CO - 2H_2O$$

$$- C \rightarrow CO_2 - 2CO$$

$$C + \tfrac{1}{2}O_2 \rightarrow CO$$

$$C - H_2 \rightarrow CO - H_2O$$

$$2C \rightarrow CH_3OH + CO - 2H_2O$$

$$C \rightarrow CH_2O - H_2O$$

The third and fourth members of the set are now discarded and the remainder rewritten to give

$$3C + 2H_2O \rightarrow CH_4 + 2CO$$

$$2CO \rightarrow CO_2 + C$$

$$2C + 2H_2O \rightarrow CH_3OH + CO$$

$$C + H_2O \rightarrow CH_2O$$

These four equations are perfectly adequate for equilibrium calculations. They are not unique since various members of the set can be combined algebraically without reducing the dimensionality $M = 4$. An equivalent set of equations with a vaguely more pleasant appearance is

$$CO_2 + C \quad \rightarrow 2CO$$

$$CH_4 + CO_2 \rightarrow CH_3OH + CO$$

$$CH_3OH + CO \rightarrow 2CH_2O$$

$$CH_2O + CO_2 \rightarrow H_2O + 2CO$$

However, this set is as arbitrary as the first. By denying the existence of elemental oxygen at equilibrium, we have precluded the possibility that the equations used to calculate the equilibrium composition will bear any semblance to plausible mechanisms for the partial oxidation of methane.

We have thus far considered thermodynamic equilibrium in homogeneous systems. When two or more phases exist, it is necessary that the requirements for reaction equilibria (i.e., Equations 4.80) be satisfied simultaneously with the requirements for phase equilibria (i.e., that the component fugacities be equal in each phase). We leave the treatment of chemical equilibria in multiphase systems to the specialized literature. However, it is useful to cite the *phase rule for reactive systems* to comprehend the possibilities:

Degrees of freedom = 2 − Number of phases

$$+ N - M - \text{Special constraints} \qquad (4.81)$$

By degrees of freedom we mean the number of variables that can be independently specified. The usual variables are temperature, pressure, and $N - 1$ mole fractions for each phase. Equation 4.81 says how many of these can be arbitrarily chosen before the remainder assume prescribed values. The special constraints typically arise when an initial condition, usually the feed composition, dictates compositions in one of the phases. Note that the phase rule variables do not determine the phase ratios, that is, the ratios of moles present in each phase to total moles. Phase ratios represent additional degrees of freedom that are not considered in Equation 4.81. The following examples illustrate these ideas.

Example 4.18 Suppose ethane is cracked at high temperatures in a vacuum. What variables determine the equilibrium composition?

Solution The system is single phase. Assume that the possible components are C_2H_6, C_2H_4, C_2H_2, CH_4, H_2, and C so that $N = 6$.[14] There are four

---

[14] The cracking chemistry may give rise to trace quantities of other hydrocarbons. For each additional compound assumed present, there will be another formation reaction; and it will remain true that $N - M = 2$.

formation reactions. None can be eliminated since elemental carbon and hydrogen are present at equilibrium. Thus $M = 4$ and

$$\text{Degrees of freedom} = 2 - 1 + 6 - 4 - S = 3 - S$$

Since ethane is specified as the feed, the carbon-to-hydrogen ratio is fixed at $\frac{1}{3}$ and $S = 1$. Thus there are two degrees of freedom, and the equilibrium composition is determined by the temperature and pressure.

The feed composition can be manipulated experimentally by choosing various amounts of C_2H_6, H_2, and so on. There appear to be $N - 1 = 5$ variables which can be chosen arbitrarily; but only one feed variable, the carbon-to-hydrogen ratio, will have any effect on the equilibrium composition.

Example 4.19   One mole of water and 1 mole of ammonia are mixed under conditions where a liquid phase and a gas phase coexist. How many degrees of freedom are there?

Solution   There are two phases, three components, and one reaction:

$$NH_3 + H_2O \rightarrow NH_4OH$$

Thus

$$\text{Degrees of freedom} = 2 - 2 + 3 - 1 - S = 2 - S$$

The constraint of a fixed feed composition does not uniquely define either the gas or liquid phase compositions. The strength of the liquid ammonia and the ratio of NH_3 to H_2O in the vapor phase are free to vary with temperature and pressure.[15] Thus $S = 0$. However, there is a phase volume constraint imposed by the initial stoichiometry. If the phase compositions are measured, the phase ratios can be calculated from a component material balance.

---

[15] The reader may choose to regard the ammonia/water interaction as a solvation process rather than a chemical reaction. Then $N = 2$, $M = 0$, $S = 0$, and there remain two degrees of freedom. Alternatively, the reaction might be regarded as

$$NH_3 + H_2O \rightarrow NH_4^+OH^-$$

Now $N = 4$, $M = 1$ but $S = 1$ since $[NH_4^+] = [OH^-]$. Still another possibility to consider two liquid phase reactions:

$$NH_3 + H_2O \rightarrow NH_4^+ + OH^-$$

$$2H_2O \rightarrow H_3O^+ + OH^-$$

so that $N = 5$, $M = 2$, and $S = 1$ since $[NH_4^+] = [OH^-] - [H_3O]$ is a constraint on the liquid phase mole fractions. There remain two degrees of freedom.

Example 4.20 Two moles of hydrogen and 1 mole of oxygen are reacted under conditions where a liquid phase and a gas phase coexist. How many degrees of freedom are there?

Solution This might appear to be a case of two phases, three components, one reaction, and no special constraints. Actually, however, the reaction equilibrium

$$H_2 + \tfrac{1}{2}O_2 \rightarrow H_2O$$

is so far to the right at conditions where liquid water exists that there will be a maximum of two components: water and whichever gas is in stoichiometric excess. Neither is in excess in the current example, so that $N = 1$. Also, $M = 0$ since the reaction goes to completion and has no further role in establishing equilibrium. Thus

$$\text{Degrees of freedom} = 2 - 2 + 1 - 0 - 0 = 1$$

so that either temperature or pressure can be selected arbitrarily. This agrees with the experimental observation that water has a fixed vapor pressure at a given temperature.

In this example there is no phase volume constraint imposed by the initial stoichiometry. To determine the phase ratio using *intensive* properties, one would have to introduce an additional thermodynamic variable such as enthalpy. Of course, we could use an *extensive* property such as the mass of one phase. The phase ratio would then be found from an overall mass balance.

Appendix 4.2

Regression Analysis

Suppose measurements of a dependent variable, for example, reaction rate, have been made at many different values of the independent variables, for example, reactant concentrations. The general goal of regression analysis is to determine a functional relationship between the dependent variable and the various independent variables, for example, to develop an equation expressing $\mathscr{R}$ as a function of $a, b, \ldots$.

For concreteness, suppose w is the single dependent variable and there are three independent variables x, y, and z. We seek a function F where

$$w = F(x, y, z) \tag{4.82}$$

Regression analysis in itself will not tell us a functional form for F. However, given an assumed form, it will tell us the best values for constants in that form. For example, suppose that w is a linear function of x, y, and z:

$$w = k + \alpha x + \beta y + \gamma z \tag{4.83}$$

Then regression analysis can be used to determine best values for k, α, β, and γ. By best we mean those values that minimize the sum of squares:

$$S_2 = \sum_{j=1}^{J} (w_j - F_j)^2$$

$$= \sum_{j=1}^{J} (w_j - k - \alpha x_j - \beta y_j - \gamma z_j)^2 \tag{4.84}$$

Equation 4.84 supposes that J experiments have been performed which gave J determinations of the dependent variable w_j for J sets of the independent variables (x_j, y_j, z_j). We now regard the experimental data as fixed and treat the **model parameters**, k, α, β, γ, as the variables. The goal is to choose these model parameters such that $S_2 \geq 0$ achieves its minimum possible value. A necessary condition for S_2 to be a minimum is that

$$\frac{\partial S_2}{\partial k} = \frac{\partial S_2}{\partial \alpha} = \frac{\partial S_2}{\partial \beta} = \frac{\partial S_2}{\partial \gamma} = 0 \tag{4.85}$$

For the assumed linear form of Equation 4.83,

$$\frac{\partial S_2}{\partial k} = 2\sum (w_j - k - \alpha x_j - \beta y_j - \gamma z_j)(-1) = 0$$

$$\frac{\partial S_2}{\partial \alpha} = 2\sum (w_j - k - \alpha x_j - \beta y_j - \gamma z_j)(-x_j) = 0$$

$$\frac{\partial S_2}{\partial \beta} = 2\sum (w_j - k - \alpha x_j - \beta y_j - \gamma z_j)(-y_j) = 0 \tag{4.86}$$

$$\frac{\partial S_2}{\partial \gamma} = 2\sum (w_j - k - \alpha x_j - \beta y_j - \gamma z_j)(-z_j) = 0$$

These may be arranged to give

$$Jk + \alpha\sum x_j + \beta\sum y_j + \gamma\sum z_j = \sum w_j$$

$$k\sum x_j + \alpha\sum x_j^2 + \beta\sum x_j y_j + \gamma\sum x_j z_j = \sum w_j x_j$$

$$k\sum y_j + \alpha\sum x_j y_j + \beta\sum y_j^2 + \gamma\sum y_j z_j = \sum w_j y_j \tag{4.87}$$

$$k\sum z_j + \alpha\sum x_j z_j + \beta\sum y_j z + \gamma\sum z_j^2 = \sum w_j z_j$$

The various sums are known from the experimental data. Thus Equations 4.87 can be solved[16] for k, α, β, and γ.

Example 4.21 Use linear regression analysis to determine α and β for the ethane iodination reaction in Example 4.7.

Solution The assumed linear form is

$$\ln \mathcal{R} = \ln k + \alpha \ln [I_2] + \beta \ln [C_2H_6] \tag{4.88}$$

The data set is the following:

$w = \ln \mathcal{R}$	$x = \ln[I_2]$	$y = \ln[C_2H_6]$
-9.56	-2.49	$-.123$
-9.99	-3.21	$-.173$
-10.29	-3.81	$-.194$

If we attempt to evaluate all three constants, k, α, and β, the following set of equations is obtained:

$$3k - 9.51\alpha - 0.49\beta = -29.84$$

$$-9.51k + 31.0203\alpha + 1.60074\beta = 95.07720 \tag{4.89}$$

$$-0.49k + 1.60074\alpha + 0.082694\beta = 4.90041$$

Solution gives $k = -8.214$, $\alpha = 0.401$, $\beta = 2.82$ so that a tentative model for this ethane iodination reaction is

$$\ln \mathcal{R} = -8.214 + 0.401 \ln [I_2] + 2.82 \ln [C_2H_6] \tag{4.90}$$

This model uses as many parameters as there are observations and thus fits the data exactly, $S_2 = 0$. One can certainly doubt the significance of such a fit. It is clear that the data are not perfect, since the material balance is not perfect. Additional data could cause large changes in the values for k, α, and β. See Problem 4.16. Certainly, the value for β seems high and could be an artifact of the limited range over which $[C_2H_6]$ was varied. Suppose we pick $\beta = 1$ on theoretical grounds. Then regression analysis can be used to find best values for the remaining parameters. The data become

$w = \ln \mathcal{R} - \ln[C_2H_6]$	$x = \ln[I_2]$
-9.44	-2.49
-9.82	-3.21
-10.10	-3.81

---

[16] No solution will exist if there are fewer observations than model parameters. The model will fit the data exactly if there are as many parameters as observations.

The least squares equations are

$$3k_0 - 9.51\alpha = -29.36$$

$$-9.51k_0 + 31.0203\alpha = 93.5088$$

(4.91)

These give $k = -9.1988$ and $\alpha = 0.5009$ so that a possible model is

$$\ln \mathscr{R} = -8.1988 + 0.5009 \ln [I_2] + \ln [C_2H_6]$$

(4.92)

Since there are now only two fitted parameters, the model does not fit the data exactly. It is a least squares fit in the normal sense, with $S_2 > 0$. It provides quite a good fit of the data:

$(\ln \mathscr{R})_{observed}$	$(\ln \mathscr{R})_{predicted}$
-9.56	-9.57
-9.99	-9.98
-10.29	-10.30

Thus original use of β as an adjustable parameter provides very little improvement in the fit, confirming that the apparent value of $\beta = 2.82$ is very likely spurious.

Regression analysis is a powerful tool for fitting models but can obviously be misused. In the above example, we used physical reasoning to avoid a spurious result. Statistical reasoning is also helpful. Confidence intervals and other statistical measures of goodness of fit can be used to judge whether or not a given parameter is significant and it if should be retained in the model. Also, statistical analysis can help in the planning of experiments so that the new data will remove a maximum amount of uncertainty in the model. See any standard text on the statistical design of experiments.

Like the linear case, *nonlinear regression* seeks to minimize the sum of squares between a set of observations and a model:

$$S_2 = \sum_{j=1}^{J} (w_j - F_j)^2$$

(4.93)

The model predictions, F_j, depend on the parameters $k, \alpha, \beta, \ldots$. The parameters are chosen so that S becomes a minimum. This minimization can be done analytically in the linear case and generates Equations 4.87. Numerical minimization is needed for the nonlinear case, and nonlinear regression can be considered a special form of multiparameter optimization.

Example 4.21 Compare the objective functions for linear and nonlinear regression analysis of Equation 4.53.

Solution In the linear case, we transform Equation 4.53 to give the linear form of Equation 4.54. The sum-of-squares objective function becomes

$$(S_2)_{\text{linear}} = \sum_{j=1}^{J} \left(\ln \mathcal{R}_j - \ln k - \alpha \ln a_j - \beta \ln b_j - \cdots \right)^2 \tag{4.94}$$

while the nonlinear objective function is

$$(S_2)_{\text{nonlinear}} = \sum_{j=1}^{J} \left(\mathcal{R}_j - ka^\alpha b^\beta \ldots \right)^2 \tag{4.95}$$

In both cases we seek values for $k, \alpha, \beta, \ldots$ that minimize S_2 for the given collection of data. However, somewhat different values may be obtained. If there is a difference, the nonlinear case is preferred since it is based directly on the observed responses $\mathcal{R}_j$ rather than on a transformation of the responses.

Example 4.22 Nonlinear regression is an option for the rate expression of Equation 4.53. There are many situations where it is mandatory. For example,

$$\mathcal{R} = \frac{ka - k'b}{1 + k_A a + k_B b}$$

is a theoretical rate expression for heterogeneous catalysis as treated in Chapter 10. No linearizing transform is available so that one must minimize

$$S_2 = \sum_{j=1}^{J} \left[\mathcal{R}_j - \frac{ka - k'b}{1 + k_A a + k_B b} \right]^2$$

to find best values for k, k', k_A, and k_B.

There is no truly robust approach to nonlinear regression, and even the best methods can cause trouble in specific instances. A review and comparison of existing techniques is given in L. T. Biegler, J. J. Damiano, and G. G. Blau, *AIChE J.*, **32** 29 (1986).

Design and Optimization Studies

This chapter, or at least the examples in it, will be written by you. We start with a common type of reaction occurring in a system having plausible physical properties. Your job is to design a reactor that is not only plausible but that will also be optimal (in some sense) for the job.

The reaction is[1]

$$A \overset{k_{\text{I}}}{\rightarrow} B \overset{k_{\text{II}}}{\rightarrow} C \tag{5.1}$$

which takes place in a liquid phase system having constant physical properties, $\rho = 800$ kg/m³, $C_p = 1.88$ kJ kg K⁻¹. The initial concentration of A is 10 percent by weight, $b_0 = 0$. Other useful data are

$$k_{\text{I}} = 10^{10} \exp\left(-\frac{10{,}000}{T}\right) \text{ s}^{-1} \qquad (\Delta H_{\mathscr{R}})_{\text{I}} = -67.7 \text{ kJ/mol}$$

$$k_{\text{II}} = 10^{12} \exp\left(-\frac{12{,}000}{T}\right) \text{ s}^{-1} \qquad (\Delta H_{\mathscr{R}})_{\text{II}} = -51.5 \text{ kJ/mol}$$

$$M_{\text{A}} = 78 \qquad M_{\text{B}} = 106 \qquad M_{\text{C}} = 134 \tag{5.2s}$$

---

[1] This is a highly proprietary process and the chemistry is secret. The reactions cannot really be elementary and first order as shown. There is an increase in molecular weight, which means that some additional component must be involved. The molecular weights suggest that the reaction is the alkylation and dialkylation of benzene with ethylene. However, the heats of reaction are slightly different than would be expected for benzene alkylation. Leave the reaction as an unknown and concentrate on the structure of the problem rather than its chemical specifics.

Profit maximization is the objective function appropriate to most industrial optimizations. How profit relates to the reactor design depends on many factors. Some are:

1. Capital and operating costs for the reactor.
2. Capital and operating costs for the upstream feed preparation and the downstream product separation and recovery equipment.
3. Values (costs and selling prices) for the raw material A and the products B and C.

It is apparent from this list that overall profit optimization must consider the total system comprising the feed, reaction, and separation steps and perhaps other factors as well. Often, however, intuition or analysis will suggest that piecewise, local optimization of individual units will yield an overall design that is at or near the global economic optimum. Suppose this is true for the present system and that our local optimization problem is to design a reactor which maximizes the outlet concentration of B. Initially, we might try to ignore capital and operating costs completely; but we would then wind up designing an infinitely large reactor operating at absolute zero temperature. Thus the volume of the reactor must be constrained by capital costs, and the allowable operating temperatures must be constrained by materials of construction limitations or by the cost of utilities. The experienced engineer works within practical design constraints using judgment based largely on experience. Primary emphasis is on finding a design that is technically feasible (it will work) and economically plausible (it should make some money). He then explores a limited set of variations about the basic design in an attempt to improve the economics.

This course does not attempt to teach plant design strategy in the global sense of profit optimization. It does aim to teach some computational aspects of reactor design which form part of the overall strategy. Thus the current design and optimization exercise contains some rather arbitrary constraints. These are imposed to avoid "absurd" answers. In real life, such constraints should always be tested. Here, we accept them as part of the exercise.

Reactor residence times will be limited to a maximum of 3 s. This limitation is meant to reflect a capital constraint. It ignores the fact some reactor designs will cost more than others even though they have the same $\bar{t}$.

The search for the best reactor requires the evaluation of many alternatives even given the constraint on $\bar{t}$. An initial question is

Batch, Semibatch, or Continuous?

The problem as stated does not give sufficient information to make this choice.

Problem 5.1   Expand the problem statement so that a continuous-flow reactor is the logical or at least a plausible choice. Under what circumstances would you choose continuous operation? Build these circumstances into the problem statement?

Assume now that the reactor is indeed continuous. A next question might be

Perfect Mixer, Piston Flow, or a Combination?

You do have enough information to answer this question, at least from the viewpoint of kinetic optimization.

Problem 5.2 Use mathematical or numerical arguments to select the best reactor type. Show enough results to convince a skeptical reader or your boss that you have the right answer. In this part of the exercise, pick the best values for $\bar{t} \leq 3s$ and $T_{isothermal}$ (unconstrained) for your reactor.

Unless the best reactor type is a single perfect mixer, operation at a constant temperature, $T_{isothermal}$, may not be optimal (or even possible). Some options are:

Isothermal, Adiabatic or Other?

Note that these possibilities exist for perfect mixers in series, for combinations of perfect mixers and piston flow reactors, or for piston flow reactors alone.

Problem 5.3 Design the best adiabatic reactor for the given reaction system. Note that your design variables are now $\bar{t} \leq 3$ and T_{in}. Assume T_{in} can be arbitrarily selected. By now you should have persuaded yourself that a piston flow reactor operated at the maximum allowable residence time optimizes the yield of B. Thus for Problem 5.3, T_{in} is the only design variable. Find this best value.

Adiabatic operation is sometimes used even when the corresponding temperature profile is suboptimal for the kinetic scheme. It may be that the necessary provisions for heat exchange are more expensive than they are worth. Given the general class of adiabatic reactors, options still exist for improving the temperature profile. Some are:

Use of Inerts, Split Feed, Interstage Heating / Cooling?

Problem 5.4 Explore each of the above possibilities for the subject reaction.

For these calculations:

(a) Assume that the inerts have the same physical properties as the reaction mass. They are added to the feed and, through dilution, mitigate any temperature changes accompanying reaction.

(b) Assume that a secondary feed stream can be mixed instantaneously with the fluid already in the reactor. The design variables (other than $\bar{t}$) for split feed are T_{in}, $\bar{t}_{split} \leq \bar{t}$, T_{split}, and Q_{split}/Q_{in}. See Figure 5.1. With four variables, you can spend a lot of time and money on computer optimization. Pick a simple case or two that has some chance of success.

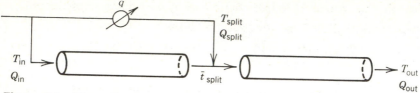

Figure 5.1   Piston flow reactor with split feed.

(c) Suppose an intermediate heat exchanger is located midway down the reactor. See Figure 5.2. What is the optimal pair of T_{in} and $T_{intermediate}$?

For these and all subsequent calculations, assume that utility limitations prevent any fluid stream from being cooled below 27°C or heated above 250°C. All intermediate temperatures are possible. Reaction temperatures higher than 250°C are possible if brought about by the reaction exotherm.

A temperature profile other than isothermal or adiabatic is often best when reaction selectivity is considered. Thus one might seek some best function, $T_{opt}(z)$, that varies continuously along the length of the reactor and that maximizes the outlet concentration of B given the fixed $\bar{t}$. Finding $T_{opt}(z)$ is a problem in *functional optimization* and is much more difficult than the problems in *parameter optimization* we have considered thus far. Also possibly difficult is the problem of physically imposing a desired axial temperature distribution on the real reactor.

Begin by considering a piston flow reactor with such good heat transfer that any desired temperature can be achieved just by setting the wall temperature. Divide the reactor into I equal-sized increments, in each of which the contents experience an age increment $\Delta\alpha = \bar{t}/I$. In a piston flow reactor the $\Delta\alpha$ values correspond to increments of length, ΔL. If all the $\Delta\alpha$'s are equal, so will be the ΔL's when $\bar{u}$ is constant. (Note that the lengths ΔL should be much longer than the Δz used in marching-ahead calculations.) We suppose now that each subreactor is maintained at a constant temperature T_i, and we wish to find those values for T_i which optimize the objective function. This is a problem in parameter optimization, the parameters being the T_i's. It is also a discrete approximation to the functional optimization discussed above, and one expects it to become a very good approximation if I is large.

Computer programs exist for solving multiparameter optimization problems. They can become quite expensive to use if I is large, and large now means something like 8. Use of large I may also be impractical or expensive to

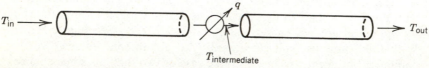

Figure 5.2   Piston flow reactor with intercooler.

implement on the real system. It is probably best to start with a small number of temperature zones, say, $I = 2$, and see if these offer much economic improvement over the single-zone case.

Problem 5.5   Determine the best two-zone temperature strategy for the consecutive reaction example. Specifically, find the isothermal zone temperatures, T_1 and T_2 that maximize b_{out}.

Usually, increasing the number of zones will increase the yield. The engineer must judge whether this improvement is worth the additional complexity in design. Often, two reaction zones will produce yields that are very close to the theoretical maximum. This is particularly true when we remove the restriction of equal ΔL and introduce the length of the first zone as a new, adjustable variable. A typical control strategy will then be $T_1 = T_{max}$ (or T_{min}) and $T_2 = T_{min}$ (or T_{max}), where T_{min} and T_{max} are extreme values for acceptable or achievable temperatures in the system. The strategy of operating at a maximum or minimum temperature for some length down the tube and then switching to the opposite extreme for the remainder of the reactor is known as **_bang-bang control_**. Very often, bang-bang control produces truly optimal results given the constraint that $T_{min} \leq T \leq T_{max}$. Suppose that the value of T_1 determined in Problem 5.5 is not physically achievable due to materials of construction or utilities limitations.

Problem 5.6   Lower T_1 from Problem 5.5 by 5°C, leave T_2 the same, and develop a bang-bang control strategy for the reactor. Specifically, pick the time $t_{switch} < \bar{t}$ at which the temperature switch should be made.

The above treatment leading to Problems 5.5 and 5.6 has presupposed the reactor configuration to be tubular. Had the configuration been N tanks in series, the temperature optimization problem would be solved by setting $I = N$ and finding the best sequence of the T_i, $i = 1, 2, \ldots, N$. For tanks in series, the temperature optimization problem is thus inherently one of parameter optimization rather than functional optimization. Had the reactor configuration been N tanks in series followed by a piston flow reactor, we would divide the piston flow section into I zones and then solve a parameter optimization problem of size $N + I$.

Optimal temperature profiles are usually monotone so that the maximum and minimum values will occur at the inlet or outlet. In constrained optimizations, the optimal profiles are usually of the simple "max–min" or "min–max" form rather than more complex forms such as "max–min–max."

We have so far ignored the fact that heat transfer coefficients are necessarily finite and that a finite length (or finite time) is needed to accomplish a temperature change. This fact is normally viewed as a hindrance to optimal design although it can sometimes be exploited to advantage. The next problem will examine such exploitation. Before doing so, however, a few caveats should be noted. Heat transfer schemes that allow feedback of downstream temperatures to upstream positions may destabilize the system and give rise to multiple steady states. Perfect mixers, in effect, allow such feedback. It can also occur in heat

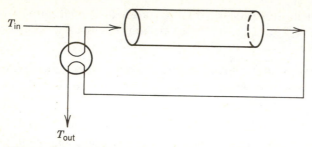

Figure 5.3 Heat integration of an exothermic reaction (dynamically unstable).

integration schemes such as that shown in Figure 5.3. The illustrated scheme seems a logical energy-saving approach for an exothermic reaction: Use the heat generated by the reaction to preheat the feed up to the reaction temperature. However, the dynamics of this system are quite bad. A disturbance in the feed stream will be propagated through the reactor and be fed back to the feed stream in the same direction as the original disturbance. Thus too hot means hotter yet, and too cold means colder yet. The desired operating condition becomes metastable with the reaction either running away or going out. The reactor can, of course, be stabilized by an independently controlled trimming preheater as illustrated in Figure 5.4.

Destabilizing positive feedback can be experienced with other temperature control schemes. Suppose the reactor is tubular and is being countercurrently heated or cooled from the shell side using a heat transfer medium of finite thermal mass (meaning finite flow rate and heat capacity). A positive temperature excursion in the reactor will cause the shell-side fluid to overheat. In countercurrent operation, the hotter-than-normal fluid will flow to upstream reactor positions, causing hotter-than-normal reactor temperatures, and so on. Again, the reactor is destabilized. Other things being equal, cocurrent heat transfer is preferred since temperature feedback to upstream positions cannot occur. However, countercurrent operation can be made stable with a suitable control system.

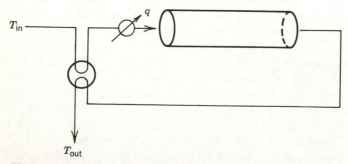

Figure 5.4 Heat integration with trim preheater (stabilized by controlling q).

Problem 5.7   Design a shell-and-tube reactor which approximates the desired temperature profile as revealed by the previous problems.

Suppose the reaction mixture can be preheated to any desired inlet temperature with negligible conversion. The reaction occurs inside tubes that have an ID of 3.8 cm and a wall thickness of 0.2 cm and are constructed from 316 stainless steel. The reactor consists of two separate zones, each 6 m long, which can be heated or cooled independently. Heating by condensation or cooling by boiling allows a constant shell-side temperature and an outside heat transfer coefficient of 5000 J m^{-2} s^{-1} K^{-1} for either heating or cooling. The inside coefficient is given by

$$\frac{h_i d_t}{\lambda} = 0.023 \, \text{Re}^{0.8} \left(\frac{C_p \mu}{\lambda} \right)^{1/3} \tag{5.3}$$

The fluid has a viscosity of 3×10^{-4} Pa s and a thermal conductivity λ of 0.10 J m^{-1} s^{-1} K^{-1}. Is the pressure drop reasonable?

The problem as stated has several design variables: T_{in}, $(T_{\text{shell}})_1$, and $(T_{\text{shell}})_2$. You are not expected to find the optimum of b_{out} with respect to all these parameters. Merely find a plausible design.

Note that Problem 5.7 does not consider utility costs in the optimization but that some values of T_{shell} would be more expensive to achieve than others. Capital costs have also been ignored. In a real problem both the tube diameter and the tube length would be design variables.

In retrospect, was there a more systematic or computationally efficient logic tree (than the one we suggested) for exploring the reactor options? If so, suggest one as:

Problem 5.8   Modify Logic Tree.

A sadistic instructor for this course would assign:

Problem 5.9   Rework Problems 5.1 through 5.7 assuming now that the first of the consecutive reactions

$$A \rightarrow B \tag{5.4}$$

is second order with rate constant $a_{\text{in}} k_I = 10^{10} \exp\left(-10,000/T \right)$ s^{-1}.

This assignment would show that design problems take far less work to assign than to solve, but your instructor is not that sadistic.

The example considered in this chapter is only one of myriad kinetic schemes. We followed a logic tree and used numerical evaluation of the branches to arrive at the "final" design. Many of the branch decisions could have been made based on qualitative, nonnumerical arguments; and such a qualitative approach is rather traditional in chemical reaction engineering. However, it is easy to devise kinetic schemes so complex that the traditional qualitative arguments become suspect. The logic trees must then be designed for the problem at hand and the branch decisions based on numerical case studies. The case study approach is the safer one to adopt even for the simple problems.

Other variants of this chapter are easy to derive. A rather fast, additional exercise would be:

Problem 5.10   Repeat, Assuming that C is the Desired Product. Obviously, the reaction should be pushed as far as possible, and the utility and mean residence time constraints quickly emerge as the key limitations. Most of the optimization problems are trivial for this case.

Problem 5.11   Interchange Rate Constants. Suppose

$$k_I = 10^{12} \exp\left(-\frac{12,000}{T}\right) \quad s^{-1}$$

$$k_{II} = 10^{10} \exp\left(-\frac{10,000}{T}\right) \quad s^{-1} \tag{5.5}$$

Then the nature of the problem is changed markedly. Qualitatively, it resembles Problem 5.10 even though B remains the desired product.

Problem 5.12   Interchange Heats of Reaction This makes a quantitative difference in the solution but qualitative aspects are similar to the original problem. Other numerical changes retain the qualitative aspects of the original problem, provided E_{II} remains greater than E_I.

Problem 5.13   Modify Reaction Mechanism These are variants on the sadistic instructor problem, Problem 5.9. The second reaction step or both reaction steps can be made second order. Again, the problem will qualitatively resemble the original one, provided $E_{II} > E_I$.

Problem 5.14   Reversible Reaction This is a more subtle variation on the theme of modified reaction mechanisms. Suppose

$$A \underset{k_{II}}{\overset{k_I}{\rightleftharpoons}} B \tag{5.6}$$

This will be a trivial optimization problem, resembling Problem 5.10 if the equilibrium favors B at high temperatures. It will be a more interesting problem if the equilibrium favors A at high temperatures. See Problem 4.10.

Chapter 5 to this point has treated reactor design and optimization for simplified kinetic schemes. Hopefully, this allows the basic concepts to be explored without imposing too great a computational burden. The following problems, although still representing a simple reaction, are more typical of industrial practice.

Problem 5.15 Methanol is to be produced from hydrogen and carbon monoxide using a commercial catalyst that is available as 7.87 mm spheres which pack to $\varepsilon = 0.4$.[2] The reaction is reversible:

$$2H_2 + CO \rightleftharpoons CH_3OH$$

with equilibrium constant

$$\log_{10} K_{equil} = \frac{3291}{T} - 7.971 \log_{10} T$$

$$+ 0.002499T - 2.953 \times 10^{-7} T^2 + 10.2$$

Laboratory kinetic studies have yielded the following rate expression:

$$\mathscr{R} = \frac{P_{H_2}^2 P_{CO} - P_{CH_3OH}/K_{equil}}{\left(k_A + k_B P_{H_2} P_{CO} + k_C P_{CH_3OH} \right)^2} \quad \text{mol m}^{-3} \text{ s}^{-1}$$

where the P values denote partial pressures in kPa and

$k_A = 0.3994 \exp(5794/T)$
$k_B = 4.7086 \times 10^{-10} \exp(7814/T)$
$k_C = 57670 - 234.5T + 0.2386T^2$

The feed gas has the following composition:

	Mole percent
CH_3OH	7
CH_4	5
CO_2	8
CO	16
H_2	64

It is available at 10.1 MPa and 473 K although heating or cooling may be done as necessary to optimize reactor performance. Assume ideal gas behavior (but see Problem 5.18) and suppose the following physical properties to be independent of temperature, pressure, and composition:

$c_p = 2.93 \text{ kJ kg}^{-1} \text{ K}^{-1}$
$\mu = 1.6 \times 10^{-5} \text{ Pa s}$
$\dfrac{\mu c_p}{\lambda} = 0.70$
$\Delta H_{\mathscr{R}} = -97.97 \text{ kJ mol}^{-1}$

It is proposed to use a multitubular reactor consisting of 38.1 mm-ID tubes

[2] These and other data were adapted from results presented at the International Workshop on Kinetic Model Development chaired by J. B. Cropley and J. M. Berty at the Annual AIChE Meeting, Chicago, 1985.

that are 12 m long. They will be heated or cooled using a boiling liquid or condensing vapor so that T_{ext} is independent of position z. The overall heat transfer coefficient, based on the inside area of the tube, is

$$U = 2800(\bar{u}_s)^{0.8} \quad \text{J m}^{-2}\,\text{s}^{-1}\,\text{K}^{-1}$$

where $\bar{u}_s$ is in m/s. Use $(\bar{u}_s)_{in} = 0.15$ m/s.

Find those values for T_{in} and T_{ext} that maximize the effluent methanol concentration. There are two operating constraints: the minimum possible pressure is atmospheric, 0.1 MPa, and the maximum allowable catalyst temperature is 553 K.

Problem 5.16   Consider the multitubular reactor of Problem 5.15. Suppose the shell side is divided to allow two independent zones of temperature control, each 6 m long. Find values for $(T_{ext})_1$ and $(T_{ext})_2$ that maximize methanol productivity.

Problem 5.17   An alternative design has been proposed for the methanol reactor of Problem 5.15. It is to be a series combination of five, identical, large-diameter packed bed reactors with intercoolers as illustrated in Figure 5.2. Assume that the reactors operate adiabatically and that the intercoolers cause no conversion and have a negligible pressure drop. There are now six design variables: T_{in}, the four intercooler exiting temperatures, and the total length of packing, $5L$. Use $(\bar{u}_s)_{in} = 0.15$ m/s. The maximum allowable temperature remains 553 K. Determine feasible values for the design variables. Is the productivity of this design, as measured in methanol production per unit reactor volume, better or worse than that for the multitubular reactor?

Problem 5.18   Reconsider the methanol synthesis reactor of Problem 5.15. The inlet pressure is high and the ideal gas law is consequently suspect. Determine through specific calculations whether this simplifying assumption had a significant effect on the results. Use the Redlich–Kwong equation of state

$$P = \frac{R_g T}{V - \beta} - \frac{\alpha}{V(V + \beta)\sqrt{T}}$$

where α and β are estimated for each of the species using critical properties:

$$\alpha_A = \frac{0.42848 R_g^2 (T_c)_A^{2.5}}{(P_c)_A}$$

$$\beta_A = \frac{0.08664 R_g (T_c)_A}{(P_c)_A}$$

Average values for the Redlich–Kwong parameters are calculated using the following mixing rules:

$$\sqrt{\alpha} = \sum y_A \sqrt{\alpha_A}$$

$$\beta = \sum y_A \beta_A$$

where y_A is the mole fraction of component A.

Problem 5.19   The temperature and pressure ranges encountered in the methanol synthesis reactor are too large to justify the assumptions of constant C_p and $\Delta H_{\mathcal{R}}$. Use literature data on the pure-component specific heats to correct for temperature and pressure effects. Determine by specific numerical examples whether these corrections will significantly affect the results of Problem 5.15.

Problem 5.20   The superficial gas velocity $(\bar{u}_s)_{in}$ was set arbitrarily at 0.15 m/s. Revisit Problem 5.15 treating $(\bar{u}_s)_{in}$ as a design variable. Use specific numerical calculations to show how methanol productivity varies with $(\bar{u}_s)_{in}$ and thus with $\bar{t}$. How will variations in $\bar{t}$ affect the design of the recovery system?

Suggestions for Further Reading

The material in this chapter is appropriately supplemented by texts on chemical engineering design and economics. A widely used text is:

M. S. Peters, and K. D. Timmerhaus, *Plant Design and Economics for Chemical Engineers*, 3rd ed., McGraw-Hill, New York, 1980.

Optimization techniques are discussed at length in:

G. S. G. Beveridge, and R. S. Schechter, *Optimization: Theory and Practice*, McGraw-Hill, New York, 1970.

Some specific results for the optimization of adiabatic reactors are given in:

K. R. Westerterp, W. P. M. van Swaaji, and A. A. Beenackers, *Chemical Reactor Design and Operation*, Wiley, New York, 1984.

6

Real Tubular Reactors in Laminar Flow

Piston flow is a *convenient* approximation of a real tubular reactor. The key to its convenience is the absence of radial or tangential variations. The dependent variables $a, b, \ldots, T, P$, are assumed to change in the axial, down-tube direction but to be completely uniform with respect to the cross section of the tube. This allows the reactor design problem to be formulated as a set of ordinary differential equations in the independent variable z. We have seen that such problems are readily solvable given the initial values $a_{in}, b_{in}, \ldots, T_{in}, P_{in}$.

Piston flow is an *accurate* approximation for many practical situations. It is usually possible to avoid tangential (θ-direction) dependence in practical reactor designs, at least for the case of premixed reactants, which we are considering throughout most of this book. It is harder but sometimes possible to avoid radial variations as well. A long, highly turbulent reactor is a typical case where piston flow will be a good approximation for most reactor design purposes. Piston flow will usually be a very bad approximation for laminar flow reactors. In these, radial variations must be considered.

6.1 Isothermal Laminar Flows with Negligible Diffusion

In laminar flow reactors, there will be a pronounced velocity gradient across the tube, with zero velocity at the wall and high velocities near the centerline. Molecules near the center will follow high-velocity streamlines and will undergo relatively little reaction. Those near the tube wall will be on low-velocity streamlines, will remain in the system for long times, and will react to near completion. Thus a gradient in composition will develop across the radius of the

165

tube. Molecular diffusion will act to alleviate this gradient but will not eliminate it, particularly in liquid phase systems with their typical molecular diffusivities of 10^{-9} to 10^{-10} m^2/s. Merrill and Hamrin[1] have found that molecular diffusion can usually be ignored in reactor design calculations if

$$\boxed{\frac{\mathscr{D}\bar{t}}{R^2} < 3 \times 10^{-3}}$$

(6.1)

where $\mathscr{D}$ is the diffusivity. This criterion is satisfied in most industrial-scale laminar flow reactors. It may not be satisfied in laboratory-scale reactors. Molecular diffusion becomes progressively less important as the size of the reactor is increased.

In the absence of diffusion, each streamline through a laminar flow reactor can be treated as if it were a piston flow reactor. Thus the system as a whole can be regarded as a large number of piston flow reactors in parallel. For the familiar case of straight streamlines and a velocity profile that depends on radial position alone, concentrations along the streamlines are given by

$$\boxed{v_z(r)\, \frac{\partial a}{\partial z} = \mathscr{R}_A}$$

(6.2)

This result is reminiscent of Equation 1.66. We have replaced the average velocity $\bar{u}$ with the velocity corresponding to a particular streamline. Equation 6.2 is written as a partial differential equation to emphasize the fact that the concentration $a = a(r, z)$ is a function of both r and z. However, given the local velocity $v_z(r)$, Equation 6.2 can be integrated as though it were an ordinary differential equation. The boundary condition at the inlet is $a(r, 0) = a_{in}$.[2] The concentration at the outlet of the tube is given by $a_{out}(r) = a(r, L)$.

Example 6.1   A first order reaction is occurring in a tubular reactor in laminar flow. Assume that $v_z(r)$ is independent of z and that mass diffusion is negligible but that the reactor is isothermal. Determine the concentration distribution of the consumed component at the reactor outlet.

Solution   For a first order reaction,

$$v_z(r)\, \frac{\partial a}{\partial z} = -ka$$

(6.3)

---

[1] L. S. Merrill, Jr., and C. E. Hamrin, Jr., *AIChE J.*, **16**, 194 (1970).
[2] It is easy to treat the more general situation where $a_{in} = a_{in}(r)$ so that the initial condition is $a(r, 0) = a_{in}(r)$. Thus we can readily accommodate a radial distribution of concentration at the inlet to the reactor.

Integrating and applying the inlet boundary condition in the usual manner gives

$$a(r, z) = a_{in} \exp\left[\frac{-kz}{v_z(r)}\right]$$ (6.4)

Evaluation at the reactor outlet gives

$$a(r, L) = a_{out}(r) = a_{in} \exp\left(\frac{-kL}{v_z(r)}\right) = a_{in} \exp(-kt)$$ (6.5)

where $t = L/v_z(r)$ is the **residence time** for the particular streamline (particular r value) being considered. Equation 6.5 gives the desired concentration distribution except that v_z has not been determined as an explicit function of r. This is a problem in fluid mechanics; and when fluid properties such as viscosity are coupled to the extent of reaction, it becomes a problem in reactor design as well. In the present example, we are assuming constant physical properties so that the fluid mechanics are uncoupled from the reactor design problem. The velocity profile has the parabolic form

$$v_z(r) = 2\bar{u}\left(1 - \frac{r^2}{R^2}\right) = u_{max}\left(1 - \frac{r^2}{R^2}\right)$$ (6.6)

so that

$$a_{out}(r) = a_{in} \exp\left\{\frac{-kL}{u_{max}[1 - (r^2/R^2)]}\right\}$$ (6.7)

Having determined $a_{out}(r)$, it remains to find the average outlet concentration when the flow from all the streamlines is mixed into a single, combined stream. This average concentration is the **convected-mean**, or **mixing-cup**, concentration. For a circular tube with a velocity profile that is a function of radius alone,

$$\bar{a}_{out} = \frac{1}{Q}\int_0^R a_{out}(r)v_z(r)2\pi r\,dr$$ (6.8)

where Q is the volumetric flow rate:

$$Q = \int_0^R v_z(r)2\pi r\,dr = \pi R^2 \bar{u}$$ (6.9)

Example 6.2 Determine $\bar{a}_{out}$ for the previous example of a first order reaction and a parabolic velocity profile.

Solution The reader is challenged to find an analytical solution by substituting Equation 6.7 into Equation 6.8. Although we are considering the simplest possible kinetic scheme and the simplest case of laminar flow, the analytical result is rather awkward to obtain and involves a tabulated function (the incomplete gamma function). Numerical integration is suggested as routine practice.

Before numerically integrating, it is useful to rewrite Equation 6.8 in dimensionless form:

$$\frac{\bar{a}_{out}}{a_{in}} = 2 \int_0^1 \frac{a_{out}(\imath)}{a_{in}} \frac{v_z(\imath)}{\bar{u}} \imath \, d\imath \qquad \text{where } \imath = \frac{r}{R} \tag{6.10}$$

For the problem at hand,

$$\frac{\bar{a}_{out}}{a_{in}} = 4 \int_0^1 \exp\left[\frac{-kL}{2\bar{u}(1 - \imath^2)}\right](1 - \imath^2)\imath \, d\imath \tag{6.11}$$

From this, it should be apparent that $\bar{a}_{out}/a_{in}$ depends only on the dimensionless rate constant $kL/\bar{u} = k\bar{t}$. The value for this constant must be known. Any convenient scheme for numerical quadrature can then be used to calculate $\bar{a}_{out}/a_{in}$, even the simple-minded scheme illustrated in Figure 2.1. However, the

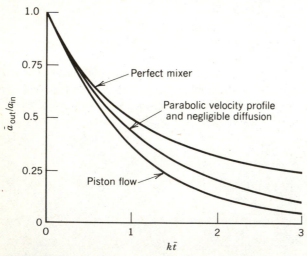

Figure 6.1 Fraction unreacted for first order kinetics in various isothermal reactors.

trapezoidal rule is recommended as being easy to implement and because it converges $0(\Delta r^2)$. Second order convergence in the radial direction will prove useful in subsequent numerical methods.

For I equally sized increments in the radial direction, the trapezoidal rule is

$$\int_0^R F(r) \, dr \approx \frac{R}{I} \left[\frac{F_0}{2} + \sum_{i=1}^{I-1} F_i + \frac{F_I}{2} \right] \qquad (6.12)$$

Applying this rule to Equation 6.11 with $k\bar{t} = 1$ gives

I	$(\bar{a}_{\text{out}}/a_{\text{in}})_I$	Δ
1	0	
		0.385063
2	0.385063	
		0.049605
4	0.434668	
		0.005714
8	0.440382	
		0.002013
16	0.442395	
		0.000617
32	0.443012	
		0.000147
64	0.443159	
Extrapolate to		
$I = \infty$	0.443208	

The extrapolation is based on the series $\frac{1}{4} + \frac{1}{16} + \frac{1}{64} + \cdots = \frac{1}{3}$. The extrapolated answer adds $\frac{1}{3}$ of the change that occurred upon the last doubling of I. Note that for the problem at hand, $k\bar{t} = 1$ requires $I = 64$ before convergence reasonably approximates $0(\Delta r^2)$; and the approximate integral for $\bar{a}_{\text{out}}/a_{\text{in}}$ has already become accurate to 3+ decimal places before the extrapolation technique could be applied.

Figure 6.1 shows how the fraction unreacted varies as a function of $k\bar{t}$. As might have been expected, the performance of the laminar flow reactor is intermediate between piston flow and perfect mixing. Do not generalize this result too far! Laminar velocity profiles exist that give reactor performance far worse than perfect mixing.

6.2 Convective Diffusion of Mass

For chemical reactions to occur, molecules must come into contact; and the mechanism for this contact is molecular motion. This is also the mechanism for diffusion. Thus diffusion is inherently important whenever reaction occurs. However, there are some reactor design problems where diffusion need not be explicitly considered, whereas others exist where a detailed accounting for molecular diffusion is critical to the analysis.

With unmixed feed, mixing occurs inside the reactor under reacting conditions. Diffusion is a slow process, and the actual rate of reaction will typically be limited by the diffusion rate rather than by the intrinsic kinetics that would prevail if the reactants were premixed. Thus diffusion is important in unmixed feed reactors unless the reaction is very slow.

With premixed reactants, molecular diffusion has already brought the reacting molecules in close proximity. In an isothermal batch reactor, various portions of the reacting mass will all react at the same rate and will thus have the same composition at any time. No concentration gradients develop, and molecular diffusion is unimportant during the reaction step of the process even though it was important during the premixing step. In a flow process, whether isothermal or not, different portions of the fluid will have different compositions. Those that have just entered the reactor will have concentrations near a_{in} whereas those near the exit will, on average at least, have concentrations closer to $\bar{a}_{out}$. Thus concentration differences exist within the system, and molecular diffusion will occur. Whether or not it is important depends on the time available for diffusion and the distance over which the diffusion must occur. The time is measured by $\bar{t}$ and the distance, by some characteristic length. For a circular tube, the characteristic length is indeed the radius, and the Merrill and Hamrin criterion, Equation 6.1, applies. For other reactor shapes, a dimensionless group such as $\mathcal{D}\bar{t}/R^2$ will still determine the importance of diffusion although the critical value will be different in, say, a triangular tube than in a circular tube.

If molecular diffusion is important, Equation 6.2, which governs reactant concentrations, must be modified. For reactions in circular tubes with one-dimensional velocity profiles, a suitable modification is

$$v_z(r)\frac{\partial a}{\partial z} = \mathcal{D}_A\left[\frac{1}{r}\frac{\partial a}{\partial r} + \frac{\partial^2 a}{\partial r^2} + \frac{\partial^2 a}{\partial z^2}\right] + \mathcal{R}_A \qquad (6.13)$$

which accounts for Fickian diffusion in both the axial and radial directions. Equation 6.13 is a form of the **convective diffusion** equation. More general forms can be found in any good textbook on mass transfer, but Equation 6.13 is sufficient for many practical situations. A simple derivation is given in Appendix 6.1.

It is useful to rewrite Equation 6.13 using dimensionless independent variables $z = z/L$ and $\imath = r/R$:

$$\frac{v_z(\imath)}{\bar{u}}\frac{\partial a}{\partial z} = \left(\frac{\mathcal{D}_A\bar{t}}{R^2}\right)\left[\frac{1}{\imath}\frac{\partial a}{\partial \imath} + \frac{\partial^2 a}{\partial \imath^2} + \left(\frac{R}{L}\right)^2\frac{\partial^2 a}{\partial z^2}\right] + \bar{t}\mathcal{R}_A \qquad (6.14)$$

The dimensionless group $\mathcal{D}\bar{t}/R^2$ is seen to arise naturally and determines whether molecular diffusion will be a significant process. The term $(R/L)^2$ determines the importance of diffusion in the radial direction compared to that in the axial direction. Normally designed tubular reactors will have $(R/L)^2 < 10^{-2}$ and typically $(R/L)^2 < 10^{-4}$. Thus axial diffusion will be negligible compared to that in the radial direction. (Some thought will show that $\partial^2 a/\partial z^2$ and $\partial^2 a/\partial \imath^2$ have comparable magnitudes.)

The version of the convective diffusion equation appropriate to most design problems is

$$\frac{v_z}{\bar{u}} \frac{\partial a}{\partial z} = \left(\frac{\mathscr{D}_A \bar{t}}{R^2} \right) \left(\frac{1}{\imath} \frac{\partial a}{\partial \imath} + \frac{\partial^2 a}{\partial \imath^2} \right) + \bar{t} \mathscr{R}_A \qquad (6.15)$$

subject to the boundary conditions that

$$a = a_{in} \qquad \text{at } z = 0 \qquad (6.16)$$

$$\frac{\partial a}{\partial \imath} = 0 \qquad \text{at } \imath = 0 \qquad (6.17)$$

$$\frac{\partial a}{\partial \imath} = 0 \qquad \text{at } \imath = 1 \qquad (6.18)$$

Equation 6.16 is our standard initial condition. It is easily generalized to $a(\imath, 0) = a_{in}(\imath)$. Equation 6.17 reflects symmetry with respect to the tube centerline. (The symmetry is a consequence of assuming no θ-direction dependence.) Equation 6.18 corresponds to assuming no mass transfer through the tube walls since there will be a diffusive flux whenever $\partial a/\partial \imath \neq 0$.

A solution to Equation 6.15 gives $a(\imath, z)$ so that the concentration of A is determined at every point in the reactor. If several reactive components are involved, a version of Equation 6.15 should be solved for each component. Unless $\mathscr{D}_A = \mathscr{D}_B = \cdots$, *convective diffusion does not preserve local stoichiometry*. Thus stoichiometric relationships like Equation 1.49 should not be used, even for single reactions.

The outlet concentration profile is given by $a(\imath, 1) = a_{out}(\imath)$. The average outlet concentration $\bar{a}_{out}$ is found using Equation 6.10.

Equation 6.15 is rarely solvable in analytical form. Analytical solutions are possible for first order reactions but they tend to be cumbersome. Thus we resort to numerical solutions.[3] Before doing so, however, we will first consider temperature effects in laminar flow reactors, which are apt to be much more important than the effects of molecular diffusion.

---

[3] Under appropriate circumstances, Equation 6.15 can also be approximated by the axial dispersion model of Chapter 7.

6.3   Temperature Profiles in Laminar Flow

Corresponding to Equation 6.15 is an analogous equation that governs the convection, conduction, and liberation of heat in a laminar flow, tubular reactor:

$$\frac{v_z}{\bar{u}}\frac{\partial T}{\partial z} = \frac{\alpha_T \bar{\imath}}{R^2}\left(\frac{1}{\imath}\frac{\partial T}{\partial \imath} + \frac{\partial^2 T}{\partial \imath^2}\right) - \frac{\Delta H_{\mathscr{R}}\mathscr{R}\bar{\imath}}{\rho C_p}.$$

(6.19)

where α_T is the thermal diffusivity. The argument that $(R/L)^2$ is small was used to justify dropping the axial conduction term just as we dropped the axial diffusion term in Equation 6.15. Note that $\Delta H_{\mathscr{R}}\mathscr{R}$ represents an implicit summation as in Equation 4.15. If the overall reaction is exothermic, $-\Delta H_{\mathscr{R}}\mathscr{R}$ will be positive so that T will be an increasing function of z.

Equation 6.19 is far from the most complex version of the energy equation, but it is adequate for the design of many tubular reactors. We shall apply it to homogeneous, laminar flow reactors in this chapter and to packed beds in Chapter 7. Two of the boundary conditions associated with Equation 6.19 are directly analogous to those for Equation 6.15:

$$T = T_{\text{in}} \quad \text{at} \quad z = 0 \quad \text{(initial condition)}$$

(6.20)

and

$$\frac{\partial T}{\partial \imath} = 0 \quad \text{at} \quad \imath = 0 \quad \text{(radial symmetry)}$$

(6.21)

The third boundary condition depends on the exact nature of the heat transfer scheme. Two common approximations are

$$\frac{\partial T}{\partial \imath} = 0 \quad \text{at} \quad \imath = 1 \quad \text{(adiabatic operation)}$$

(6.22)

and

$$T = T_{\text{wall}} \quad \text{at} \quad \imath = 1 \quad \text{(constant wall temperature)}$$

(6.23)

The assumption of constant wall temperature may be generalized to include a specified temperature profile along the wall, $T = T_{\text{wall}}(z)$ at $\imath = 1$, which is a form useful for studies in functional optimization.

Solutions to Equation 6.19 give the internal temperature distribution $T(\imath, z)$, the outlet temperature profile $T(\imath, 1) = T_{\text{out}}(\imath)$, and the mixing-cup average outlet temperature

$$\overline{T}_{\text{out}} = 2 \int_0^1 T_{\text{out}}(\imath) \frac{v_z(\imath)}{\overline{u}} \imath \, d\imath \tag{6.24}$$

Analytical solutions are generally unavailable except when the heat of reaction is negligible. Even then, the analytical solutions are computationally cumbersome. Thus we will be concerned with numerical solutions. Note that the reaction rate $\mathscr{R} = \mathscr{R}(a, b, \ldots, T)$ couples Equations 6.15 and 6.19 so that the pair must be solved simultaneously. This is no great chore. Solving the pair together is little more difficult than solving a single partial differential equation. Similarly, there is little advantage to dropping the molecular diffusion terms (thereby solving Equation 6.19 simultaneously with Equation 6.2) even when the Merrill and Hamrin criterion is satisfied. The radial diffusion terms may as well be retained. This is not true for the axial diffusion or conduction terms. Retaining them would greatly increase the difficulty of obtaining a solution, for reasons that will be apparent in Chapter 7.

6.4 Numerical Solution Techniques

Many techniques have been developed for the numerical solution of partial differential equations. The best method depends on the specific equations being solved and, to a lesser extent, on the geometry of the system. Partial differential equations having the form of Equation 6.15 or Equation 6.19 are known as parabolic PDEs and are among the easiest to solve. We give here the simplest possible method of solution, one that is directly analogous to the marching-ahead technique we have used for ordinary differential equations. The reader is warned that the method is computationally inefficient and that other techniques should be considered if repetitive solutions are needed for optimization and case studies. The method we shall use is based on finite difference approximations for the partial derivatives. Finite element methods will occasionally give better performance; and even within the class of finite difference techniques, alternate methods such as the "method of lines" should be considered if the computing cost becomes significant. However, for a solution done only once, programming ease is usually more important than computational efficiency.

A finite difference approximation for the partial derivative of temperature in the axial direction is

$$\frac{\partial T}{\partial z} \approx \frac{T(r, z + \Delta z) - T(r, z)}{\Delta z} \tag{6.25}$$

This approximation is called a *forward difference* since it involves the points z and $z + \Delta z$. (See Appendix 6.2 for a discussion of finite difference approximations.) In the radial direction we prefer to use a second order, *central difference* approximation for the first partial derivative:

$$\frac{\partial T}{\partial r} \approx \frac{T(r + \Delta r, z) - T(r - \Delta r, z)}{2\,\Delta r} \tag{6.26}$$

which is seen to involve the $r - \Delta r$ and $r + \Delta r$ points. For the second radial derivative we use

$$\frac{\partial^2 T}{\partial r^2} = \frac{\partial}{\partial r}\left(\frac{\partial T}{\partial r}\right) \approx \frac{T(r + \Delta r, z) - 2T(r, z) + T(r - \Delta r, z)}{\Delta r^2} \tag{6.27}$$

These approximations (or similar ones written in terms of the reduced variables $\imath$ and z) are substituted into the governing PDE, Equation 6.19. The result is then solved for the temperature at the $z + \Delta z$ position:

$$T(\imath, z + \Delta z) = \left[1 - 2\left(\frac{\bar{u}\alpha_T \imath}{v_z R^2}\right)\frac{\Delta z}{\Delta \imath^2}\right]T(\imath, z)$$

$$+ \left(\frac{\bar{u}\alpha_T \imath}{v_z R^2}\right)\frac{\Delta z}{\Delta \imath^2}\left[1 + \frac{\Delta \imath}{2\imath}\right]T(\imath + \Delta \imath, z) \tag{6.28}$$

$$+ \left(\frac{\bar{u}\alpha_T \imath}{v_z R^2}\right)\frac{\Delta z}{\Delta \imath^2}\left[1 - \frac{\Delta \imath}{2\imath}\right]T(\imath - \Delta \imath, z) - \frac{\Delta H_{\mathscr{R}}\mathscr{R}\imath\bar{u}\,\Delta z}{\rho C_P v_z}$$

Equation 6.28 allows temperatures at the "new" axial position $z + \Delta z$ to be calculated given values at the "old" position z. Three different radial positions, $\imath - \Delta \imath$, $\imath$, and $\imath + \Delta \imath$, at the old axial location are used in the calculations. See Figure 6.2. The three circled points at axial position z are used to calculate the new value at the point $z + \Delta z$. The dotted lines in Figure 6.2 show how the radial position $\imath$ can be changed to determine temperatures as a function of radius. The use of values at axial position z to determine values at position $z + \Delta z$ is repeated for all values of $\imath$. Thus the complete radial profile at $z + \Delta z$ can be determined from knowledge of the profile at z. Colloquially, this solution technique can be called marching ahead with a sideways shuffle.

The $\Delta H_{\mathscr{R}}\mathscr{R}/\rho C_P$ term in Equation 6.28 will depend on temperature and composition. It is evaluated using temperature and composition values at the point $(\imath, z)$.

Special versions of Equation 6.28 are required at the tube wall, $\imath = 1$ $(r = R)$, and by the centerline, $\imath = 0$ $(r = 0)$. When the wall temperature is specified, as in Equation 6.23, this temperature is used directly and Equation 6.28 is not evaluated at $\imath = 1$. If the zero flux condition, Equation 6.22, is appropriate, we estimate T_{wall} from the fact that $\partial T/\partial \imath$ is zero at the wall. There are several

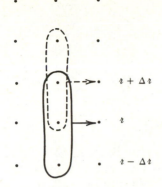

$\imath + \Delta\imath$

$\imath$

$\imath - \Delta\imath$

$z - \Delta z \quad z \quad z + \Delta z$

Figure 6.2 Computational scheme for marching-ahead solution.

ways of doing this. A good one is

$$T_{\text{wall}}(z + \Delta z) = \frac{4T(1 - \Delta\imath, z + \Delta z) - T(1 - 2\,\Delta\imath, z + \Delta z)}{3} \tag{6.29}$$

which converges $0(\Delta\imath^2)$. This result comes from approximating T as a quadratic in $\imath$, $T = C_1 + C_2\imath + C_3\imath^2$, in the vicinity of the wall. The constants C_1, C_2, and C_3 are found by fitting the quadratic to temperatures at the points $\imath = 1 - \Delta\imath$ and $\imath = 1 - 2\,\Delta\imath$, and by forcing $\partial T/\partial\imath$ to zero at $\imath = 1$. Equivalently, Equation 6.29 is obtained by using a second order, forward difference approximation for the derivative at $\imath = 1$. (See Appendix 6.2, Equation 6.90.)

Application of radial symmetry, Equation 6.21, is done by considering a fictitious point at location $\imath = -\Delta\imath$ and by assigning the value $T(-\Delta\imath) = T(\Delta\imath)$ at this point. This gives a zero first derivative according to the central difference approximation of Equation 6.26. However, Equation 6.28 cannot be used directly at $\imath = 0$ due to the appearance of the radius in the denominator. Instead, we must re-examine Equation 6.19 in the limit of $\imath \to 0$. L'Hospital's rule may be used to show that

$$\lim_{\imath \to 0} \frac{1}{\imath}\frac{\partial T}{\partial\imath} = \lim_{\imath \to 0} \frac{\partial T/\partial\imath}{\imath} = \frac{\partial^2 T}{\partial\imath^2}\bigg|_{\imath=0} \tag{6.30}$$

Thus Equation 6.19 becomes

$$\frac{v_z}{\bar{u}}\frac{\partial T}{\partial z} = \frac{2\alpha_T\bar{\imath}}{R^2}\left(\frac{\partial^2 T}{\partial\imath^2}\right) - \frac{\Delta H_{\mathscr{R}}\mathscr{R}\bar{\imath}}{\rho C_p} \qquad \text{at} \qquad \imath = 0 \tag{6.31}$$

The finite difference approximations, Equations 6.25 and 6.27, are substituted into Equation 6.31 to obtain

$$T(0, z + \Delta z) = \left[1 - 4\left(\frac{\bar{u}\alpha_T \imath}{v_z(0) R^2}\right)\frac{\Delta z}{\Delta \imath^2}\right] T(0, z)$$

$$+ 4\left(\frac{\bar{u}\alpha_T \imath}{v_z(0) R^2}\right)\frac{\Delta z}{\Delta \imath^2} T(\Delta \imath, z) - \frac{\Delta H_{\mathscr{R}} \mathscr{R} \bar{\imath} \bar{u} \Delta z}{\rho C_p v_z(0)} \tag{6.32}$$

where we have set $T(-\Delta \imath, z) = T(+\Delta \imath, z)$.

Given initial values $T_{\text{in}}(\imath)$, we can now march down the tube. Equation 6.28 is used for all interior points, $\Delta \imath \leq \imath \leq 1 - \Delta \imath$. Equation 6.32 is used for $\imath = 0$. For $\imath = 1$, $T(1, z) = T_{\text{wall}}(z)$ will either be known directly or calculated from Equation 6.29 after the values at $\imath = 1 - \Delta \imath$ and $\imath = 1 - 2\Delta \imath$ have been determined.

A marching-ahead solution to a parabolic partial differential equation is seen to be conceptually straightforward and directly analogous to the marching-ahead method we have used for solving ordinary differential equations. The difficulties associated with the numerical solution are the familiar ones of accuracy and stability. Stability considerations will be treated first.

Were it not for the $\Delta H_{\mathscr{R}} \mathscr{R}$ term, Equation 6.19 would be linear. For linear equations, a marching-ahead solution will be stable provided the coefficients on the $T(\imath, z)$ terms in Equations 6.28 and 6.32 are non-negative. This criterion gives

$$\frac{\Delta z}{\Delta \imath^2} \leq \frac{v_z(\imath) R^2}{2 \bar{u} \alpha_T \imath} \qquad 0 < \imath \leq 1 \tag{6.33}$$

and

$$\frac{\Delta z}{\Delta \imath^2} \leq \frac{v_z(0) R^2}{4 \bar{u} \alpha_T \imath} \qquad \imath = 0 \tag{6.34}$$

for all $\imath$, $0 \leq \imath \leq 1$. The more restrictive of these equations must be used to calculate a maximum value for $\Delta z/\Delta \imath^2$. We pick some suitably small value of $\Delta \imath$ (the size of $\Delta \imath$ being determined by accuracy considerations) and then pick Δz so that the maximum is not exceeded. Note that Δz will vary as $\Delta \imath^2$ by this approach so that the allowable Δz may be very small indeed. Such is the price paid for the simplicity of the marching-ahead solution.

Which of Equations 6.33 and 6.34 is the more restrictive depends on the shape of the velocity profile. If v_z were flat[4] (a condition which can be approached with strong heating at the walls or with a highly non-Newtonian

---

[4] The situation of laminar flow with $v_z(r) = \bar{u}$ is sometimes called **toothpaste flow**. If you have ever used Stripe™ toothpaste, you recognize that toothpaste flow is quite different than piston flow. There is no mixing in the radial direction. Concentration gradients in the radial direction may exist; assuming radial symmetry, they are governed by Equation 6.15.

fluid), then $v_z = \bar{u}$. The centerline condition, Equation 6.34, is the more demanding in this circumstance and

$$\frac{\Delta z}{\Delta z^2} \le \frac{R^2}{4\alpha_T \bar{t}} \tag{6.35}$$

For real velocity profiles, $v_z(1) = 0$ by the zero-slip condition of hydrodynamics. This appears to cause problems at the tube wall, but we do not face the stability problem exactly at $z = 1$ since T_{wall} will be calculated from special formulas. However, the point $z = 1 - \Delta z$ must be considered, and Equation 6.33 evaluated at $z = 1 - \Delta z$ will usually give a more restrictive value for Δz than Equation 6.34. When the velocity profile is parabolic, $v_z(z = 1 - \Delta z) = 2\bar{u}(2\,\Delta z - \Delta z^2)$. Then Equation 6.33 gives

$$\frac{\Delta z}{\Delta z^2} \le \frac{(2\,\Delta z - \Delta z^2)R^2}{\alpha_T \bar{t}} \tag{6.36}$$

while Equation 6.34 gives

$$\frac{\Delta z}{\Delta z^2} \le \frac{R^2}{2\alpha_T \bar{t}} \tag{6.37}$$

The near-wall result (Equation 6.36) will be more restrictive than the centerline result (Equation 6.37) if $\Delta z \le 0.25$ (four or more increments in the radial direction). With strong cooling at the wall or with a bulk polymerization, $v_z/\bar{u}$ at $z = 1 - \Delta z$ will be even lower than for the parabolic profile. Thus the constraint of Equation 6.33 evaluated at $z = 1 - \Delta z$ will be even more restrictive. The situation can be reversed with strong heating at the wall or with a highly non-Newtonian fluid. As a general rule, the stability criterion should be checked using Equation 6.34 at $z = 0$ and Equation 6.33 at $z = 1 - \Delta z$. The smaller value for z is then used in the marching-ahead calculations.

Example 6.3 Calculate the minimum number of axial steps needed for stability, $J_{\min}$, assuming $v_z = 2\bar{u}(1 - r^2)$ and $I = 4, 8, 16, 32,$ and 64. Note that $\Delta z \equiv 1/I$ and that $J_{\min}$ is the smallest integer for which $\Delta z = 1/J$ satisfies the stability criterion. Assume $\alpha_T \bar{t}/R^2 = 1.0$ for these calculations.

Solution Equation 6.36 provides the stability criterion for this case. For $\alpha_T \bar{t}/R^2 = 1.0$ it becomes $\Delta z \le 2\,\Delta z^3 - \Delta z^4$. This gives the following results:

I	$\Delta z_{\min}$	$J_{\min}$
4	2.734×10^{-2}	37
8	3.662×10^{-3}	274
16	4.730×10^{-4}	2115
32	6.008×10^{-5}	16645
64	7.570×10^{-6}	132105

The rapid increase in $J_{\min}$ demonstrates a basic weakness of the marching-ahead technique. A single run at $I = 64$ might take days on a personal computer. Multiple runs, as needed for optimization studies, would tax most people's budget on a mainframe computer. Fortunately, the assumed value for $\alpha_T \bar{t}/R^2$ is relatively large, and this gives large values for J_{min}. Assuming $\alpha_T \bar{t}/R^2 = 0.1$ will decrease J_{min} by a factor of 10. Even so, a more sophisticated technique such as the method of lines is recommended whenever the reactor model becomes reasonably complex. Software for the method of lines and for other PDE solvers is available for most mainframe computers.

Except in trivial cases, Equation 6.19 will be nonlinear due to the temperature and concentration dependence of $\Delta H_{\mathscr{R}} \mathscr{R}$. Thus there is no guarantee of stability even when Δz is determined by the above method. The linear stability criterion should be regarded as a necessary condition for computational stability. Practical experience indicates that it is usually a sufficient condition as well.

The finite difference approximations of Equations 6.26 and 6.27 and the boundary condition of Equation 6.29 all converge $0(\Delta \imath^2)$. The finite difference approximation for $\partial T/\partial z$ converges $0(\Delta z)$, but when the stability criterion is given by Equation 6.34, Δz will be proportional to $\Delta \imath^2$. Thus one expects the entire computation to converge $0(\Delta \imath^2)$. This includes the integration of Equation 6.24 when performed using the trapezoidal rule. When the stability criterion is determined from Equation 6.33, Δz will vary approximately as $\Delta \imath^3$. This gives slower convergence in the radial direction than in the axial direction. The slowest converging step determines the behavior of the solution as both $\Delta \imath$ and Δz approach zero. Overall convergence will again be $0(\Delta \imath^2)$.

Example 6.4   Find the temperature distribution in a laminar flow, tubular heat exchanger having uniform inlet temperature T_{in} and constant wall temperature T_{wall}. Ignore the temperature dependence of viscosity so that the velocity profile can be assumed parabolic. Use $\alpha_T \bar{t}/R^2 = 0.4$ and report your results in terms of the dimensionless temperature

$$\tau = \frac{T - T_{in}}{T_{wall} - T_{in}} \tag{6.38}$$

Solution   This problem has no heat generated by reaction and is therefore linear. The dimensionless temperature ranges from $\tau = 0$ at the inlet to $\tau = 1$ at the walls. Since no heat is generated, $0 \leq \tau \leq 1$ at every point in the heat exchanger. The solution, $\tau(\imath, z)$, depends only on the value of $\alpha_T \bar{t}/R^2$ and is the same for all values of T_{in} and T_{wall}. Known as the Graetz problem, an analytical solution based on eigenfunction expansions was found in the late 19th century. This solution is awkward to evaluate, but numerical values are easily calculated by the marching-ahead technique.

Use $\Delta \imath = 0.25$. The stability criterion at the centerline gives

$$\frac{\Delta z}{\Delta \imath^2} \leq \frac{u_{max} R^2}{4 \bar{u} \alpha_T \bar{t}} = \frac{R^2}{2 \alpha_T \bar{t}} = 1.25 \tag{6.39}$$

and at $\imath = 1 - \Delta\imath = 0.75$ the criterion is

$$\frac{\Delta z}{\Delta\imath^2} \frac{(2\,\Delta\imath - \Delta\imath^2)R^2}{\alpha_T \bar{\imath}} = 1.094 \tag{6.40}$$

which is the more restrictive. Since $\Delta\imath = 0.25$, $\Delta z < 0.0684$. We would like $(\Delta z)^{-1}$ to be an integer: 15 is the minimum possible integer, but we will choose 16 so that $\Delta z = 0.0625$.

Since $v_z = v_z(\imath)$, the marching-ahead equations will depend on r. For $\Delta\imath = 0.25$ and $\Delta z = 0.0625$, these are

$$\tau(1, z + \Delta z) = 1.0$$

$$\tau(0.75, z + \Delta z) = 0.0857\tau(0.75, z) + 0.5333\tau(1, z) + 0.3810\tau(0.5, z)$$

$$\tau(0.5, z + \Delta z) = 0.4667\tau(0.5, z) + 0.3333\tau(0.75, z) + 0.2000\tau(0.25, z)$$

$$\tau(0.25, z + \Delta z) = 0.5733\tau(0.25, z) + 0.3200\tau(0.5, z) + .1067\tau(0, z)$$

$$\tau(0, z + \Delta z) = 0.2000\tau(0, z) + 0.8000\tau(0.25, z) \tag{6.41}$$

Note that the coefficients of the reduced temperatures sum to 1.0 in each equation. This is necessary because the asymptotic solution ($z \gg 1$) must give $\tau = 1$ for all $\imath$. Had there been a source term, $-\Delta H_{\mathscr{R}}\mathscr{R}/\rho C_P$, the coefficients would be unchanged but an additional generation term would be added to each equation.

The marching-ahead technique gives the following results:

z	$\imath = 0$	$\imath = .25$	$\imath = .50$	$\imath = .75$	$\imath = 1.0$
0	0	0	0	0	1.0000
0.0625	0	0	0	0.5333	1.0000
0.1250	0	0	0.1778	0.5790	1.0000
0.1875	0	0.0569	0.2760	0.6507	1.0000
0.2500	0.0455	0.1209	0.3571	0.6942	1.0000
0.3125	0.1058	0.1884	0.4222	0.7289	1.0000
0.3740	0.1719	0.2544	0.4777	0.7567	1.0000
0.4375	0.2379	0.3171	0.5260	0.7802	1.0000
0.5000	0.3013	0.3755	0.5690	0.8006	1.0000
0.5625	0.3607	0.4295	0.6075	0.8187	1.0000
0.6250	0.4157	0.4791	0.6423	0.8349	1.0000
0.6875	0.4664	0.5246	0.6739	0.8496	1.0000
0.7500	0.5129	0.5661	0.7026	0.8629	1.0000
0.8125	0.5555	0.6041	0.7287	0.8749	1.0000
0.8750	0.5944	0.6388	0.7525	0.8859	1.0000
0.9375	0.6299	0.6705	0.7743	0.8960	1.0000
1.0000	0.6624	0.6994	0.7941	0.9051	1.0000

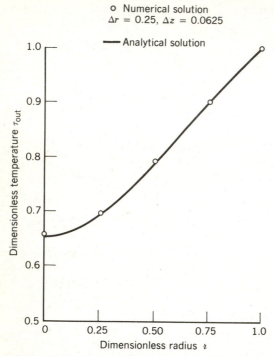

Figure 6.3 Numerical versus anlytical solutions to the Graetz problem with $\alpha_T \bar{t}/R^2 = 0.4$.

Figure 6.3 shows these results for $\tau(\imath, 1) = \tau_{out}(\imath)$ and compares them to the analytical solution. The accuracy is quite good even for a course grid like $\Delta \imath = 0.25$ and $\Delta z = 0.0625$.

Thus far we have discussed only the solution to the temperature distribution problem and not that for composition. However, comparison of Equations 6.15 and 6.19 shows that the marching-ahead solution will have the same form for concentration as for temperature. We need only replace α_T by $\mathcal{D}_A$ and $-\Delta H_{\mathcal{R}}\mathcal{R}/\rho C_p$ by $\mathcal{R}_A$. Then Equations 6.28, 6.29, and 6.32 can be used to calculate $a(\imath, z + \Delta z)$ in the same manner as they are used to calculate $T(\imath, z + \Delta z)$. Note that the boundary condition of zero flux at the wall, Equation 6.29, is used for concentration.

Example 6.5 Solve the convective diffusion equation for an isothermal, first order reaction occurring in a laminar flow reactor. Specifically, find $\bar{a}_{out}/a_{in}$ as a function of $D_A\bar{t}/R^2$ for $k\bar{t} = 1$. The velocity profile is parabolic.

Solution It happens that an analytical solution is known for this case of first order reaction. The mathematically inclined reader may choose to find the solution in the literature. We choose to illustrate a numerical approach which is easily adaptable to nonlinear kinetics and multiple reactants. We seek to solve

$$2(1 - i^2)\frac{\partial a}{\partial z} = \frac{D_A \bar{t}}{R^2}\left(\frac{1}{i}\frac{\partial a}{\partial i} + \frac{\partial^2 a}{\partial i^2}\right) - k\bar{t}a \qquad (6.42)$$

subject to

$$a(i,0) = a_{\text{in}} \qquad \frac{\partial a}{\partial r} = 0 \text{ at } i = 0 \text{ and } i = 1$$

The marching-ahead equation at the centerline is the concentration analog of Equation 6.32:

$$a(0, z + \Delta z) = \left[1 - \frac{4\bar{u}}{v_z(0)}\left(\frac{D_A \bar{t}}{R^2}\right)\frac{\Delta z}{\Delta i^2}\right]a(0, z)$$

$$+ \frac{4\bar{u}}{v_z(0)}\left(\frac{D_A \bar{t}}{R^2}\right)\frac{\Delta z}{\Delta i^2}a(\Delta i, z) - \frac{\bar{u}k\bar{t}a(0, z)\Delta z}{v_z(0)} \qquad (6.43)$$

For an interior point we use the analog of Equation 6.28:

$$a(i, z + \Delta z) = \left[1 - \frac{2\bar{u}}{v_z}\left(\frac{D_A \bar{t}}{R^2}\right)\frac{\Delta z}{\Delta i^2}\right]a(i, z)$$

$$+ \frac{\bar{u}}{v_z}\left(\frac{D_A \bar{t}}{R^2}\right)\frac{\Delta z}{\Delta i^2}\left[1 + \frac{\Delta i}{2i}\right]a(i, +\Delta i, z)$$

$$+ \frac{\bar{u}}{v_z}\left(\frac{D_A \bar{t}}{R^2}\right)\frac{\Delta z}{\Delta i^2}\left[1 - \frac{\Delta r}{2i}\right]a(i - \Delta i, z) - \frac{k\bar{t}a(0, z)\Delta z}{(v/\bar{u})}$$

$$(6.44)$$

where $v_z = 2(1 - i^2)$. The zero flux condition at the wall is expressed analogously to Equation 6.29:

$$a_{\text{wall}}(z + \Delta z) = \frac{4a(1 - \Delta i, z + \Delta z) - a(1 - 2\Delta i, z + \Delta z)}{3} \qquad (6.45)$$

The stability criterion is obtained from Equation 6.36:

$$\frac{\Delta z}{\Delta i^2} \le \frac{(2\Delta i - \Delta i^2)R^2}{D_A \bar{t}} \qquad (6.46)$$

The inlet boundary condition gives $a(\imath, 0) = a_{in} = 1$ for all $\imath$. The convected mean exit concentration is found using Equation 6.8 and the trapezoidal rule

$$\bar{a}_{out} = \frac{1}{I} \sum_{i=1}^{I-1} 4\imath (1 - \imath^2) a(i\,\Delta\imath, 1) \tag{6.47}$$

Note that neither $a(0, 1)$ nor $a(1, 1)$ contributes to the convected mean since $2\pi r v_z$ vanishes at $r = 0$ and $r = R$.

Having set up all the working equations, it remains only to program the computer and choose appropriate values for $\Delta\imath$ and Δz. Pick $\Delta\imath = 1/I$, starting at a small value for I and doubling it until the required accuracy is achieved. For each integer I, $J = 1/\Delta z$ is chosen as an integer which satisfies the stability criterion. The following results show possible choices for the case of $\mathscr{D}_A \bar{t}/\imath^2 = 0.1$. Also shown are calculated values for the average outlet concentration.

I	Δz_{max}	J	$\bar{a}_{out}/a_{in}$
2	1.8750	1	0.7500
4	0.2734	4	0.4756
8	0.0366	28	0.4165
16	0.0047	212	0.4064
32	0.0006	1665	0.4044

Figure 6.4 shows a plot of $\bar{a}_{out}/a_{in}$ as a function of the diffusion parameter $\mathscr{D}_A \bar{t}/R^2$. The laminar flow reactor behaves as a piston flow reactor for high values of $\mathscr{D}_A \bar{t}/R^2$, but diffusion becomes negligible at about $\mathscr{D}_A \bar{t}/R^2 = 0.001$.

Using the marching-ahead method with a parabolic velocity profile forces a very fine grid in the axial direction. Even with modern computers, this can cause

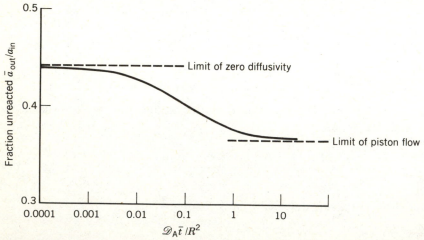

Figure 6.4   First order reaction with $k\bar{t} = 1$ in a tube with a parabolic velocity profile.

problems when highly accurate results are needed. Appendix 6.3 describes alternative finite difference approximations which eliminate stability concerns. Thus, for example, algorithms exist that allow $\Delta z \approx \Delta \imath$ so that both Δz and $\Delta \imath$ can be halved together, and the number of computations increases by a factor of only 4. The price for this is greater complexity in the individual calculations.

The equations governing the convective diffusion of heat and mass will generally be coupled through the temperature and composition dependence of $\mathscr{R}$. The marching-ahead approach allows their simultaneous solution in a straightforward manner. If molecular diffusion is negligible, concentration can be marched ahead as an ordinary differential equation rather than as a partial differential equation. Thus $a(\imath, z + \Delta z)$ would be calculated using just $a(\imath, z)$ rather than the three-radial-point method illustrated in Figure 6.2. However, this saves little in computation, and radial diffusion of mass may as well be retained even though $\mathscr{D}_A$ is small. Since $\mathscr{D}_A \ll \alpha_T$, the stability criterion for $\Delta z/\Delta \imath^2$ will be determined by the temperature equation rather than the composition equation.

6.4.1   Slit Flow and Rectangular Coordinates

Results to this point have been confined to cylindrical coordinate systems and to tubular reactors. Tubes are an extremely practical geometry which is widely used for chemical reactors. Less common is slit flow as occurs between parallel plates. Suppose the plates are very wide so that sidewall effects are negligible. Then the analog of Equation 6.15 in rectangular coordinates is

$$
\frac{v_z}{\bar{u}}\frac{\partial a}{\partial z} = \frac{\mathscr{D}_A \bar{\imath}}{R^2}\frac{\partial^2 a}{\partial y^2} + \bar{\imath}\mathscr{R}_A
\tag{6.48}
$$

The rectangular coordinate y plays the role of $\imath$. For uniformity of notation, the distance between plates can be denoted as $2R$. See Figure 6.5 for the case of stationary plates with fluid flow driven by a pressure gradient in the z-direction.

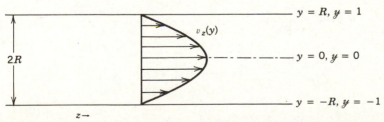

Figure 6.5   Pressure drive flow between parallel plates with both plates stationary.

The velocity profile for a Newtonian fluid of constant viscosity is

$$v_z(y) = 1.5\bar{u}\left(\frac{1-y^2}{R^2}\right) = 1.5\bar{u}(1-y^2) \tag{6.49}$$

The boundary conditions associated with Equation 6.48 are analogous to those for Equation 6.15:

$$a = a_{in} \quad \text{at } z = 0$$

$$\frac{\partial a}{\partial y} = 0 \quad \text{at } y = 0$$

$$\frac{\partial a}{\partial y} = 0 \quad \text{at } y = 1 \tag{6.50}$$

The mathematics are seen to be entirely analogous to those for tubular geometries considered previously. The marching-ahead equations assume a similar but simpler form.

Example 6.6 Determine the marching-ahead equations corresponding to Equations 6.48 through 6.50.

Solution Proceeding as for cylindrical coordinates, $\partial a/\partial z$ is approximated by a first order, forward difference

$$\frac{\partial a}{\partial z} \approx \frac{a(y, z + \Delta z) - a(y, z)}{\Delta z} \tag{6.51}$$

and $\partial^2 a/\partial y^2$ is approximated using a second order difference

$$\frac{\partial^2 a}{\partial y^2} \approx \frac{a(y + \Delta y, z) - 2a(y, z) + a(y - \Delta y, z)}{\Delta y^2} \tag{6.52}$$

Substituting these into Equation 6.48 and solving for $a(y, z + \Delta z)$ gives

$$a(y, z + \Delta z) = \left[1 - 2\left(\frac{\mathscr{D}_A t}{R^2}\right)\frac{\bar{u}}{v_z}\frac{\Delta z}{\Delta y^2}\right]a(y, z)$$

$$+ \left[\frac{\mathscr{D}_A t}{R^2}\frac{\bar{u}}{v_z}\frac{\Delta z}{\Delta y^2}\right][a(y + \Delta y, z) + a(y - \Delta y, z)] \tag{6.53}$$

The centerline causes no difficulty since Equation 6.48 does not contain a term like $(1/y)\,\partial a/\partial y$. Equation 6.53 can be used at $y = 0$ just by setting $a(\Delta y, z) = a(-\Delta y, z)$. A special form is required at the wall. Analogously to Equation 6.29 we have

$$a_{\text{wall}}(z + \Delta z) = \frac{4a(1 - \Delta y, z + \Delta z) - a(1 - 2\,\Delta y, z + \Delta z)}{3} \tag{6.54}$$

The stability criterion associated with this marching-ahead scheme is

$$\frac{\Delta z}{\Delta y^2} \le \frac{v_z(y)R^2}{2\bar{u}\alpha_T i} \qquad 0 \le i < 1 \tag{6.55}$$

Example 6.7 Assume $v_z(y) = \bar{u}$ for the slit flow geometry of Figure 6.5. Demonstrate the instability of the marching-ahead equations when the stability criterion is violated. Assume $\mathscr{R}_A = 0$.

Solution Suppose

$$2\left(\frac{\mathscr{D}_A i}{\mathscr{R}^2}\frac{\bar{u}}{v_z}\frac{\Delta z}{\Delta y^2}\right) = 2 \tag{6.56}$$

so that Equation 6.53 becomes

$$a(y, z + \Delta z) = -a(y, z) + a(y + \Delta y, z) + a(y - \Delta y, z) \tag{6.57}$$

This equation conserves mass and admits the steady state solution $a(y, z) = \text{const} = a_{\text{in}}$. However, this solution is numerically unstable in the presence of small errors. Observe the following calculations which are done for interior points:

$\uparrow$ y						
	0	0	0	0	0	+0.1
	0	0	0	0	+0.1	−0.4
Exact	0	0	0	+0.1	−0.3	+1.0
solution	0	0	+0.1	−0.2	+0.6	−1.6
	0	+0.1	−0.1	+0.3	−0.7	+1.9
Original	0	0	+0.1	−0.2	+0.6	−1.6 Propagated
error	0	0	0	+0.1	−0.3	+1.0 error
	0	0	0	0	+0.1	−0.4
	0	0	0	0	0	+0.1

$$z \rightarrow$$

The original error ($+0.1$ instead of 0) grows exponentially in magnitude and oscillates in sign. The marching-ahead scheme is clearly unstable in the presence of small blunders or round-off errors.

Example 6.8 Repeat the previous example for a case which satisfies the stability criterion.

Solution The largest value that satisfies the stability criterion is

$$2\left(\frac{\mathcal{D}_A t}{R^2}\frac{\bar{u}}{v_z}\frac{\Delta z}{\Delta y^2}\right) = 1 \tag{6.58}$$

This gives the following marching-ahead equation:

$$a(y, z + \Delta z) = 0.5a(y + \Delta y, z) + 0.5a(y - \Delta y, z) \tag{6.59}$$

Introduction of a small error at an interior point gives the following results:

y ↑

0	0	0	0	0	0.00625
0	0	0	0	0.0125	0
0	0	0	0.025	0	0.0250
0	0	0.05	0	0.0375	0
0	0.1	0	0.05	0	0.0375
0	0	0.05	0	0.0375	0
0	0	0	0.025	0	0.0250
0	0	0	0	0.0125	0
0	0	0	0	0	0.00625

Original error

Propagated error

$z \rightarrow$

Here, the magnitude of the error decreases at downstream points. Suppose the original error represents injection of a small amount of nonreactive tracer. Then the local tracer concentrations decrease in the downstream direction and gradually spread out due to diffusion in the y-direction. However, the total quantity of injected tracer (0.1 in this case) remains constant. This is exactly how we would expect the real system to behave. The marching-ahead scheme is now both conservative and stable.

Figure 6.6 shows another flow geometry for which rectangular coordinates are useful. The bottom plate is stationary, but the top plate moves at velocity V_0.

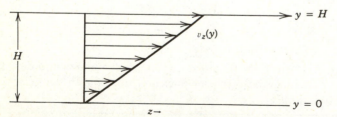

Figure 6.6 Drag flow between parallel plates with the upper plate in motion and no axial pressure drop.

The velocity profile is linear:

$$v_z(y) = \frac{V_0 y}{H} \tag{6.60}$$

Mass transfer with reaction is governed by Equation 6.48. Appropriate boundary conditions are

$$\frac{\partial a}{\partial y} = 0 \quad \text{at } y = 0, H \tag{6.61}$$

An analogous equation holds for heat transfer but possible boundary conditions now include

$$T = T_{\text{upper}} \quad \text{at } y = H$$

$$T = T_{\text{lower}} \quad \text{at } y = 0 \tag{6.62}$$

The drag flow of Figure 6.6 is used to model polymer-processing equipment. Sometimes, a pressure-driven flow is superimposed on the drag flow.

6.5 Variable-Viscosity Problems

Real fluids have viscosities that are functions of temperature and composition. This means that the viscosity will vary across the radius of a tubular reactor and that the velocity profile will be something other than parabolic. If the viscosity is lower near the wall, as in heating, the velocity profile will be flattened compared to the parabolic distribution. If the viscosity is higher near the wall, as in cooling, the velocity profile will be elongated. These phenomena can be important in laminar flow reactors, affecting performance and even operability. Generally speaking, a flattened velocity profile will improve performance by more closely approaching piston flow. Conversely, an elongated profile will usually hurt performance.

This section gives a simplified method for including the effects of variable viscosity in a reactor design problem. It is restricted to low Reynolds numbers, $Re < 100$, and is used mainly for reactions involving polymers. It illustrates the coupling of the equations of mass, energy, and momentum transport in a reactor design problem of practical significance. Increasingly, elaborate computer codes are being devised which recognize this coupling in complex flow geometries. These codes are being verified and are becoming design tools for the reaction engineer. The present example is representative of a general class of single phase, variable-viscosity, variable-density problems, yet avoids complications in mathematical or numerical analysis.

Two assumptions are needed to treat the variable-viscosity problem in its simplest form:

(i) The momentum of the fluid is negligible compared to viscous forces.

(ii) The radial velocity component v_r is negligible compared to the axial component v_z.

The first of these assumptions drops the momentum terms from the equations of motion, giving a condition known as *creeping flow*. This assumption leaves v_r and v_z coupled through a pair of simultaneous, partial differential equations. The pair can be solved when circumstances warrant, but the second assumption allows much greater simplification. It allows v_z to be given by a single, ordinary differential equation:

$$0 = \frac{-dP}{dz} + \frac{1}{r}\frac{d}{dr}\left(\mu r \frac{dv_z}{dr}\right) \tag{6.63}$$

This equation is subject to the boundary conditions of radial symmetry, $dv_z/dr = 0$ at $r = 0$, and zero slip at the wall, $v_z = 0$ at $r = R$. Solution gives

$$v_z = \frac{1}{2}\left(\frac{-dP}{dz}\right)\int_r^R \frac{r\,dr}{\mu} = u_0 \frac{\int_r^R \frac{r}{\mu}\,dr}{\int_0^R \frac{r}{\mu}\,dr} \tag{6.64}$$

The quantities u_0 and dP/dz appear as constants, at least local constants. These can be found from knowledge of mass flow rates:

$$\int_0^R 2\pi r v_z \rho\,dr = Q\bar{\rho} = \pi R^2 \bar{u}\bar{\rho} \tag{6.65}$$

Equation 6.65 expresses the overall material balance constraint of constant mass flow down the tube. It also gives constant volumetric flow if ρ is constant. Combining Equations 6.64 and 6.65 gives

$$\frac{v_z(r)}{\bar{u}} = \frac{R^2}{2}\frac{\int_r^R \frac{r}{\mu}\,dr}{\int_0^R r\frac{\rho}{\bar{\rho}}\int_r^R \frac{r}{\mu}\,dr\,dr} \tag{6.66}$$

When only the dimensionless profile is needed, the computational scheme is especially simple. Rewrite Equation 6.64 as

$$v_z = \frac{u_0 I(r)}{\int_0^R \frac{r}{\mu}\,dr} \tag{6.67}$$

where

$$I(r) = \int_r^R \frac{r\,dr}{\mu} \tag{6.68}$$

and define

$$\Omega = \frac{1}{R^2} \int_0^R \frac{\rho}{\bar{\rho}} rI(r)\,dr \tag{6.69}$$

Then Equation 6.56 becomes

$$\frac{v_z(r)}{\bar{u}} = \frac{1}{2} \frac{I(r)}{\Omega} \tag{6.70}$$

Example 6.9   Suppose a marching-ahead solution for the equations of mass and energy has given the radial composition and temperature profiles at some point z in a tubular reactor. Physical property correlations have then been used to estimate the viscosity profile:

$\imath = \dfrac{r}{R}$	$\mu(r)$ Pa s
1.0	148
0.8	55
0.6	20
0.4	7
0.2	3
0	1

Use these data to estimate the fully developed velocity profile at this axial position. Assume constant density.

Solution   We first calculate the value of $I(r)$ for each radial position using a dimensionless version of Equation 6.68:

$$I(\imath) = \int_\imath^1 \frac{\imath\,d\imath}{\mu} \tag{6.71}$$

The integration can be done using the trapezoid rule and marching from the tube wall toward the center:

$$I(1) = 0 \tag{6.72}$$

$$I(\imath) = I(\imath + \Delta\imath) + \left[\frac{\imath + \Delta\imath}{\mu(\imath + \Delta\imath)} + \frac{\imath}{\mu(\imath)} \right] \frac{\Delta\imath}{2} \tag{6.73}$$

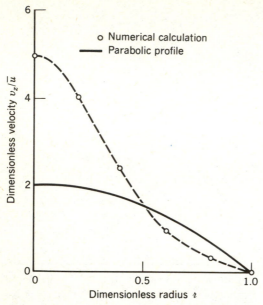

Figure 6.7 Elongated velocity profile typical of strong cooling or polymerization at the wall.

When these calculations are done, we integrate again to obtain

$$\Omega = \int_0^1 \imath I(\imath)\, d\imath \tag{6.74}$$

The trapezoidal rule is suitable for this integral as well. Finally, we can obtain the dimensionless velocity profile from

$$\frac{v_z(\imath)}{\bar{u}} = \frac{1}{2}\frac{I(\imath)}{\Omega} \tag{6.75}$$

The computational scheme is illustrated below:

$\imath$	μ	$I(\imath)$	$v_z/\bar{u}$
1	148	0	0
0.8	55	0.00213	0.31
0.6	20	0.00658	0.95
0.4	7	0.01530	2.21
0.2	3	0.02768	4.00
0	1	0.03435	4.96

where we have determined that $\Omega = 0.00346$. See Figure 6.7.

The marching-ahead equations for $T(r, z + \Delta z)$ and $a(r, z + \Delta z)$ require that $v_z/\bar{u}$ be known at the "old" axial location. We will know $\mu(r)$ at the old axial location and can thus find $v_z/\bar{u}$ using Equation 6.66. Each time we take a step in the axial direction, we must first recalculate the marching-ahead coefficients using the local values for $v_z(r)/\bar{u}$. This is time consuming but straightforward provided Δz is chosen small enough. The stability criterion should be satisfied at all points down the tube. It is entirely possible that a downstream velocity profile, distorted because of temperature or composition changes, will impose a more demanding limit on Δz than the inlet profile $v_z(r, 0)$.

Equation 6.66 allows v_z to be calculated as a function of radius. It will also be true that μ and thus v_z are functions of axial position z. This will cause no difficulty provided the changes in the axial direction are slow. The formulation of Equation 6.63 treats the fully developed velocity profile $v_z(r)$, which corresponds to the local values of $\mu(r)$ without regard to upstream or downstream conditions. Changes in $\mu(r)$ must be gradual enough that the adjustment from one axial velocity profile to another requires only small velocities in the radial direction. We have assumed v_r is small enough that it does not affect the equation of motion for v_z. This does not mean that v_r is zero. In fact, values for v_r can be calculated from the fluid continuity equation

$$\frac{1}{r}\frac{\partial}{\partial r}(rv_r\rho) + \frac{\partial(v_z\rho)}{\partial z} = 0 \tag{6.76}$$

which can be integrated to give

$$v_r = -\frac{1}{r\rho}\int_0^r r\rho\frac{\partial v_z}{\partial z}\,dr \tag{6.77}$$

If fluid streamlines are required, they can be calculated from

$$\int_0^{r_0} rv_z(r, 0)\rho\,dr = \int_0^r r'v_z(r', z)\rho\,dr' \tag{6.78}$$

This mass balance equation shows that material that is at radial position r_0 when $z = 0$ will move to radial position r for some $z > 0$.

Radial motion of fluid can have a significant, cummulative effect on the convective diffusion equations even when v_r has a negligible effect on the equation of motion for v_z. Thus Equation 6.66 can give an accurate approximation for v_z even though Equations 6.15 and 6.19 need to be modified to account for radial convection:

$$v_z\frac{\partial a}{\partial z} + v_r\frac{\partial a}{\partial r} = \mathcal{D}_A\left(\frac{1}{r}\frac{\partial a}{\partial r} + \frac{\partial^2 a}{\partial r^2}\right) + \mathcal{R}_A \tag{6.79}$$

and

$$v_z \frac{\partial T}{\partial z} + v_r \frac{\partial T}{\partial r} = \alpha_T \left(\frac{1}{r} \frac{\partial T}{\partial r} + \frac{\partial^2 T}{\partial r^2} \right) - \Delta H_{\mathscr{R}} \mathscr{R} \qquad (6.80)$$

The boundary conditions are unchanged. Marching-ahead equations based on these PDEs are considered in Problem 6.8. Examples of their solutions are given in Section 12.3.

Suggestions for Further Reading

The convective diffusion equations for mass and energy are given detailed treatments in most texts on transport phenomena. The modern classic is:

R. B. Bird, W. E. Stewart, and E. N. Lightfoot, *Transport Phenomena*, Wiley, New York, 1960.

Numerical methods useful in their solution are described in:

M. E. Davis, *Numerical Methods and Modeling for Chemical Engineers*, Wiley, New York, 1984.

B. A. Finlayson, *Nonlinear Analysis in Chemical Engineering*, McGraw-Hill, New York, 1980.

M. Kubicek and V. Hlavacek, *Numerical Solution of Nonlinear Boundary Value Problems with Applications*, Prentice-Hall, Englewood Cliffs, NJ, 1983.

L. Lapidus and G. F. Pinder, *Numerical Solution of Partial Differential Equations in Science and Engineering*, Wiley, New York, 1982.

Practical applications to laminar flow reactors are still mainly in the research literature. The first good treatment of a variable-viscosity reactor is:

S. Lynn and J. E. Huff, "Polymerization in a Tubular Reactor", *AIChE J.*, **17**, 475 (1971).

A detailed model of an industrially important reaction, the styrene polymerization, is given by

C. E. Wyman and L. F. Carter, "A Numerical Model for Tubular Polymerization Reactors", *AIChE Symp. Ser.*, **72**, 1 (1976).

See also Chapter 12.

The appropriateness of neglecting radial flow in the axial momentum equation yet of retaining it in the convective diffusion equation is described in:

H. S. McLaughlin, R. Mallikarjun, and E. B. Nauman, "The Effect of Radial Velocities on Laminar Flow, Tubular Reactor Models," *AIChE J.*, **32**, 419 (1986).

Problems

6.1 Polymerizations often give such high viscosities that laminar flow is inevitable. A typical monomer diffusivity in a polymerizing mixture is 10^{-10} m^2/s (the diffusivity of the polymer will be much lower). A pilot-scale reactor might have a radius of 1 cm. How long can the mean residence time be before molecular diffusion is important? What about a production-scale reactor with $R = 10$ cm?

6.2 The velocity profile for laminar, non-Newtonian flow in a pipe can sometimes be approximated as

$$u = u_0\left[1 - \left(\frac{r}{R}\right)^{\eta+1/\eta}\right]$$

where η is called the flow index, or power law constant. The case $\eta = 1$ corresponds to a Newtonian fluid and gives a parabolic velocity profile. Find $\bar{a}_{out}/a_{in}$ for a first order reaction given $k\bar{t} = 1.0$ and $\eta = 0.5$. Assume negligible diffusion.

6.3 An unreconstructed cgs'er messed up your viscosity correlation by reporting his results in centipoise rather than Pascal seconds. How does this affect the sample velocity profile calculated in Section 6.5?

6.4 Suppose you are marching down the infamous tube and at step j have determined the temperature and composition at each radial point. A correlation is available to calculate viscosity, and it gives the results tabulated below. Assume constant density and Re = 0.1. Determine the axial velocity profile. Plot your results and compare them to the parabolic distribution.

$\imath = r/R$	Case 1, Isothermal (μ/μ_{in})	Case 2, Cooling (μ/μ_{in})	Case 3, Heating (μ/μ_{in})
1.000	1.0	54.6	0.018
0.875	1.0	33.1	0.030
0.750	1.0	20.1	0.050
0.625	1.0	12.2	0.082
0.500	1.0	7.4	0.135
0.375	1.0	4.5	0.223
0.250	1.0	2.7	0.368
0.125	1.0	1.6	0.607
0	1.0	1.0	1.000

6.5 Show that, for constant density, Equation 6.66 becomes

$$\frac{v_z(r)}{\bar{u}} = \frac{R^2 \int_r^R \frac{r}{\mu} \, dr}{\int_0^R \frac{r^3}{\mu} \, dr}$$

6.6 Duplicate Figure 6.4 for $k\bar{t} = 2$.

6.7   Free-radical polymerizations tend to be highly exothermic. The following data are representative of the thermal (i.e., spontaneous) polymerization of styrene:

$\lambda = 0.13$ J m^{-1} s^{-1} K^{-1}

$\mathscr{D}_A = 10^{-9}$ m^2/s

$\Delta H = -8 \times 10^4$ J/mol

$c_p = 1.9 \times 10^3$ J kg^{-1} K^{-1}

$\rho = 950$ kg/m^3

$a_{in} = 9200$ mol/m^3

$L = 7$ m

$\bar{t} = 1$ hr

$k = 10^{10} \exp(-10{,}000/T)$ hr^{-1}

$T_{in} = 60°C$

$T_{wall} = 140°C$

Assume laminar flow and a parabolic velocity distribution. Calculate the temperature and composition profiles in the reactor using $\Delta t = 0.2$. Consider two cases: (a) $R = 0.01$ m; (b) $R = 0.20$ m.

6.8   Develop a marching-ahead technique for solving Equation 6.79. Use a second order, central difference approximation for $\partial a/\partial t$ wherever this derivative occurs. Note that this allows convergence $0(\Delta t^2)$ and does not change the stability criterion. Note also that v_r vanishes at $t = 0$ and 1. This means that previously derived results for the wall and centerline still hold. Only one new equation, that for interior points, is required.

6.9   Stepwise condensation polymerizations can be modeled as a second order decay of the unreacted fraction of functional groups, $1 - X$, where X is the conversion of functional groups. Neglecting diffusion,

$$v_z \frac{\partial(1-X)}{\partial z} + v_r \frac{\partial(1-X)}{\partial r} = -k(1-X)^2$$

where $X = 0$ at $z = 0$ and $\partial x/\partial r = 0$ at $r = 0$. The following viscosity relationship

$$\frac{\mu}{\mu_0} = 1 + 100 X^3$$

is representative of condensation polymerizations in solvents. Determine $X(r, z)$ and $v_z(r, z)$ for a tubular reactor with $k\bar{t} = 2$.

6.10 Show how the pressure drop can be calculated for a variable-viscosity, laminar flow reactor such as that in Problem 6.9.

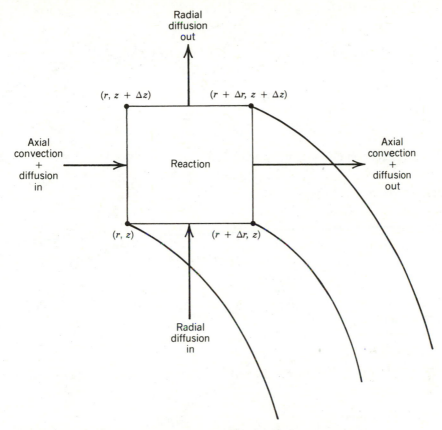

Figure 6.8 Differential volume element in cylindrical coordinates.

Appendix 6.1

The Convective Diffusion Equation

This section derives a simple version of the convective diffusion equation, applicable to tubular reactors with a one-dimensional velocity profile, $v_z(r)$. The starting point is Equation 1.2 applied to the differential volume element shown in Figure 6.8. The volume element is located at point (r, z) and is in the shape of a ring. Note that θ dependence is ignored so that the results will not be applicable to low-viscosity systems having significant natural convection.

Component A is transported by diffusion and axial convection. The diffusive flux is governed by Fick's law. Thus

$$\text{Radial diffusion in} = -\mathscr{D}_A \frac{\partial a}{\partial r}\bigg|_r [2\pi r \Delta z]$$

$$\text{Axial diffusion in} = -\mathscr{D}_A \frac{\partial a}{\partial z}\bigg|_z [2\pi r \Delta r]$$

Axial convection in $= v_z a|_z [2\pi r \Delta r]$

Radial diffusion out $= -\mathscr{D}_A \dfrac{\partial a}{\partial r}\bigg|_{r+\Delta r} [2\pi (r + \Delta r) \Delta z]$

Axial diffusion out $= -\mathscr{D}_A \dfrac{\partial a}{\partial z}\bigg|_{z+\Delta z} [2\pi r \Delta r]$

Axial convection out $= v_z a|_{z+\Delta z} [2\pi r \Delta r]$

Additional terms needed for Equation 1.2 are

Formation of A by reaction $= \mathscr{R}_A [2\pi r \Delta r \Delta z]$

Accumulation $= \dfrac{\partial a}{\partial t} [2\pi r \Delta r \Delta z]$

Applying Equation 1.2,

$$-\mathscr{D}_A \frac{\partial a}{\partial r}\bigg|_{r} [2\pi r \Delta z] - \mathscr{D}_A \frac{\partial a}{\partial z}\bigg|_{z} [2\pi r \Delta r] + v_z a|_z [2\pi r \Delta r] + \mathscr{R}_A [2\pi r \Delta r \Delta z]$$

$$= -\mathscr{D}_A \frac{\partial a}{\partial r}\bigg|_{r+\Delta r} [2\pi (r + \Delta r) \Delta z] - \mathscr{D}_A \frac{\partial a}{\partial z}\bigg|_{z+\Delta z} [2\pi r \Delta r] \qquad (6.81)$$

$$+ v_z a|_{z+\Delta z} [2\pi r \Delta r] + \frac{\partial a}{\partial t} [2\pi r \Delta r \Delta z]$$

This result can be divided through by $2\pi r \Delta r \Delta z$ and rearranged to give

$$\frac{\partial a}{\partial t} + \frac{v_z a|_{z+\Delta z} - v_z a|_z}{\Delta z} = \frac{\mathscr{D}_A \dfrac{\partial a}{\partial z}\bigg|_{z+\Delta z} - \mathscr{D}_A \dfrac{\partial a}{\partial z}\bigg|_{z}}{\Delta z}$$

$$(6.82)$$

$$+ \frac{\mathscr{D}_A \dfrac{\partial a}{\partial r}\bigg|_{r+\Delta r} - \mathscr{D}_A \dfrac{\partial a}{\partial r}\bigg|_{r}}{\Delta r} + \mathscr{D}_A \frac{\partial a}{\partial r}\left(\frac{1}{r}\right) + \mathscr{R}_A$$

The limit is now taken as $\Delta r \to 0$ and $\Delta z \to 0$. The result is

$$\frac{\partial a}{\partial t} + \frac{\partial (v_z a)}{\partial z} = \frac{\partial \left(\mathscr{D}_A \dfrac{\partial a}{\partial z} \right)}{\partial z} + \frac{\partial \left(\mathscr{D}_A \dfrac{\partial a}{\partial r} \right)}{\partial r}$$

$$+ \frac{\mathscr{D}_A}{r} \frac{\partial a}{\partial r} + \mathscr{R}_A \tag{6.83}$$

Common simplifications are steady state operation, v_z independent of z, and constant diffusivity. Then

$$v_z \frac{\partial a}{\partial z} = \mathscr{D}_A \left[\frac{\partial^2 a}{\partial z^2} + \frac{1}{r} \frac{\partial a}{\partial r} + \frac{\partial^2 a}{\partial r^2} \right] + \mathscr{R}_A \tag{6.84}$$

which is the same as Equation 6.13.

Appendix 6.2

Finite Difference Approximations

This section describes a number of finite difference approximations useful for solving second order partial differential equations, that is, equations containing terms such as $\partial^2 f/\partial x^2$. The basic idea is to approximate f as a polynomial in x and then to differentiate the polynomial to obtain estimates for derivatives such as $\partial f/\partial x$ and $\partial^2 f/\partial x^2$. The polynomial approximation is a local one that applies to some region of space centered about point x. When the point changes, the polynomial approximation will usually change as well.

We begin by fitting a quadratic to the three points depicted below.

The quadratic has the form

$$f = A + Bx + Cx^2 \tag{6.85}$$

Writing it for the three points gives

$$f_- = A - B\Delta x + C\Delta x^2$$

$$f_0 = A \tag{6.86}$$

$$f_+ = A + B\Delta x + C\Delta x^2$$

These equations are solved for A, B, and C to give

$$f = f_0 + \left(\frac{f_+ - f_-}{2\,\Delta x}\right)x + \left(\frac{f_+ - 2f_0 + f_-}{2\,\Delta x^2}\right)x^2 \tag{6.87}$$

This is a **second order approximation** and can be used to obtain derivatives up to the second. Equation 6.87 is differentiated to give

$$\frac{\partial f}{\partial x} = \left(\frac{f_+ - f_-}{2\,\Delta x}\right) + \left(\frac{f_+ - 2f_0 + f_-}{\Delta x^2}\right)x \tag{6.88}$$

and

$$\frac{\partial^2 f}{\partial x^2} = \frac{f_+ - 2f_0 - f_1}{\Delta x^2} \tag{6.89}$$

The value of the first derivative depends on the position at which it is evaluated. Setting $x = +\Delta x$ gives a **second order, forward difference**:

$$\left.\frac{\partial f}{\partial x}\right|_+ \approx \frac{3f_+ - 4f_0 + f_-}{2\,\Delta x} \tag{6.90}$$

Setting $x = 0$ gives a **second order, central difference**:

$$\left.\frac{\partial f}{\partial x}\right|_0 \approx \frac{f_+ - f_-}{2\,\Delta x} \tag{6.91}$$

Setting $x = -\Delta x$ gives a **second order, backward difference**:

$$\left.\frac{\partial f}{\partial x}\right|_- \approx \frac{-f_+ + 4f_0 - 3f_-}{2\,\Delta x} \tag{6.92}$$

The second derivative is constant (independent of x) for this second order approximation. We consider it to be a central difference:

$$\left.\frac{\partial^2 f}{\partial x^2}\right|_0 \approx \frac{f_+ - 2f_0 + f_-}{\Delta x^2} \tag{6.93}$$

All higher derivatives are zero. Obviously, to obtain a nontrivial approximation to an nth derivative requires at least an nth order polynomial. The various nontrivial derivatives obtained from an nth order polynomial will converge $0(\Delta x^n)$.

Example 6.10   Suppose $f(x) = x \exp(x)$. Then $f_+ = \Delta x \exp(\Delta x)$, $f_0 = 0$, $f_- = -\Delta x \exp(-\Delta x)$. The various derivative approximations are

$$\left. \frac{\partial f}{\partial x} \right|_+ \approx \frac{3e^{\Delta x} - e^{-\Delta x}}{2}$$

$$\left. \frac{\partial f}{\partial x} \right|_0 \approx \frac{e^{\Delta x} + e^{-\Delta x}}{2}$$

(6.94)

$$\left. \frac{\partial f}{\partial x} \right|_- \approx \frac{-e^{\Delta x} + 3e^{-\Delta x}}{2}$$

$$\left. \frac{\partial^2 f}{\partial x^2} \right|_0 \approx \frac{e^{\Delta x} - e^{-\Delta x}}{\Delta x}$$

Evaluating them as a function of Δx gives the following:

Δx	$\left.\frac{\partial f}{\partial x}\right|_+$	Δ	$\left.\frac{\partial f}{\partial x}\right|_0$	Δ	$\left.\frac{\partial f}{\partial x}\right|_-$	Δ	$\left.\frac{\partial^2 f}{\partial x^2}\right|_0$	Δ
1	3.893		1.543		−0.807		2.350	
		1.723				−0.892		0.266
$\frac{1}{2}$	2.170		1.128		0.805		2.084	
		0.633		0.096		−0.441		0.063
$\frac{1}{4}$	1.537		1.031		0.526		2.021	
		0.279		0.024		−0.231		0.016
$\frac{1}{8}$	1.258		1.008		0.757		2.005	
		0.131		0.006		−0.120		0.004
$\frac{1}{16}$	1.127		1.002		0.877		2.001	
		0.064		0.001		−0.061		0.001
$\frac{1}{32}$	1.063		1.000		0.938		2.000	
∞	1		1		1		2	

It is apparent that the central difference approximations converge $0(\Delta x^2)$. The forward and backward approximations to the first derivative converge $0(\Delta x)$.

This is because they are really approximating the derivatives at the points $x = \pm \Delta x$ rather than at $x = 0$.

For a *first order approximation*, we fit a straight line between the points $x = 0$ and $x = \Delta x$ to get the *first order, forward difference approximation*

$$\left. \frac{\partial f}{\partial x} \right|_{+\Delta x/2} \approx \frac{f_+ - f_0}{\Delta x} \tag{6.95}$$

and between the points $x = -\Delta x$ and $x = 0$ to get the *first order, backward difference* approximation

$$\left. \frac{\partial f}{\partial x} \right|_{-\Delta x/2} \approx \frac{f_0 - f_-}{\Delta x} \tag{6.96}$$

These both converge $0(\Delta x)$.

Appendix 6.3

Implicit Differencing Schemes

The marching-ahead method for solving parabolic PDEs is called an *explicit* method because the "new" value $T(r, z + \Delta z)$ is given as an explicit function of the "old" values $T(r, z)$, $T(r - \Delta r, z), \ldots$. See, for example, Equation 6.28. This explicit scheme is obtained by using a first order, forward difference approximation for the axial derivative. See, for example, Equation 6.25. Other approximations for $\partial T/\partial z$ are given in Appendix 6.2. These usually give rise to *implicit* methods where $T(r, z + \Delta z)$ is not found directly but is given as one member of a set of simultaneous algebraic equations. The simplest implicit scheme is known as *backward differencing* and is based on a first order, backward difference approximation for $\partial T/\partial z$:

$$\frac{\Delta T}{\Delta z} \approx \frac{T(r, z) - T(r, z - \Delta z)}{\Delta z} \tag{6.97}$$

Instead of Equation 6.28 we obtain

$$\left[1 + \frac{2\bar{u}\alpha_T \iota}{v_z R^2} \frac{\Delta z}{\Delta \iota^2} \right] T(\iota, z) - \frac{\bar{u}\alpha_T \iota}{v_z R^2} \frac{\Delta z}{\Delta \iota^2} \left[1 + \frac{\Delta \iota}{2\iota} \right] T(\iota + \Delta \iota, z)$$

$$- \frac{2\bar{u}\alpha_T \iota}{v_z R^2} \frac{\Delta z}{\Delta \iota^2} \left[1 - \frac{\Delta \iota}{2\iota} \right] T(\iota - \Delta \iota, z) \tag{6.98}$$

$$= T(\iota, z - \Delta z) - \frac{\Delta H_{\mathscr{R}}}{\rho C_p} \frac{\mathscr{R} \bar{\iota} \bar{u} \Delta z}{v_z}$$

Here, the temperatures on the left-hand side are the new, unknown values while that on the right is the previous, known value. Note that the heat sink/source term is evaluated at the previous location, $z - \Delta z$.

Equation 6.98 cannot be solved directly since there are three unknowns. However, if a version of Equation 6.98 is written for every interior point, $0 < \imath < 1$, and if appropriate special forms are written for $\imath = 0$ and $\imath = 1$, then as many equations are obtained as there are unknown temperatures. The resulting algebraic equations are linear and can be solved by matrix inversion.

The backward-differencing scheme is stable for all $\Delta \imath$ and Δz so that $I = 1/\Delta \imath$ and $J = 1/\Delta z$ can be picked independently. This avoids the need for extremely small Δz values as was encountered in the second example of Section 6.4. The method converges $0(\Delta \imath^2, \Delta z)$.

Example 6.11 Use the backward differencing method to solve the heat transfer problem of Example 6.4. Select $\Delta \imath = 0.25$ and $\Delta z = 0.0625$.

Solution Equations 6.41 are the forward-marching equations for $\Delta \imath = 0.25$ and $\Delta z = 0.0625$. The backward-differencing equations can be determined from them by comparing Equation 6.98 with Equation 6.28. The result is

$$\tau(1, z) = 1.0$$

$$- 0.5333\tau(1, z + 1.9143\tau(0.75, z) - 0.3810\tau(0.5, z) = \tau(0.75, z - \Delta z)$$

$$- 0.3333\tau(0.75, z) + 1.5333\tau(0.5, z) - 0.2000\tau(0.25, z) = \tau(0.5, z - \Delta z)$$

$$- 0.3200\tau(0.5, z) + 1.4267\tau(0.25, z) - 0.10676\tau(0, z) = \tau(0.25, z - \Delta z)$$

$$- 0.8000\tau(0.25, z) + 1.8000\tau(0, z) = \tau(0, z - \Delta z)$$

$$\text{(6.99)}$$

In matrix form

$$\begin{bmatrix} 1 & 0 & 0 & 0 & 0 \\ -0.5333 & 1.9143 & -0.3810 & 0 & 0 \\ 0 & -0.3333 & 1.5333 & -0.2000 & 0 \\ 0 & 0 & -0.3200 & 1.4267 & -0.1067 \\ 0 & 0 & 0 & -0.8000 & 1.8000 \end{bmatrix} \begin{bmatrix} \tau(1, z) \\ \tau(0.75, z) \\ \tau(0.5, z) \\ \tau(0.25, z) \\ \tau(0, z) \end{bmatrix}$$

$$= \begin{bmatrix} 1.0 \\ \tau(0.75, z - \Delta z) \\ \tau(0.5, z - \Delta z) \\ \tau(0.25, z - \Delta z) \\ \tau(0, z - \Delta z) \end{bmatrix} \qquad \text{(6.100)}$$

This system of equations is solved for each z beginning with the inlet boundary condition:

$$
\begin{bmatrix}
1.0 \\
\tau(0.75,\, z - \Delta z) \\
\tau(0.5,\, z - \Delta z) \\
\tau(0.25,\, z - \Delta z) \\
\tau(0,\, z - \Delta z)
\end{bmatrix}
=
\begin{bmatrix}
1 \\
0 \\
0 \\
0 \\
0
\end{bmatrix}
\qquad \text{(6.101)}
$$

Results are

z	$\imath = 0$	$\imath = .25$	$\imath = .5$	$\imath = .75$	$\imath = 1.0$
0	0	0	0	0	1.0000
0.0625	0.0067	0.0152	0.0654	0.2916	1.0000
0.1250	0.0241	0.0458	0.1487	0.4605	1.0000
0.1875	0.0525	0.0880	0.2313	0.5652	1.0000
0.2500	0.0901	0.1372	0.3068	0.6349	1.0000
0.3125	0.1345	0.1901	0.3737	0.6846	1.0000
0.3740	0.1832	0.2440	0.4326	0.7223	1.0000
0.4375	0.2338	0.2972	0.4844	0.7523	1.0000
0.5000	0.2848	0.3486	0.5303	0.7771	1.0000
0.5625	0.3349	0.3975	0.5712	0.7982	1.0000
0.6250	0.3832	0.4437	0.6079	0.8166	1.0000
0.6875	0.4293	0.4869	0.6410	0.8327	1.0000
0.7500	0.4728	0.5271	0.6710	0.9471	1.0000
0.8125	0.5135	0.5645	0.6982	0.8601	1.0000
0.8750	0.5515	0.5991	0.7230	0.8718	1.0000
0.9375	0.5869	0.6311	0.7456	0.8824	1.0000
1.0000	0.6196	0.6606	0.7664	0.8921	1.0000

The backward-differencing method requires the solution of $I + 1$ simultaneous equations to find the radial temperature profile. The solution is still marched ahead in the axial direction. *Fully implicit* schemes exist where $(J + 1)(I + 1)$ equations are solved simultaneously, one for each grid point in the total system. Fully implicit schemes may be used for problems where axial conduction or diffusion are important so that second derivatives in the axial direction, $\partial^2 T/\partial z^2$ or $\partial^2 a/\partial z^2$, must be retained in the partial differential equation. An alternative approach for this case is the shooting method described in Chapter 7. When applied to partial differential equations, shooting methods are usually implemented using an implicit technique in the radial direction. This gives rise to a tridiagonal matrix which must be inverted at each step in axial marching. The Thomas algorithm is a simple and efficient way of performing this inversion. See Davis[5] for a description. Davis also discusses finite difference approximations which combine forward and backward differencing. One of these, *Crank–Nicholson*, is widely used. It is implicit, unconditionally stable (at least for the linear case), and converges $0(\Delta r^2, \Delta z^2)$.

[5] M. E. Davis, *Numerical Methods and Modeling for Chemical Engineers*, Wiley, New York, 1984.

7

Real Tubular Reactors in Turbulent Flow

Turbulent flow reactors are modeled quite differently from laminar flow reactors. In a turbulent flow field, nonzero velocity components exist in all three coordinate directions, and they fluctuate with time. Statistical methods must be used to obtain time average values for the various components and to characterize the instantaneous fluctuations about these averages. We write

$$\mathbf{V} = v + \psi \tag{7.1}$$

where ψ represents the fluctuating velocity and v is a time average:

$$v = \lim_{t \to \infty} \frac{1}{t} \int_0^t \mathbf{V} \, d\theta \tag{7.2}$$

For turbulent flow in long, empty pipes, the time average velocities in the radial and tangential directions are zero since there is no net flow in these directions:

$$v_r = v_\theta = 0 \tag{7.3}$$

The axial velocity component will have a time average profile $v_z(r)$, which is a function of radial position. This profile is considerably flatter than the parabolic profile of laminar flow, but a profile nevertheless exists. The zero-slip boundary condition still applies and forces $v_z(R) = 0$. The time average velocity changes very rapidly near the tube wall. The region over which the change occurs is known as the **hydrodynamic boundary layer**. Sufficiently near the wall, flow in the boundary layer will be laminar, with attendant limitations on heat and mass transfer. Outside the boundary layer—meaning closer to the center of the tube—the time average velocity profile is approximately flat. Flow in this region

203

is known as *core turbulence*. Here, the fluctuating velocity components are high and give rapid rates of heat and mass transfer in the radial direction. Thus turbulent flow reactors are often modeled as having no composition or temperature gradients in the radial direction. This is not quite the same as assuming piston flow. At high Reynolds numbers, the boundary layer thickness becomes small and a situation akin to piston flow is approached. At lower Reynolds numbers, a more sophisticated model may be needed.

7.1   The Axial Dispersion Model

Suppose a small pulse of an ideal, nonreactive tracer is injected into a tube at the center. An ideal tracer is identical to the bulk fluid in terms of flow properties but is distinguishable in some nonflow aspect that is detectable with suitable instrumentation. Typical tracers are dyes, radioisotopes, and salt solutions.

The first and most obvious thing that happens to the tracer is movement downstream at a rate equal to the time average axial velocity. If we are dealing with a stationary coordinate system (called an *Eulerian* coordinate system), the injected pulse will soon be gone from view. This approach resolves the immediate question of what happens to the tracer but does not provide much help in the analysis of tubular reactors. Shift to a moving (*Lagrangian*) coordinate system that translates down the tube with the same velocity as the fluid. In this coordinate system, the center of the injected pulse remains stationary; but individual tracer particles spread out from the center due to the combined effects of molecular diffusion and the fluctuating velocity components. If the time average velocity profile were truly flat, the tracer concentration would soon become uniform in the radial and tangential directions but would spread indefinitely in the axial direction. This kind of mixing has not been encountered in our previous discussions. Axial mixing is disallowed in the piston flow model. In laminar flow, the only mechanism for random mixing is molecular diffusion; and this was neglected in the axial direction because axial concentration or temperature gradients are so much smaller than radial gradients. In turbulent flow, effective diffusivities are enhanced to the point of virtually eliminating the radial gradients and of causing a significant amount of mixing in the axial direction.

At high Reynolds numbers, boundary layers will be thin and the core velocity profile flat. Piston flow with superimposed axial mixing might be a plausible model for this situation. The piston flow model is

$$\bar{u}\frac{da}{dz} = \mathscr{R}_A \tag{7.4}$$

We add a term for axial mixing by a diffusionlike mechanism:

$$\bar{u}\frac{da}{dz} = D\frac{d^2a}{dz^2} + \mathscr{R}_A \tag{7.5}$$

The parameter D is known as the *axial dispersion coefficient*. At very high Reynolds numbers, it governs a type of mixing that is due solely to fluctuating velocities in the axial direction. These fluctuating axial velocities cause mixing by a random process that is conceptually similar to molecular diffusion except that the fluid elements being mixed are much larger than molecules.

At lower Reynolds numbers, the axial velocity profile will not be flat; and it might seem that another correction must be added to Equation 7.5. It turns out, however, that Equation 7.5 remains a good model for real turbulent reactors (and even some laminar ones) given suitable values for D. These suitable values are determined experimentally using transient experiments with nonreactive tracers. A correlation for D is shown in Figure 7.1. The dimensionless group $D/\bar{u}d_t = D/2\bar{u}R$ depends on the Reynolds number and on molecular diffusivitiy as measured by the Schmidt number $Sc = \mu/\rho\mathcal{D}_A$.

For the laminar region, D is known analytically:

$$D = \mathcal{D}_A + \frac{\bar{u}^2 R^2}{48\mathcal{D}_A} \tag{7.6}$$

Note, however, that the axial dispersion model is a reasonable approximation to laminar flow reactors only when they are sufficiently long:

$$\frac{L}{R} > \frac{\bar{u}R}{8\mathcal{D}_A} \tag{7.7}$$

The axial dispersion model also provides a reasonably good description of fluid mixing in packed-bed reactors. Figure 7.2 gives a correlation from which $\bar{u}L/D$ can be estimated. For design purposes, assuming $\bar{u}d_p/D = 1$, where d_p is the packing diameter, is fairly conservative in the sense that it will overestimate the effects of axial dispersion.

The axial dispersion model lumps the combined effects of fluctuating velocity components, nonflat velocity profiles, and molecular diffusion into the single parameter D. At very close levels of scrutiny, real systems behave differently than predicted by the axial dispersion model; but the model is amazingly accurate for most purposes.

The boundary conditions associated with Equation 7.5 are known as Danckwerts, or *close* boundary, conditions. They are obtained from flux balances across the inlet and outlet planes of the reactor. The flux entering at $z = 0$ is $\bar{u}a_{in} = \bar{u}a(0 -)$ and is due solely to convection. This must be matched by the flux leaving the plane at $z = 0$ since no reaction occurs in a plane of zero volume. The flux leaving the plane at $z = 0$ is $\bar{u}a(0 +)$ due to convection and $-Dda/dz|_{0+}$ due to diffusion. Thus

$$\bar{u}a_{in} = \bar{u}a(0^+) - D\frac{da}{dz}\Big|_{0^+} \tag{7.8}$$

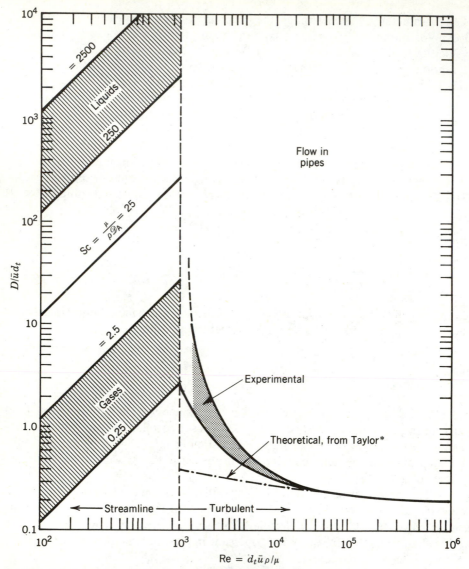

Figure 7.1 Correlation for the dispersion of fluids flowing in pipes. [adapted from O. Levenspiel, *Ind. Eng. Chem.*, **50**, 343 (1958).]

A similar balance is made at $z = L$, but we also use the fact that $a(L-) = a(L+) = a_{\text{out}}$. This gives

$$\left.\frac{da}{dz}\right|_{L} = 0 \tag{7.9}$$

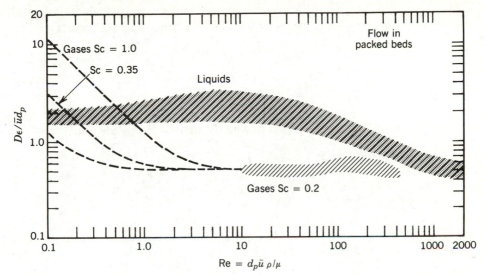

Figure 7.2 Correlation of dispersion to Reynolds number. [O. Levenspiel, *Chemical Reaction Engineering*, 2nd ed., Wiley, 1972, p. 282.]

at the outlet. These specific conditions assume that $D = 0$ for the inlet and outlet piping to the reactor. It turns out, however, that reaction conversions predicted using the axial dispersion model are independent of D in the inlet and outlet regions provided no reaction occurs in these regions.

Equation 7.5 is an ODE, and an analytical solution is readily obtained when the reactor is isothermal and the reaction is first order. The general solution to Equation 7.5 with $\mathscr{R}_A = -ka$ is

$$a(z) = \exp\left(\frac{\text{Pe}}{2}\frac{z}{L}\right)\left[C_1\exp\left(+s\frac{\text{Pe}}{2}\frac{z}{L}\right) + C_2\exp\left(-s\frac{\text{Pe}}{2}\frac{z}{L}\right)\right] \tag{7.10}$$

where $\text{Pe} = \bar{u}L/D$ is the **Peclet number** and

$$s = \sqrt{1 + \frac{4k\bar{t}}{\text{Pe}}} \tag{7.11}$$

The constants C_1 and C_2 are evaluated using the boundary conditions, and the outlet concentration is found by setting $z = L$. This gives

$$\boxed{\frac{a_{\text{out}}}{a_{\text{in}}} = \frac{4s\exp\left(\dfrac{\text{Pe}}{2}\right)}{(1+s)^2\exp\left(\dfrac{s\,\text{Pe}}{2}\right) - (1-s)^2\exp\left(\dfrac{-s\,\text{Pe}}{2}\right)}} \tag{7.12}$$

Conversions predicted from Equation 7.12 depend only on the values of $k\bar{t}$ and Pe. They are smaller than those for piston flow but larger than for perfect mixing. In fact,

$$\lim_{\text{Pe} \to \infty} \frac{a_{\text{out}}}{a_{\text{in}}} = e^{-k\bar{t}} \tag{7.13}$$

so that the model approaches piston flow in the limit of high Peclet numbers (low D). Also,

$$\lim_{\text{Pe} \to 0} \frac{a_{\text{out}}}{a_{\text{in}}} = \frac{1}{1 + k\bar{t}} \tag{7.14}$$

so that the axial dispersion model approaches perfect mixing at low Peclet numbers (high D). However, use of this model is not recommended—or it should be used with considerable caution—for Peclet numbers less than 8. It is also not recommended for laminar flow reactors except for rough estimates. For Pe > 8, Equation 7.12 becomes, to an excellent approximation,

$$\frac{a_{\text{out}}}{a_{\text{in}}} = \frac{4s \exp\left[\dfrac{\text{Pe}}{2}(1 - s)\right]}{(1 + s)^2} \tag{7.15}$$

For fully turbulent reactors, $\bar{u}R/D$ will be in the range of 0.8 to 2.5, so that Pe = $\bar{u}L/D$ will be $0.8L/R$ to $2.5L/R$. This means that piston flow, Equation 7.13, will be an excellent approximation to all but very short reactors operated at relatively high values of $k\bar{t}$. Figure 7.3 shows the errors in conversion which would result from assuming piston flow, Equation 7.13, rather than using the axial dispersion model, Equation 7.12. The difference is negligible for most industrial systems. For an open tube in well-developed turbulent flow (Re > 10,000), the assumption of piston flow is normally quite accurate.

7.1.1 Numerical Solutions

For reactions other than first order, Equation 7.5 must be integrated numerically. Numerical integration is also required when the rate constant k varies with position down the tube as in nonisothermal reactors or reactors with catalyst deactivation.

The numerical solution of Equation 7.5 is rather more complicated than for the first order ODEs we have encountered in analyzing piston flow reactors. In some ways, it is even more complicated than for the PDEs encountered in laminar flow systems with radial gradients. The reason for the difficulty is the second derivative in the axial direction, d^2a/dz^2. Converting to the dimension-

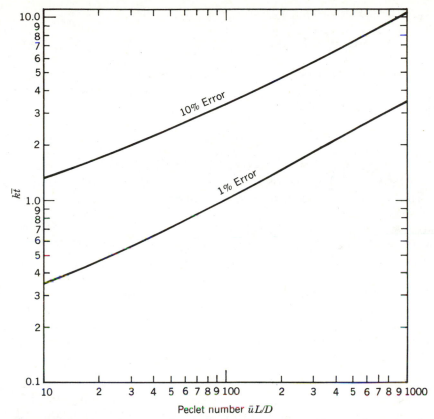

Figure 7.3 Error in conversions calculated for a first order reaction using the piston flow model rather than axial dispersion.

less length variable $z = z/L$ gives

$$\frac{da}{dz} = \frac{1}{\text{Pe}}\left(\frac{d^2a}{dz^2}\right) + \mathscr{R}_A \bar{t} \tag{7.16}$$

Using a backward difference approximation for da/dz and the usual central difference for d^2a/dz^2 gives

$$a_{j+1} = (2 + \text{Pe}\,\Delta z)a_j - (1 + \text{Pe}\,\Delta z)a_{j-1} - \text{Pe}\,\mathscr{R}_A \bar{t}\,\Delta z^2 \tag{7.17}$$

Thus the value for the next, marching-ahead point requires knowledge of *two* previous points. To calculate a_2, we need to know both a_0 and a_1. The boundary conditions, Equations 7.8 and 7.9, give us neither of these directly. In finite difference form, the inlet boundary condition is

$$a_1 = (1 + \text{Pe}\,\Delta z)a_0 - \text{Pe}\,\Delta z a_{\text{in}} \tag{7.18}$$

Thus if a_0 is known, a_1 can be calculated. One possible approach to solving Equation 7.5 is to guess a_0, to calculate a_1, and then to march down the tube. The outlet boundary condition is

$$a_{J+1} = a_J \tag{7.19}$$

where J is the number of steps in the axial direction, that is $J\Delta z = 1$. If Equation 7.19 is satisfied, the correct value for a_0 was used. Otherwise, guess a new a_0. This approach is known as **forward shooting**.

The shooting method may not always work in practice. What we have done is to convert a **boundary value** problem into an easier-to-solve **initial value** problem. Unfortunately, the conversion gives a numerical computation which is **ill-conditioned**. Extreme precision is needed at the inlet of the tube to get reasonable accuracy at the outlet. The phenomenon is akin to problems which arise in the numerical inversion of matrices and Laplace transforms.

Example 7.1 Use forward shooting to solve Equation 7.16 for a first order reaction with Pe = 16 and $k\bar{t} = 2$. Compare the result to the analytical solution, Equation 7.12.

Solution Set $\Delta z = 1/32$ so that $\text{Pe}\,\Delta z = 0.5$ and $k\bar{t}\,\text{Pe}\,\Delta z^2 = 0.03125$. Set $a_{\text{in}} = 1$ so that dimensionless or normalized concentrations are determined. Equations 7.18 and 7.17 become

$$a_1 = 1.5a_0 - 0.5$$

$$a_{j+1} = 2.53125a_j - 1.5a_{j-1} \tag{7.20}$$

Results for a succession of guesses for a_0 give

$\dfrac{a_0}{a_{\text{in}}}$	$\dfrac{a_{32}}{a_{\text{in}}} = \dfrac{a_{\text{out}}}{a_{\text{in}}}$	$\dfrac{a_{33}}{a_{\text{in}}}$
0.90342	−20.8	−33.0
0.90343	+0.93	+1.37
0.903429	−1.24	−2.06
0.9034296	+0.0630	+0.0004
0.90342965	+0.1715	+0.1723
0.903429649	+0.1693	+0.1689
0.9034296493	0.1699	0.1699

The answer by the shooting method is $a_{\text{out}}/a_{\text{in}} = 0.17$. The analytical result is

0.1640. Note that the shooting method requires extreme precision on guesses of a_0 to obtain an answer of limited accuracy. Double precision is needed on 16-bit computer. Better accuracy with the numerical approach could be achieved with a smaller step size ($J = 64$ gives $a_0 = 0.9012498065$ and $a_{out} = 0.1670$) or a more sophisticated integration routine. Note, however, that a more sophisticated integration technique would only give a more accurate value for a_{out} once the right guess for a_0 has been made. It would not eliminate the ill-conditioning inherent in forward shooting.

The best solution to such numerical difficulties is to change methods. For the axial dispersion model, integration in the reverse direction eliminates most of the difficulty. Let $x = 1 - z$. Then the model becomes

$$\frac{da}{dx} = -\frac{1}{Pe}\frac{d^2a}{dx^2} - \mathscr{R}_A \bar{t} \tag{7.21}$$

$$\frac{da}{dx} = 0 \qquad \text{at } x = 0 \tag{7.22}$$

$$a + \frac{1}{Pe}\frac{da}{dx} = a_{in} \qquad \text{at } x = 1 \tag{7.23}$$

The marching-ahead equation for *reverse shooting* is obtained from Equation 7.21 using a backward difference for da/dx and a central difference for d^2a/dx^2

$$a_{j+1} = (2 - Pe\,\Delta x)a_j - (1 - Pe\,\Delta x)a_{j-1} - \mathscr{R}_A \bar{t}\, Pe\,\Delta x^2 \tag{7.24}$$

It is solved by guessing a_0 (which is now the outlet concentration) and setting $a_1 = a_0$ to satisfy Equation 7.22. Equation 7.24 is then marched ahead. The right guess for a_0 was made when Equation 7.23 is satisfied. In finite difference form, we seek a_0 such that

$$a_J + \frac{a_J - a_{J-1}}{Pe\,\Delta x} = a_{in} \tag{7.25}$$

Example 7.2   Repeat Example 7.1 using reverse shooting.

Solution   For $J = 32$, Equation 7.24 gives

$$a_{j+1} = 1.53125a_j - 0.5a_{j-1}$$

Guess $a_0 = a_1$; calculate a_{31} and a_{32}; and then use Equation 7.25 to calculate

a_{in}. Some results are

$$a_0 = a_{out} \qquad a_{in} = a_J + \frac{a_J - a_{J-1}}{Pe\,\Delta x}$$

0.16	1.0073
0.15	0.9444
0.159	1.0010

Thus we obtain $a_{out} = 0.16$ for a step size of $\Delta x = 0.03125$. The ill-conditioning problem has been solved, but the solution remains inaccurate due to the simple integration scheme and large step size.

The reverse shooting method is particularly well suited to Runge–Kutta integration. See Example 7.3.

7.1.2 Nonisothermal Axial Dispersion

The axial dispersion model is readily applied to nonisothermal reactors. The turbulent mixing that leads to flat concentration profiles will also give flat temperature profiles. Thus $T(r, z) = T(z)$. Equation 7.5 remains valid for nonisothermal reactors except that the reaction rate $\mathcal{R}_A$ will be a function of T and thus of z.

An expression for the axial dispersion of heat can be written in direct analogy to Equation 7.5:

$$\bar{u}\frac{dT}{dz} = E\frac{d^2T}{dz^2} + \mathcal{S} \tag{7.26}$$

where E is the axial dispersion coefficient for heat and $\mathcal{S}$ represents the sum of all source terms. For well-developed turbulence, the thermal Peclet number $(Pe)_{thermal} = \bar{u}L/E$ should be identical to the mass Peclet number $Pe = \bar{u}L/D$. At lower Reynolds numbers, one would expect $\bar{u}L/E$ to depend on a thermal Schmidt number,[1] $(Sc)_{thermal} = \mu/\rho\alpha_T$, just as $\bar{u}L/D$ depends on $\mu/\rho\mathcal{D}_A$.

The usual source terms associated with Equation 7.26 are heat transfer through the tube wall and heat of reaction:

$$\bar{u}\frac{dT}{dz} = E\frac{d^2T}{dz^2} - \frac{2U}{\rho C_p}\frac{(T - T_{ext})}{R} - \frac{\Delta H_{\mathcal{R}}\mathcal{R}}{\rho C_p} \tag{7.27}$$

The boundary condition at the inlet is

$$\bar{u}T_{in} = \bar{u}T(0^+) - E\frac{dT}{dz}\bigg|_{0^+} \tag{7.28a}$$

---

[1] This dimensionless group is more conventionally called the Prandtl number, defined as $\mu C_p/\lambda_T$, where λ_T is the thermal conductivity.

and

$$\frac{dT}{dz}\bigg|_L = 0 \tag{7.28b}$$

at the outlet. Equation 7.27 must be solved simultaneously with Equation 7.5; and when $\mathscr{R}$ has an Arrhenius temperature dependence, the solution must be numerical. As is the case with axial dispersion of mass, axial dispersion of heat is negligible in many industrial reactors. Piston flow is normally an excellent approximation for tubular reactors at $Re > 10,000$.

Example 7.3   Explore the consequences of ignoring axial dispersion in tubular reactors operating at a moderate Reynolds number, $Re = 10,000$. Consider a highly exothermic reaction and suppose that operation is near a point of thermal instability where a small change in operating conditions can give a large change in the hotspot temperature.

Solution   We begin by applying a piston flow model to a simple, first order reaction with constant physical properties.

$$\frac{da}{dz} = -k_0 \bar{t} \exp\left(\frac{-E}{R_g T}\right) a \tag{7.29}$$

$$\frac{dT}{dz} = -\frac{2U\bar{t}}{\rho C_p R}(T - T_{\text{ext}}) + \left(\frac{-\Delta H_{\mathscr{R}} a_{\text{in}}}{\rho C_p}\right) k_0 \bar{t} \exp\left(\frac{-E}{R_g T}\right) \frac{a}{a_{\text{in}}} \tag{7.30}$$

The following parameter values give rise to a near runaway:

$$k_0 \bar{t} = 2 \times 10^{11} \text{ (dimensionless)}$$

$$\frac{E}{R_g} = 10,000 \text{ K}$$

$$\left(-\frac{\Delta H_{\mathscr{R}} a_{\text{in}}}{\rho C_p}\right) = 200 \text{ K}$$

$$\frac{2U\bar{t}}{\rho C_p R} = 10 \text{ (dimensionless)}$$

$$T_{\text{in}} = T_{\text{ext}} = 373 \text{ K}$$

Integration of Equations 7.29 and 7.30 gives $a_{\text{out}}/a_{\text{in}} = 0.209$, $T_{\text{out}} = 376$ K, and a hotspot, $T_{\text{max}} = 403$ K, occurring at $z = 0.47$.

Turn now to the axial dispersion model. The appropriate dispersion term, $Pe^{-1} d^2a/dz^2$ or $(Pe)_T^{-1} d^2T/dz^2$, is added to the right-hand sides of Equations 7.29 and 7.30. Plausible values at $Re = 10,000$ are

$$\frac{D}{\bar{u}d_t} = 0.45 \qquad \frac{D}{\bar{u}L} = 0.045$$

$$\frac{E}{\bar{u}d_t} = 0.6 \qquad \frac{E}{\bar{u}L} = 0.060$$

where we have assumed an extremely low aspect ratio, $L/d_t = 10$, to magnify the effects of axial dispersion.

We will use the reverse shooting method. Substituting $x = 1 - z$ gives

$$\frac{da}{dx} = -\frac{D}{\bar{u}L}\frac{d^2a}{dx^2} + k_0\bar{t}a \exp\left(\frac{-E}{R_gT}\right) \tag{7.31}$$

$$\frac{dT}{dx} = -\frac{E}{\bar{u}L}\frac{d^2T}{dx^2} + \frac{2U\bar{t}}{\rho C_p R}(T - T_{ext}) - \left(\frac{-\Delta H_{\mathscr{R}}a_{in}}{\rho C_p}\right)k_0\bar{t}\exp\left(\frac{-E}{R_gT}\right)\frac{a}{a_{in}} \tag{7.32}$$

The boundary conditions are

$$a = a_{out} \quad \text{and} \quad T = T_{out} \qquad \text{at } x = 0 \tag{7.33}$$

$$\frac{da}{dx} = \frac{dT}{dx} = 0 \qquad \text{at } x = 0 \tag{7.34}$$

where a_{out} and T_{out} must be guessed. The guesses are confirmed when

$$a + \frac{D}{\bar{u}L}\frac{da}{dx} = a_{in} \qquad \text{at } x = 1 \tag{7.35}$$

and

$$T + \frac{E}{\bar{u}L}\frac{dT}{dx} = T_{in} \qquad \text{at } x = 1 \tag{7.36}$$

We are now ready to solve Equations 7.31 through 7.36 by any convenient means. The simple marching-ahead scheme is, of course, one means. However, these equations are also well suited to Runge–Kutta integration.

The Runge–Kutta scheme (Appendix 2.1) applies only to first order, ordinary differential equations. To use it here, Equations 7.31 and 7.32 must be converted to an equivalent set of first order equations. Let

$$a' = \frac{da}{dx} \quad \text{and} \quad T' = \frac{dT}{dx} \tag{7.37}$$

Then Equations 7.31 and 7.32 can be rewritten as

$$\frac{da}{dx} = a' = 0 \qquad a = a_0 \text{ at } x = 0 \tag{7.38}$$

$$\frac{D}{\bar{u}L}\frac{da'}{dx} + a' - k_0\bar{t}\exp\left(\frac{-E}{R_gT}\right)a = 0 \qquad a' = 0 \text{ at } x = 0 \tag{7.39}$$

$$\frac{dT}{dx} = T' = 0 \qquad T = T_0 \text{ at } x = 0 \tag{7.40}$$

$$\frac{E}{\bar{u}L}\frac{dT'}{dx} + T' - \frac{2U\bar{t}}{\rho C_p R}(T - T_{\text{ext}}) + \left(\frac{-\Delta H_{\mathscr{R}}a_{\text{in}}}{\rho C_p}\right)k_0\bar{t}\exp\left(\frac{-E}{R_gT}\right)\frac{a}{a_{\text{in}}} = 0$$

$$T' = 0 \text{ at } x = 0 \tag{7.41}$$

There are four equations in four dependent variables, a, a', T, and T'. They can now be integrated using the Runge–Kutta method as outlined in Appendix 2.1.

A double trial-and-error procedure is needed to determine $a_0 = a_{\text{out}}$ and $T_0 = T_{\text{out}}$. If done only once, this is probably best done by hand. If repeated evaluations are necessary, a two-dimensional Newton's method can be used. Define

$$F(a_0, T_0) = a + \frac{D}{\bar{u}L}a' - a_{\text{in}}$$

$$\tag{7.42}$$

$$G(a_0, T_0) = T + \frac{E}{\bar{u}L}T' - T_{\text{in}}$$

and use the methodology of Appendix 3.1 to find a_0 and T_0 such that $F = G = 0$.
Following is a comparison of results with and without axial dispersions:

	Piston Flow	Axial Dispersion
$D/\bar{u}L$	0	0.045
$E/\bar{u}L$	0	0.060
$a_{\text{out}}/a_{\text{in}}$	0.209	0.340
T_{out}	376 K	379 K
T_{max}	403 K	392 K
$z(T_{\text{max}})$	0.47	0.63

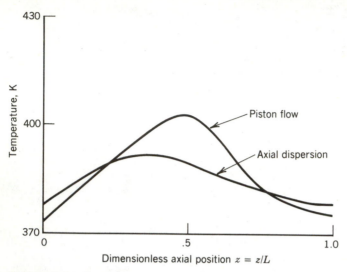

Figure 7.4 Comparison of piston flow and axial dispersion models at conditions near thermal runaway.

This example was chosen to be sensitive to axial dispersion, yet its effects are fairly modest. As would be expected, conversions are lower and the hotspots are colder when axial dispersion is considered. See Figure 7.4.

A more dramatic comparison of piston flow and axial dispersion models is shown in Figure 7.5. Input parameters are the same as above except that $T_{in} = 374$. The piston model predicts ***parametric sensitivity***, where a small change in T_{in} leads to a large change in T_{max}. A classical runaway—which is characterized by a positive second derivative, $d^2T/dz^2 > 0$, at locations just in front of the hotspot—occurs in the piston flow case.

7.2   Packed-Bed Reactors

Packed-bed reactors are very widely used, particularly for solid-catalyzed heterogeneous reactions in which the packing serves as the catalyst. The velocity profile is, of course, quite complex in a packed bed. When measured at a distance from the surface of the packing, velocities are found to be approximately uniform except near the tube wall. Random packing gives higher voidages and thus higher velocities at the wall. However, the velocity profile is almost invariably modeled as being flat. This is not meant to imply piston flow in the sense that radial gradients in composition and temperature are negligible. The packing limits radial mixing to the point that quite large differences can develop across the tube. Radial concentration and temperature profiles can be modeled using an effective

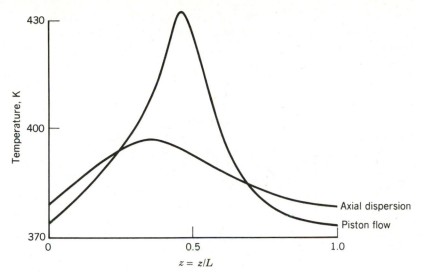

Figure 7.5 Comparison of piston flow and axial dispersion models under conditions of a predicted runaway.

radial diffusivity. For the case of mass transfer,

$$\bar{u}_s \frac{\partial a}{\partial z} = D_r \left(\frac{1}{r} \frac{\partial a}{\partial r} + \frac{\partial^2 a}{\partial r^2} \right) + \varepsilon \mathcal{R}_A \tag{7.43}$$

This is similar to the convective diffusion equation used to analyze laminar flow reactors. Now, however, the velocity profile is assumed to be flat, and D_r is an empirically determined parameter instead of a molecular diffusivity. Note that Equation 7.43 uses the superficial velocity, $\bar{u}_s = Q/A_c$, and that $\varepsilon \mathcal{R}_A$ is the reaction rate per total volume (fluid and packing) while $\mathcal{R}_A$ by itself is the rate per fluid-phase volume. Except for the appearance of the void fraction ε, there is no overt sign that the reactor is a packed bed. This model is ***pseudohomogeneous***. It ignores the detailed interactions between the packing and the fluid, these effects being lumped in the value for D_r. The concentration, a is the fluid-phase concentration, and the rate expression, $\mathcal{R}_A(a, b, \dots)$, is based on fluid-phase concentrations. This approach is obviously satisfactory when the reaction is truly homogeneous. For heterogeneous, solid-catalyzed reactions, the rate is presumably governed by surface concentrations (see Chapter 10), but the use of pseudohomogeneous kinetic expressions is very common.

The boundary conditions associated with Equation 7.43 are

$$a = a_{in} \quad \text{at } z = 0$$

$$\frac{\partial a}{\partial r} = 0 \quad \text{at } r = 0 \quad \text{and} \quad r = R \tag{7.44}$$

Equation 7.43 can be solved by the explicit marching-ahead procedure used for laminar flow reactors. The stability criterion is most demanding when evaluated at the centerline:

$$\frac{\Delta z}{\Delta \ell^2} \leq \frac{R^2}{4D_r \ell} \tag{7.45}$$

This result follows directly from Equation 6.35 but is now written for mass transfer, with D_r replacing α_T.

Analogous to Equation 7.43 is the corresponding heat balance

$$\bar{u}_s \frac{\partial T}{\partial z} = E_r \left(\frac{1}{r}\frac{\partial T}{\partial r} + \frac{\partial^2 T}{\partial r^2} \right) - \frac{\Delta H_{\mathscr{R}}\varepsilon\mathscr{R}}{\rho C_p} \tag{7.46}$$

where E_r is the effective thermal diffusivity in the radial direction. This is related to the effective thermal conductivity λ_r in the usual manner:

$$E_r = \frac{\lambda_r}{\rho C_p} \tag{7.47}$$

The boundary conditions associated with Equation 7.46 are

$$T = T_{\text{in}} \qquad \text{at } z = 0 \tag{7.48}$$

$$\frac{\partial T}{\partial r} = 0 \qquad \text{at } r = 0 \tag{7.49}$$

$$h_r(T - T_{\text{wall}}) = -\lambda_r \frac{\partial T}{\partial r} \qquad \text{at } r = R \tag{7.50}$$

The wall boundary condition has a form not previously encountered. It accounts for the especially high resistance to heat transfer that is observed near the wall in packed-bed reactors. Most of the heat transfer within a packed bed is by fluid convection in the axial direction and by conduction through the solid packing in the radial direction. The high voidage near the wall lowers the effective conductivity. As before, T_{wall} is the inside temperature of the tube, but this may now be significantly different than the fluid temperature just a short distance in from the wall. In essence, the system is modeled as if there were a thin thermal boundary layer across which heat is transferred at a rate proportional to the temperature difference $T(r = R) - T_{\text{wall}}$. This rate of transfer is matched to transfer within the bed by a mechanism of effective radial conduction; thus follows Equation 7.50.

In finite difference form, Equation 7.50 can be written as

$$T(R, z) = \frac{h_r \Delta r T_{\text{wall}} + \lambda_r T(R - \Delta r, z)}{h_r \Delta r + \lambda_r} \tag{7.51}$$

which has a truncation error of $0(\Delta r)$. An alternative result which converges as

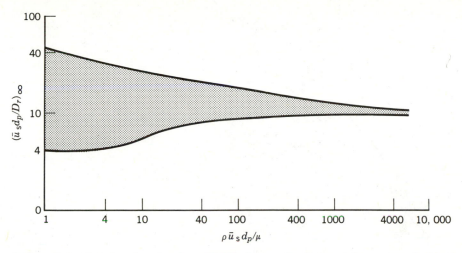

Figure 7.6 Existing data for the radial Peclet number in large-diameter packed beds.

Δr^2 can be obtained by approximating $\partial T / \partial r$ with a second order, forward difference. The result is

$$T(R, z) = \frac{2 h_r \Delta r T_{\text{wall}} + 4\lambda_r T(R - \Delta r, z) - \lambda_r T(R - 2\Delta r, z)}{2 h_r \Delta r + 3 \lambda_r} \tag{7.52}$$

Equations 7.43 and 7.46 are widely employed to model the steady state behaviour of packed-bed reactors.

The parameters D_r, λ_r, and h_r have historically been found by fitting the model to actual operating data from pilot plants or full-scale reactors. The resulting correlations tended to have a limited range of validity and were seldom published. Fortunately, generally useful correlations—some with good theoretical bases—are beginning to emerge.[2]

Figure 7.6 shows the range of existing data for $\bar{u}_s d_p / D_r$. The indicated data apply to large-diameter tubes where $d_t / d_p \to \infty$. Many practical designs have $d_t / d_p < 10$, and the results in Figure 7.6 should be corrected using

$$\frac{\bar{u}_s d_p}{D_r} = \frac{\left(\dfrac{\bar{u}_s d_p}{D_r} \right)_\infty}{1 + 19.4 \left(\dfrac{d_p}{d_t} \right)^2} \tag{7.53}$$

Correlations for λ_r are shown in Figure 7.7. Data in this figure include both

[2] Summaries of extant correlations and references to the original literature are provided by G. F. Froment and K. B. Bischoff, *Chemical Reaction Analysis and Design*, Wiley, New York, 1979, pp. 532–538, and by A. G. Dixon and D. L. Cresswell, *AIChE J.*, **25**, 663 (1979).

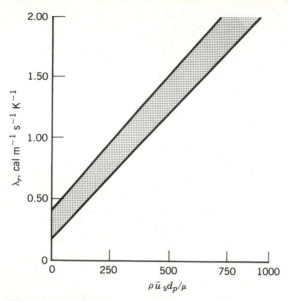

Figure 7.7 Existing correlations for the effective radial conductivity in packed beds.

ceramic and metallic packings, and more accurate results can be obtained from semitheoretical predictions.[3] The wall heat transfer coefficient h_r can be predicted with good accuracy using

$$\frac{h_r d_p}{\lambda_r} = 3 \left(\frac{\rho d_p \bar{u}_s}{\mu} \right)^{-0.25} \tag{7.54}$$

Example 7.4 The oxidation of *o*-xylene to phthalic anhyride is conducted in a multitubular reactor using air at approximately atmospheric pressure as the oxidant:

$$\tag{7.55}$$

---

[3]A. G. Dixon and D. L. Cresswell, *AIChE J.*, **25**, 663 (1979).

Side reactions including complete oxidation are important but will not be considered here. The o-xylene concentration is low, $a_{in} = 44$ g/m^3, to stay under the explosive limit. Due to the large excess of oxygen, the reaction is pseudo-first order in o-xylene concentration with $\ln \varepsilon k = 19.837 - 13{,}636/T$, where k is in s^{-1}. The tube is packed with 3-mm pellets consisting of V$_2$O$_5$ on potassium-promoted silica. It has an ID of 50 mm, is 5 m long, is operated with a superficial velocity of 1 m/s, and has a wall temperature of 623 K. Use $\rho = 1.29$ kg/m^3, $\mu = 3 \times 10^{-5}$ Pa s, $C_p = 0.237$ cal gm^{-1} K^{-1}, and $\Delta H = -307$ kcal/mol.

Use the two-dimensional, radial dispersion model to estimate the hotspot temperature and location for $T_{in} = 623$ K.

Solution We first need to estimate D_r, λ_r, E_r, and h_r. The particle Reynolds numbers, $\rho \bar{u}_s d_p / \mu$, is 130. Figure 7.6 suggests $(\bar{u} d_p / D_r)_\infty \approx 10$. A small correction, giving $\bar{u} d_p / D_r = 8$, is obtained from Equation 7.53 so that $D_r = 3.8 \times 10^{-4}$ m^2/s. Figure 7.7 gives $\lambda_r \approx 0.4$ cal m^{-1} s^{-1} K^{-1} so that $E_R = \lambda_r / \rho C_p = 1.3 \times 10^{-3}$ m^2/s. Equation 7.54 gives $h_r = 120$ cal m^{-2} s^{-1} K^{-1}.

A marching-ahead scheme will be used to solve Equations 7.43 and 7.46. The centerline stability criterion for the temperature equation is the most demanding. Thus

$$\frac{\Delta z}{\Delta \imath^2} \leq \frac{R^2}{4 E_r \bar{t}_s} = 0.024 \tag{7.56}$$

where $\bar{t}_s = L / \bar{u}_s$. Choose $\Delta \imath = 0.2$, $I = 5$. Then the stability criterion is satisfied by $\Delta z = 4 \times 10^{-4}$, $J = 1200$. The marching-ahead equations for concentration are

$$a_{0,j+1} = (1 - 4G_A) a_{0,j} + 4G_A a_{1,j} - \varepsilon k \bar{t}_s a_{0,j} \Delta z$$

$$a_{i,j+1} = (1 - 2G_A) a_{i,j} + G_A \left(1 + \frac{0.5}{i}\right) a_{i+1,j}$$

$$+ G_A \left(1 - \frac{0.5}{i}\right) a_{i-1,j} - \varepsilon k \bar{t}_s a_{i,j} \Delta z \qquad 1 \leq i \leq I - 1$$

$$a_{I,j+1} = \frac{4 a_{I-1,j+1} - a_{I-2,j+1}}{3} \tag{7.57}$$

where $G_A = (D_r \bar{t}_s / R^2) \Delta z / \Delta \imath^2 = 0.0633$. The marching-ahead equations for temperature are

$$T_{0,\,j+1} = (1 - 4G_T)T_{0,\,j} + 4G_T T_{1,\,j} + \left(\frac{-\Delta H_{\mathscr{R}} a_{\text{in}}}{\rho C_p}\right)\left(\frac{\varepsilon k \bar{\imath}_s a_{0,\,j}}{a_{\text{in}}}\right)\Delta z$$

$$T_{i,\,j+1} = (1 - 2G_T)T_{i,\,j} + G_T\left(1 + \frac{0.5}{i}\right)T_{i+1,\,j}$$

$$+ G_T\left(1 - \frac{0.5}{i}\right)T_{i-1,\,j} + \left(\frac{-\Delta H_{\mathscr{R}} a_{\text{in}}}{\rho C_p}\right)\varepsilon k \bar{\imath}_s a_{i,\,j}\,\Delta z$$

$$1 \le i \le I - 1$$

$$T_{I,\,j+1} = \frac{2h_r R\,\Delta\imath\, T_{\text{wall}} + 4\lambda_r T_{I-1,\,j+1} - \lambda_r T_{I-2,\,j+1}}{2h_r R\,\Delta\imath + 3\lambda_r} \qquad \textbf{(7.58)}$$

where $G_T = (E_r \bar{\imath}_s / R^2)\,\Delta z/\Delta\imath^2 = 0.2167$.

The hotspot is located at $\imath = 0$ and $z = 0.107$. $T_{\text{max}} = 632.7$ K. Figure 7.8 shows the radial temperature and concentration profiles at the hotspot.

The example uses high gas velocities and small-diameter catalyst particles. This gives a high heat transfer coefficient at the tube wall and mitigates possible temperature and composition gradients within the catalyst particles. It also

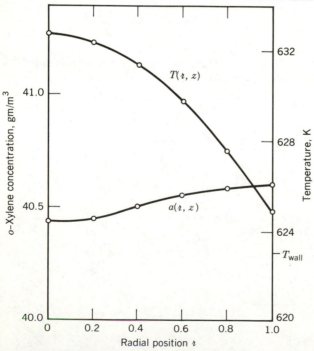

Figure 7.8 Radial temperature and composition profiles at a hotspot.

causes large pressure drops. See Problems 7.4 through 7.6 for further discussion of phthalic anhydride processes.

Axial dispersion terms $D(\partial^2 a/\partial z^2)$ and $E(\partial^2 T/\partial z^2)$ can be added to Equations 7.43 and 7.46. This converts the initial value problem to a *two-point boundary value problem* in the axial direction. The resulting equations can be solved by applying the shooting method to a PDE. However, axial dispersion is usually negligible since $\bar{u}L/D$ and $\bar{u}L/E$ are typically much greater than 100 in a packed-bed reactor.

A more important embellishment to the foregoing treatment of packed-bed reactors is to allow for temperature and concentration gradients *within* catalyst pellets.

The intrapellet diffusion of heat and mass is governed by partial differential equations that are about as complex as those governing the bulk properties of the bed. A set of simultaneous PDEs must be solved to estimate the extent of reaction and conversion occurring within a single pellet. These local values are then substituted into Equations 7.43 and 7.46 so that we need to solve a set of PDEs that are embedded within a set of PDEs. The resulting system truly reflects the complexity of heterogeneous reactors, and solutions still tax the fastest computers. Most industrial reactors are designed on the basis of pseudohomogeneous models as in Equations 7.43 and 7.46. In fact, radial gradients are sometimes neglected even in single-tube calculations for multitubular designs. This forces actual plant designs to be somewhat more conservative than they would otherwise be. Reasonable correlations for radial heat transfer now exist and should be used.

Suggestions for Further Reading

An excellent introduction to the axial dispersion model and to methods for measuring dispersion coefficients is given by:

O. Levenspiel, *Chemical Reaction Engineering*, 2nd ed., Wiley, New York, 1972.

A more advanced account is given in:

E. B. Nauman and B. A. Buffham, *Mixing in Continuous Flow Systems*, Wiley, New York, 1983.

The heat and mass transfer phenomena associated with packed bed reactors is described in:

G. F. Froment and K. B. Bischoff, *Chemical Reaction Analysis and Design*, Wiley, New York, 1979.

Correlations for heat transfer in packed beds are still being developed. The current state of the art is represented by:

A. G. Dixon and D. L. Cresswell, "Theoretical Prediction of Effective Heat Transfer Parameters in Packed Beds", *AIChE J.*, **25**, 663–676 (1979).

See Chapter 10 for a discussion of heterogeneous catalysis in packed beds and for additional references.

Problems

7.1 Nerve gas is to be thermally decomposed by oxidation in a 50 cm-diameter tubular reactor that is approximately isothermal at 620°C. The entering concentration of nerve gas is 1% by volume. The outlet concentration must be less than 1 part in 10^{12} by volume. The observed half-life for the reaction is 0.2 s. How long should you make the tube given an inlet velocity of 20 m/s? What will be the pressure drop given an atmospheric discharge?

7.2 Determine the yield of a second order reaction in an isothermal, tubular reactor governed by the axial dispersion model with Pe = 16 and $a_{in}k\bar{t} = 2$.

7.3 The marching equation for reverse shooting, Equation 7.24, was developed using a first order, backward difference approximation for da/dx even though a second order approximation was necessary for d^2a/dx^2. Since the locations $j - 1, j, j + 1$ are involved anyway, would it not be better to use a second order, central difference approximation for da/dx? Would this allow convergence $0(\Delta x^2)$ for the reverse shooting method?

7.4 Example 7.4 on the partial oxidation of *o*-xylene used a pseudo-first order kinetic scheme. For this to be justified, the oxygen concentration must be approximately constant, which in turn requires a low oxygen consumption and a low pressure drop. Are these assumptions reasonable for the reactor in Example 7.4? Specifically, estimate the total change in oxygen concentration given atmospheric discharge pressure and $a_{out} = 21$ gm/m³. Assume $\varepsilon = 0.4$.

7.5 Phthalic anhydride will, in the presence of the V_2O_5 catalyst of Example 7.4, undergo complete oxidation:

$$O + 7.5O_2 \rightarrow 8CO_2 + 2H_2O \qquad (I)$$

where $\Delta H_{\mathscr{R}} = -760$ kcal/mol. Suppose this reaction, like the primary reaction, is pseudo-first order, and that $\ln k_I = 12.300 - 10,000/T$. Repeat the hotspot calculation of Example 7.4, now taking into account the second reaction.

7.6 An alternative route to phthalic anhydride is the partial oxidation of naphthalene:

$$+ 4.5O_2 \rightarrow \quad O + 2CO_2 + 2H_2O \qquad (II)$$

where $\Delta H_{\mathscr{R}} = -430$ kcal/mol. This reaction is performed using a promoted V_2O_5 catalyst on silica much like that considered in Example 7.4. Suppose $\ln K_{II} = 31.6800 - 19{,}100/T$ for the above reaction and that the subsequent, complete oxidation of phthalic anhydride follows the kinetics of Problem 7.5. Suppose it is desired to use the same reactor as in Example 7.4 for naphthalene oxidation but now with $a_{in} = 53$ gm/m^3. Determine values for T_{in} and T_{wall} that maximize the production of phthalic anhydride from naphthalene.

7.7 Methanol is produced from syngas according to the following reaction:

$$2H_2 + CO \rightleftharpoons CH_3OH$$

Problem 5.15 gave rate, equilibrium, and physical property data and studied the reaction in 38.1 mm-ID tubes each 12 m long. There, a lumped parameter model was used which ignored radial variations in composition and temperature. Did this simplifying assumption have a major influence on the design calculations? Resolve this question by using the distributed parameter model of Section 7.2. Treat the case of $T_{in} = 473$ K and $T_{wall} = 485$ K.

7.8 Klier et al., *J. Catalysis*, **74**, 343 (1982), have reported experimental results on a 30/70 Cu/ZnO catalyst used for the methanol synthesis of Problems 5.15 and 7.7. It is available as $\frac{3}{16}$ by $\frac{3}{16}$ extruded cylindrical pellets which have a bulk density of 75 lb/ft^3 and a voidage of 40%. Rate data have been fit to the following form:

$$\mathscr{R} = k \left(\frac{K'' P_{CO_2}}{P_{CO} + K'' P_{CO_2}} \right)^3 \frac{K_{CO} K H_2^2 \left[P_{CO} P_{H_2}^2 - \left(P_{CH_2OH}/K' \right) \right]}{\left(1 + K_{CO} P_{CO} + K_{CO_2} P_{CO_2} + K_{H_2} P_{H_2} \right)^3}$$

where all partial pressures are measured in atm and $\mathscr{R}$ is in mol hr^{-1} gm catalyst^{-1}. The various constants, in units consistent with the above, have been determined as follows:

T (°C)	K_{CO}	K_{H_2}	K_{CO_2}	K''	K
225	12.52	1.77	39.62	158.2	1.064
235	8.58	1.40	21.52	125.4	1.253
250	5.00	1.00	9.00	90.0	1.584

The equilibrium constant K' is given as follows:

$$K' = \frac{3.27 \times 10^{-13} \exp(11{,}678/T)}{1 - P(1.95 \times 10^{-4}) \exp(1703/T)}$$

where P is in atm and T is in K. Suppose you have been given the job of evaluating competing claims by various researchers and catalyst manufacturers:

(a) Is there a significant difference between the Klier catalyst and the Berty–Cropley catalyst of Problem 5.15?

(b) If you were designing a methanol reactor, which catalyst would you specify? Assume the same cost and operating life.

(c) If you had a methanol reactor designed for the Berty–Cropley catalyst, how would it perform if the Klier catalyst were retrofitted?

8

Unsteady Reactors

The general material balance of Section 1.1 contains an accumulation term which enables its use for unsteady state reactors. Except for some brief examples in Chapter 1, we have neglected this term, and it is true that the great majority of chemical reactors are designed for steady state operation. However, even steady state reactors must occasionally start up and shut down. Also, an understanding of process dynamics is necessary to design the control systems needed to handle upsets and to enable operation at steady states that would otherwise be unstable.

Unsteady mass and energy balances consider three kinds of accumulation:

$$\text{Total mass} \qquad \frac{d(V\rho)}{d\theta}$$

$$\text{Component moles} \qquad \frac{d(Va)}{d\theta}$$

$$\text{Enthalphy} \qquad \frac{d(V\rho H)}{d\theta}$$

These accumulation terms are added to the appropriate steady state balances to convert them to unsteady balances. We begin with a lumped system, the now familiar stirred tank reactor.

8.1 Unsteady Stirred Tanks

The steady state balance for total mass is

$$Q_{in}\rho_{in} - Q_{out}\rho_{out} = 0 \tag{8.1}$$

227

A perfect mixer has the same composition and physical properties within the system as at the outlet. Thus the unsteady state balance for total mass is

$$\frac{d(V\rho_{out})}{d\theta} = Q_{in}\rho_{in} - Q_{out}\rho_{out} \tag{8.2}$$

Liquid phase systems with approximately constant density are common. Thus the usual simplification of Equation 8.2 is

$$\frac{dV}{d\theta} = Q_{in} - Q_{out} \tag{8.3}$$

The component balance for a general case is

$$\frac{d(Va_{out})}{d\theta} = Q_{in}a_{in} - Q_{out}a_{out} + V\mathcal{R}_A \tag{8.4}$$

Time-dependent volumes and flow rates are explicitly allowed in this equation and must be used to analyze most startup and shutdown transients. Some control strategies can be effectively analyzed with the constant-volume, constant-flow rate version of Equation 8.4:

$$\bar{t}\frac{da_{out}}{d\theta} = a_{in} - a_{out} + \bar{t}\mathcal{R}_A \tag{8.5}$$

The enthalpy balance for a fairly general case is

$$\frac{d(V\rho_{out}H_{out})}{d\theta} = Q_{in}\rho_{in}H_{in} - Q_{out}\rho_{out}H_{out} - V\Delta H_{\mathcal{R}}\mathcal{R} \\ + UA_{ext}(T_{ext} - T_{out}) \tag{8.6}$$

A still more general case is discussed in Problem 8.10. Typical simplifications are constant volume and flow rate, constant density, and replacement of enthalpy with $C_p(T - T_0)$. This gives

$$\bar{t}\frac{dT_{out}}{d\theta} = T_{in} - T_{out} - \frac{\bar{t}\Delta H_{\mathcal{R}}\mathcal{R}}{\rho C_p} + \frac{UA_{ext}}{V\rho C_p}(T_{ext} - T_{out}) \tag{8.7}$$

Being a lumped system, the transient performance of a perfectly mixed reactor is governed by a set of ODEs. The minimum set is just Equation 8.5, which governs the reaction of a single component with time-varying inlet concentration, $a_{in} = a_{in}(\theta)$. The maximum set has separate ODEs for each of the variables V, H_{out}, a_{out}, b_{out}, ... The ODEs must be supplemented by a set of initial conditions and by any thermodynamic relations needed to determine dependent properties such as density and enthalpy. The maximum set will consist of Equations 8.2 and 8.6 and N versions of Equation 8.4, where N is the number of components in the system. The maximum dimensionality is thus $2 + N$. This can always be reduced to 2 plus the number of independent reactions using the reaction coordinate method of Chapter 3. However, such reductions are seldom necessary from a computational viewpoint.

8.1.1 Transients in Isothermal Perfect Mixers

If the system is isothermal with $T_{in} = T_{out}$, Equation 8.6 is unnecessary. Unsteady behavior in an isothermal perfect mixer is governed by a maximum of $N + 1$ ordinary differential equations. Except for highly complicated reactions such as polymerizations (where N is theoretically infinite), solutions are usually straightforward. Numerical methods are similar to those used for steady state, piston flow reactors; and analytical solutions are usually possible for first order kinetics.

Example 8.1 Consider a first order reaction occurring in a perfect mixer where the inlet concentration of reactant has been held constant at a_0 for $\theta < 0$. At time $\theta = 0$, the inlet concentration is changed to a_1. Thus $a_{in} = a_0$ for $\theta < 0$ and $a_{in} = a_1$ for $\theta > 0$. Find the outlet response for $\theta > 0$ assuming isothermal, constant-volume, constant-density operation.

Solution This situation requires Equation 8.5 only:

$$\bar{t}\frac{da_{out}}{d\theta} = a_1 - a_{out} - k\bar{t}a_{out} \qquad \theta > 0 \tag{8.8}$$

A general solution is

$$a_{out} = \frac{a_1}{1 + k\bar{t}} + C\exp\left(-\frac{1 + k\bar{t}}{\bar{t}}\theta\right) \tag{8.9}$$

as may be verified by differentiation. The constant C depends on an initial condition which has the form

$$a_{out} = a_{initial} \qquad \text{at } \theta = 0 \tag{8.10}$$

For the problem at hand, $a_{initial}$ is equal to the steady state response of the

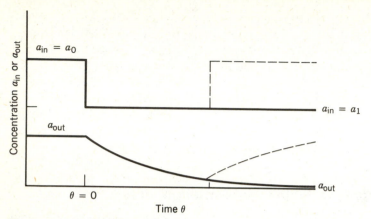

Figure 8.1 Dynamic response of a perfectly mixed reactor to changes in inlet concentration.

reactor to an input of a_0:

$$a_{\text{initial}} = \frac{a_0}{1 + k\bar{t}} \tag{8.11}$$

Applying this initial condition to Equation 8.9 gives

$$a_{\text{out}} = \frac{a_1 + (a_0 - a_1)\exp\left[-(1 + k\bar{t})\theta/\bar{t}\right]}{1 + k\bar{t}} \tag{8.12}$$

as the desired solution. Figure 8.1 illustrates Equation 8.12 and also shows the effect of restoring a_{in} to its original value at some time $\theta > 0$.

Example 8.1 shows how an isothermal perfect mixer with first order reaction responds to an abrupt change in inlet concentration. The outlet concentration moves from an initial steady state to a final one in a gradual fashion. If the inlet concentration is returned to its original value, the outlet concentration returns to its original value. If the time period for an input disturbance is small, the outlet response is small. The magnitude of the outlet disturbance will never be larger than the magnitude of the inlet disturbance. The system is stable. Indeed, it is **open-loop stable**, which means that steady state operation can be achieved without resort to a feedback control system.

Example 8.2   Analyze a startup transient for an isothermal, constant-density stirred tank reactor. Suppose the tank is initially empty and is filled at rate $Q_{\text{in}} = Q$. A first order reaction begins immediately. When the tank is full, discharge flow commences at rate $Q_{out} = Q$. Find a_{out} as a function of time.

Solution The steady state flow rate is $Q_{in} = Q_{out} = Q$. Filling starts at time $\theta = 0$ and the tank becomes full at time $\theta = \bar{t}$, where $\bar{t} = V_{full}/Q$. For $0 < \theta < \bar{t}$, Equation 8.3 gives

$$\frac{dV}{d\theta} = Q \tag{8.13}$$

or

$$V = Q\theta \qquad 0 < \theta < \bar{t} \tag{8.14}$$

Equation 8.4 becomes

$$\frac{d(Va)}{d\theta} = \frac{V\,da}{d\theta} + a\frac{dV}{d\theta} = Qa_0 - Vka \tag{8.15}$$

where a denotes the concentration within the reactor. Note that $a_{out} = a$ if there were any discharge flow. Substituting for $V = V(\theta)$ gives

$$Q\theta\frac{da}{d\theta} + Qa = Qa_0 - Q\theta ka \tag{8.16}$$

or

$$\theta\frac{da}{d\theta} + (1 + \theta k)a = a_0 \qquad 0 < \theta < \bar{t} \tag{8.17}$$

The solution of this first order ODE is left to the reader. The initial condition is $a = a_0$ at $\theta = 0$. The steady state concentration will be $a_{out} = a_0/(1 + k\bar{t})$ and will be achieved when $\theta \gg \bar{t}$. Does the concentration of A overshoot or undershoot the steady state value during startup, $0 < \theta < \bar{t}$?

During the startup transient of the previous example, material was fed to the reactor during the period $0 < \theta < \bar{t}$ but there was no discharge. Operation during this period can be classified as semibatch or, more particularly, as *fed-batch*. Fed-batch reactions, with or without a subsequent phase of continuous operation, are commonly used for fermentations.

There are many ways to start up a stirred tank reactor; the fed-batch method of Example 8.2 is only one of many possibilities. It is generally desired to begin continuous operation only when the vessel is full and when the concentration within the vessel has reached its steady state value. This can be done with fed-batch. If Q_{in} is low during the fed-batch period, the vessel fills slowly and the reactant concentration will be low when the vessel is finally full. If Q_{in} is high, the vessel fills quickly and the reactant concentration will be near a_{in} when the vessel is full. Obviously, an intermediate filling rate exists in which the vessel is full and the steady state outlet concentration is reached simultaneously. See Problem 8.2. Perhaps the most common way of starting a stirred tank reactor is to use an initial period of fed-batch operation with large Q_{in} to quickly fill the vessel. This is followed by a period of true batch operation, during which time the reactant concentration achieves its steady state value. Finally, the input and discharge flows are started with $Q_{in} = Q_{out} = Q$.

Just as there are many ways of starting a reactor, there are many ways of shutting it down. The next examples illustrate two possibilities.

Example 8.3   A perfect mixer has been operating at steady state. It is desired to shut it down. Suppose this is done by setting $Q_{in} = 0$ while continuing $Q_{out} = Q$ until the reactor is empty. Assume isothermal, constant-volume operation with first order reaction.

Solution   It is perhaps obvious that stopping the input flow will cause the system to behave as a batch reactor. The initial concentration of the batch is the steady state value $a_0/(1 + k\bar{t})$, and the concentration decreases with time as the vessel discharges:

$$a_{out} = \frac{a_0 \exp(-k\theta)}{1 + k\bar{t}} \qquad \theta > 0 \tag{8.18}$$

where we have assumed the shutdown transient starts at time $\theta = 0$. The transient lasts until the vessel is empty which occurs at time $\theta = \bar{t}$ for the special case $Q_{in} = 0$, $Q_{out} = Q$ for $\theta > 0$.

The rather intuitive approach used to obtain Equation 8.18 should be reinforced by more formal mathematics. Equation 8.4 gives

$$V\frac{da_{out}}{d\theta} + a_{out}\frac{dV}{d\theta} = -Q_{out}a_{out} + V\mathcal{R}_A \tag{8.19}$$

and Equation 8.3 gives

$$\frac{dV}{d\theta} = -Q_{out} \tag{8.20}$$

Substituting Equation 8.20 into Equation 8.19 cancels all Q_{out} and V terms giving the batch reaction equation

$$\frac{da_{out}}{d\theta} = -ka_{out} \tag{8.21}$$

the solution to which is Equation 8.18.

Example 8.4 A shutdown strategy known as *constant RTD control* sets Q_{out} slightly above its steady state value, say, $Q_{out} = 1.1Q$, while setting Q_{in} proportional to the volume in the vessel:

$$Q_{in} = \frac{V}{\bar{t}} = \frac{V(\theta)}{\bar{t}} \tag{8.22}$$

where $\bar{t}$ is the steady state value V_0/Q. Explore the consequences of this strategy for an isothermal, constant-density perfect mixer.

Solution Equation 8.3 becomes

$$\frac{dV}{d\theta} = \frac{V}{\bar{t}} - 1.1Q \tag{8.23}$$

subject to the initial condition that $V = V_0$ at $\theta = 0$. Solution gives

$$V = 1.1V_0 - 0.1V_0 \exp\left(+\frac{\theta}{\bar{t}}\right) \tag{8.24}$$

The vessel goes empty at the time

$$\theta = \bar{t} \ln \frac{1.1}{0.1} = 2.4\bar{t} \tag{8.25}$$

During the interval $0 < \theta < 2.4\bar{t}$, the volume is gradually decreasing while the outlet concentration is governed by Equation 8.4. For the given shutdown strategy, Equation 8.4 becomes

$$a_{out}\frac{dV}{d\theta} + V\frac{a_{out}}{d\theta} = \frac{Va_{in}}{\bar{t}} - 1.1Qa_{out} + V\mathscr{R}_A \tag{8.26}$$

Replacing $dV/d\theta$ by Equation 8.23 and canceling where possible gives

$$\bar{t}\frac{da_{out}}{d\theta} = a_{in} - a_{out} + \bar{t}\mathscr{R}_A \tag{8.27}$$

which is identical to Equation 8.5. However, Equation 8.5 was derived for a constant-volume reactor while Equation 8.27 was derived for the special case of variable-volume operation now being considered. Thus the reactor dynamics for shutdown using constant RTD control are the same as if the reactor were operating at constant volume. Specifically, $da_{out}/d\theta$ will be zero and a_{out} will remain at its steady state value provided only that a_{in} remains constant. Thus we have the rather surprising result that the reactor shutdown does not alter the outlet concentration. This result is valid for arbitrary reaction kinetics and for any $Q_{out} > 0$ provided only that a steady state existed prior to shutdown and that Q_{in} is kept proportional to V according to Equation 8.22. See Nauman and Carter[1] for additional details and an experimental verification.

Consider now a complex set of reactions occurring in a perfect mixer operated at steady state. In Chapter 3 we determined the outlet composition by solving a set of simultaneous algebraic equations. The composition can also be determined by solving a set of simultaneous ODEs. Suppose the reactor is instantaneously filled at time $\theta = 0$ and that the initial concentrations are $a_0, b_0, \ldots$. Typically, but not inevitably, these initial concentrations are set equal to the inlet concentrations used during steady state operation. The outlet concentrations for time $\theta > 0$ are governed by Equation 8.5 written for each component:

$$\bar{t}\frac{da_{out}}{d\theta} = a_{in} - a_{out} + \bar{t}\mathscr{R}_A$$

$$\bar{t}\frac{db_{out}}{d\theta} = b_{in} - b_{out} + \bar{t}\mathscr{R}_B \tag{8.28}$$

$$\vdots \qquad \vdots$$

subject to the initial conditions that $a_{out} = a_0$, $b_{out} = b_0, \ldots$ at $\theta = 0$. These equations are coupled through the concentration dependence of reaction rate $\mathscr{R}(a, b, \ldots)$. They may be solved by any convenient method suitable for first order, simultaneous ODEs. The solution will show the evolution of outlet concentrations from the initial values to the steady state values, provided that the steady state exists and is stable.

Example 8.5 Suppose the competing, elementary reactions

$$A + B \xrightarrow{k_I} C$$

$$A \xrightarrow{k_{II}} D$$

[1] E. B. Nauman and K. Carter, *I & EC Proc. Des. Dev.*, **13**, 275 (1974).

occur in a perfect mixer. Use the transient technique to determine the steady state outlet composition and to explore system stability. Suppose $k_I a_{in} \bar{t} = 3$, $k_{II} \bar{t} = 1.0$, $b_{in} = 1.5 a_{in}$, $c_{in} = 0$, and $d_{in} = 0.1 a_{in}$.

Solution Equations 8.28 are written for each component. They can be expressed in finite difference form as

$$\left(\frac{a}{a_{in}}\right)_{j+1} = \left(\frac{a}{a_{in}}\right)_j + \left[1 - (1 + k_{II}\bar{t})\left(\frac{a}{a_{in}}\right)_j - k_I a_{in}\bar{t}\left(\frac{a}{a_{in}}\right)_j\left(\frac{b}{a_{in}}\right)_j\right]\Delta\tau$$

$$\left(\frac{b}{a_{in}}\right)_{j+1} = \left(\frac{b}{a_{in}}\right)_j + \left[\frac{b_{in}}{a_{in}} - \left(\frac{b}{a_{in}}\right)_j - k_I a_{in}\bar{t}\left(\frac{a}{a_{in}}\right)_j\left(\frac{b}{a_{in}}\right)_j\right]\Delta\tau$$

$$\left(\frac{c}{a_{in}}\right)_{j+1} = \left(\frac{c}{a_{in}}\right)_j + \left[\frac{c_{in}}{a_{in}} - \left(\frac{c}{a_{in}}\right)_j + k_I a_{in}\bar{t}\left(\frac{a}{a_{in}}\right)_j\left(\frac{b}{a_{in}}\right)_j\right]\Delta\tau \qquad (8.29)$$

$$\left(\frac{d}{a_{in}}\right)_{j+1} = \left(\frac{d}{a_{in}}\right)_j + \left[\frac{d_{in}}{a_{in}} - \left(\frac{d}{a_{in}}\right)_j + k_{II}\left(\frac{a}{a_{in}}\right)_j\right]\Delta\tau$$

where $\tau = \theta/\bar{t}$ is a dimensionless time. Equations 8.29 are directly suitable for marching-ahead calculations, although they can be written more compactly as

$$\alpha_{j+1} = \alpha_j + (1 - 2\alpha_j - 3\alpha_j\beta_j)\,\Delta\tau \qquad \alpha_0 = 1$$

$$\beta_{j+1} = \beta_j + (1.5 - \beta_j - 3\alpha_j\beta_j)\,\Delta\tau \qquad \beta_0 = 1.5$$

$$\gamma_{j+1} = \gamma_j + (0 - \gamma_j + 3\gamma_j\beta_j)\,\Delta\tau \qquad \gamma_0 = 0 \qquad (8.30)$$

$$\delta_{j+1} = \delta_j + (0.1 - \delta_j + \alpha_j)\,\Delta\tau \qquad \delta_0 = 0.1$$

where $k_I a_{in}\bar{t}$ and $k_{II}\bar{t}$ have been replaced with their numerical values.

Figure 8.2 illustrates the solution to Equations 8.30 for the indicated case where the initial concentrations have been set equal to the inlet concentrations. Numerical values for the steady state concentrations are also indicated on Figure 8.2. Had only these values been required, they could have been found more easily by using the steady state equations. These can be obtained from Equations 8.30 by supposing $\alpha_{j+1} = \alpha_j$, and so on. This gives

$$0 = 1 - 2\alpha - 3\alpha\beta$$

$$0 = 1.5 - \beta - 3\alpha\beta$$

$$0 = -\gamma + 3\alpha\beta \qquad (8.31)$$

$$0 = 0.1 - \delta + \alpha$$

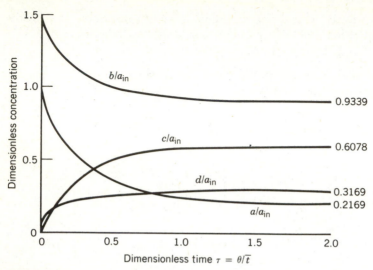

Figure 8.2 Transient approach to a stable steady state in a perfectly mixed stirred tank reactor.

which are readily solvable in the present example. Solution of Equations 8.31 gives the same outlet concentrations as are achieved asymptotically in the transient solution. The transient solution shows the steady state to be stable and achievable from the inlet composition.

The algebraic equations of Chapter 3 (or Equations 8.31 in Example 8.5) can always be solved to find one or more steady states. However, they provide no direct information about stability. If the transient technique of this section indicates a steady state, then that steady state is stable and achievable from the initial composition used in the calculations. If the same steady state is found for all possible initial compositions, then that steady state is unique and globally stable. This is the usual case for isothermal reactions in a perfect mixer. However, nonlinear kinetic schemes have been found that exhibit multiple steady states or that never achieve a steady state. See Problem 8.7. Multiple steady states are common in nonisothermal reactors although at least one of them is usually stable.

8.1.2   Nonisothermal Stirred Tank Reactors

Nonisothermal stirred tanks have been defined as having T_{out} different than T_{in}. More to the point, the temperature in a nonisothermal stirred tank is governed by an enthalpy balance that contains the heat of reaction as a significant term. If the heat of reaction is unimportant so that a desired T_{out} can be imposed on the system regardless of the extent of reaction, then the reactor dynamics can be analyzed by the methods of the previous section even though the reactor is technically nonisothermal. Thus we focus on situations where Equation 8.6 must be explicitly considered as part of the design.

Example 8.6   The styrene polymerization example of Section 4.2.1 showed three steady states. Explore the stability of the middle one, which has $a_{out}/a_{in} = 0.738$ and $T_{out} = 404$ K.

Solution   Accumulation terms must be added to the component mass balance of Equation 4.19 and to the heat balance of Equation 4.21. This gives

$$\bar{t}\frac{da_{out}}{d\theta} = a_{in} - a_{out} - a_{out}k_0\bar{t}\exp\left(\frac{-E}{R_gT_{out}}\right) \tag{8.32}$$

and

$$\bar{t}\frac{dT_{out}}{d\theta} = T_{in} - T_{out} - \frac{a_{in}\Delta H_{\mathscr{R}}}{\rho C_p}\left(\frac{a_{out}}{a_{in}}\right)k_o\bar{t}\exp\left(\frac{-E}{R_gT_{out}}\right) \tag{8.33}$$

Using the values specified in Section 4.2.1. gives

$$2\frac{d(a_{out}/a_{in})}{d\theta} = 1 - \left(\frac{a_{out}}{a_{in}}\right)\left[1 + 2\times10^{10}\exp\left(\frac{-10,000}{T_{out}}\right)\right] \tag{8.34}$$

$$2\frac{dt_{out}}{d\theta} = 300 - T_{out} + 8\times10^{12}\left(\frac{a_{out}}{a_{in}}\right)\exp\left(\frac{-10,000}{T_{out}}\right) \tag{8.35}$$

Substitution of the exact steady state solution will make the right-hand sides of Equation 8.34 and 8.35 vanish. Then both time derivatives will be zero. However, neither experimental nor numerical approximations can be exact. Suppose continuous operation is started at initial conditions of $a_{out}/a_{in} = 0.74$ and $T_{out} = 405$ K, Figure 8.3 shows concentration and temperature trajectories determined by numerical solution of Equations 8.34 and 8.35. Even though the initial conditions were quite close to the metastable state, the reaction runs away, heading toward the upper steady state. Other starting conditions will generate similar runaways or will cause extinctions. Long-term operation at or near the metastable state is impossible without some form of control system.

The styrene polymerizer of the previous example is ***open-loop unstable*** with respect to the middle steady state. The other two steady states are stable but undesirable. Either the material balance, Equation 8.32, or the energy balance, Equation 8.33, must be changed to operate the system at the middle steady state. Usually, it is the heat balance that is altered since reactor dynamics tend to be more sensitive to temperature than composition. Autorefrigeration (see Section

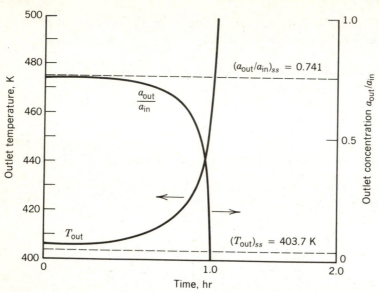

Figure 8.3 Behavior of a styrene polymerizer with initial conditions near a metastable state.

4.2.1) is an excellent means of stabilizing the operation. Another approach is to allow for external heat exchange, supplementing the heat balance by a $UA_{ext}(T_{ext} - T_{out})$ term. In laboratory-scale equipment, UA_{ext} may be so large that setting T_{ext} effectively sets the internal operating temperature. In industrial-scale equipment, UA_{ext} will be limited in size, the temperature difference $\Delta T = T_{ext} - T_{out}$ will be significant in magnitude, and some form of feedback control on T_{ext} will be necessary to stabilize operation at or near the desired point. When T_{out} is higher than its desired value, T_{ext} must be lowered and vice versa. This can be accomplished with an automatic controller.

Example 8.7 Devise an automatic control system based on external heat exchange for the styrene polymerizer of the previous examples.

Solution Equation 8.7 provides the energy balance. Parameters are identical to those of the previous example but the dimensionless group $UA_{ext}/V\rho C_p$ must also be specified. For operation at an adiabatic steady state, as in the current example, there is no net load on the heat exchanger. It will alternately supply and remove heat to maintain T_{out} at its desired value, and UA_{ext} need be sized only to handle disturbances. Alternative heating and cooling is generally undesirable. A better choice is to raise T_{in} so that the reaction always has a positive exotherm at the desired operating point. The exchanger will be sized somewhat larger and will always remove heat. Assume $T_{in} = 330$ K and

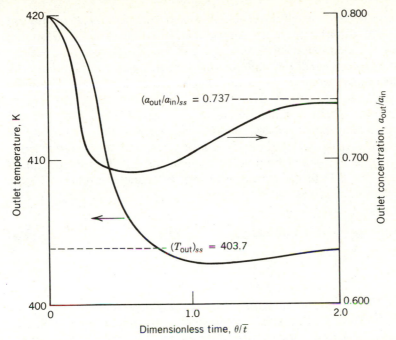

Figure 8.4 Approach to a metastable steady state stabilized by feedback control.

$UA/V\rho C_p = 0.5$. Then Equation 8.7 becomes

$$2\frac{dT_{out}}{d\theta} = 330 - T_{out} + 8 \times 10^{12}\left(\frac{a_{out}}{a_{in}}\right)\exp\left(-\frac{10\,000}{T_{out}}\right) + 0.5(T_{ext} - T_{out})$$

(8.36)

Suppose we would like the middle steady state to remain at its old location: $a_{out}/a_{in} = 0.741$, $T_{out} = 403.7$ K. Substituting these values into Equation 8.36, setting $dT_{out}/d\theta = 0$, and solving for T_{ext} gives $T_{ext} = 343.7$. Consider a control action of the form

$$T_{ext} = 343.7 + G(403.7 - T_{out})$$ (8.37)

where G is a constant of proportionality. When T_{out} is higher than the desired value, Equation 8.37 will lower T_{ext} and vice versa.

Figure 8.4 shows the effect of combining the control action of Equation 8.37 with Equations 8.32 and 8.36. This result uses $G = 20$ and a larger initial disturbance, $a_{out}/a_{in} = 0.80$ and $T_{out} = 410$ K, than in the previous example, yet the system achieves the desired steady state.

The controller represented by Equation 8.37 is rudimentary. It would not handle an initial disturbance much larger than the one given it. Had the calculation for the metastable value of $T_{ext} = 343.7$ K been wrong, the reactor would have stabilized at conditions different than the desired ones. The controller has only **proportional control**. It lacks the **integral** (reset) action and **derivative** action used in most controllers. Nevertheless, it does illustrate the use of feedback control to operate at a metastable state. In actual practice, T_{out} would be measured by a temperature sensor. The measurement would be compared to a desired or setpoint value (403.7 K in the above case), and the resulting error signal $\varepsilon = T_{desired} - T_{out}$ would be used to manipulate T_{ext}. This manipulation occurs via Equation 8.37 in the current example. In actual controllers, a more complicated algorithm such as

$$T_{ext} = G_0 + G_1\varepsilon + G_2\int_0^\theta \varepsilon\, d\theta' + G_3\frac{d\varepsilon}{d\theta} \tag{8.38}$$

would be used.

Example 8.8 This example cites a real study of a laboratory CSTR that exhibits complex dynamics and limit cycles in the absence of a feedback controller. We cite the work of Vermeulen and Fortuin, who studied the acid-catalyzed hydration of 2,3-epoxy-1-propanol to glycerol:

$$
\begin{array}{c}
\text{H}\quad\text{H}\quad\text{H} \\
|\quad\ \ |\quad\ \ | \\
\text{H}-\text{C}-\text{C}-\text{C}-\text{H} + \text{H}_2\text{O} \xrightarrow{\text{H}_2\text{SO}_4} \\
|\quad\ \ \backslash\!/ \\
\text{OH}\quad\text{O}
\end{array}
\qquad
\begin{array}{c}
\text{H}\quad\text{H}\quad\text{H} \\
|\quad\ \ |\quad\ \ | \\
\text{H}-\text{C}-\text{C}-\text{C}-\text{H} \\
|\quad\ \ |\quad\ \ | \\
\text{OH}\ \text{OH}\ \text{OH}
\end{array}
\tag{8.39}
$$

The reactor has separate feed streams for an aqueous solution of the epoxy and for an aqueous solution of the acid. Startup begins with the vessel initially full of acid.

The chemistry seems fairly simple. The water concentration is high and approximately constant so that the reaction is pseudo-first order with respect to the epoxy. The rate is also proportional to the hydrogen ion concentration h. Thus

$$R = k_0 \exp\left(-\frac{E}{R_g T}\right)eh \tag{8.40}$$

where e is the epoxy concentration. The sulfuric acid dissociates in two equi-

---

[2] D. P. Vermeulen and J. M. H. Fortuin, Paper presented at ISCRE, Philadelphia, 1986.

librium steps:

$$H_2SO_4 = H^+ + HSO_4^- \qquad K_1 = \frac{[H^+][HSO_4^-]}{[H_2SO_4]}$$

$$HSO_4^- = H^+ + SO_4^{2-} \qquad K_2 = \frac{[H^+][SO_4^{2-}]}{[HSO_4^-]} \tag{8.41}$$

The hydrogen ion concentration can be found from

$$\frac{h_{out}^3}{K_1} + h_{out}^2 + (K_2 - s_{out})h_{out} - 2K_2 s_{out} = 0 \tag{8.42}$$

where s is the total sulfate concentration.

There are three ODEs which govern the system. For sulfate, which is not consumed,

$$\bar{t}\frac{ds_{out}}{d\theta} = s_{in} - s_{out} \qquad s_{out} = s_0 \quad \text{at} \quad \theta = 0 \tag{8.43}$$

For the epoxy,

$$\bar{t}\frac{de_{out}}{d\theta} = e_{in} - e_{out} - k_0\bar{t}\exp\left(-\frac{E}{R_g T_{out}}\right)e_{out}h_{out}$$

$$e_{out} = 0 \quad \text{at} \quad \theta = 0 \tag{8.44}$$

For temperature,

$$\left(\rho V C_p + m_R C_R\right)\frac{dT_{out}}{d\theta} = \rho Q(C_p)_{in}T_{in} - Q\rho C_p T_{out} + UA_{ext}(T_{ext} - T_{out})$$

$$+ q + (-\Delta H_{\mathscr{R}})\rho V k_0 \exp\left(-\frac{E}{R_g T_{out}}\right)e_{out}h_{out}$$

$$T_{out} = T_0 \quad \text{at} \quad \theta = 0 \tag{8.45}$$

This heat balance contains two terms not seen before: $m_R C_R$ represents the mass times specific heat of the agitator and vessel walls and q represents the energy input by the agitator. Although the model is nominally for constant physical properties, Vermeulen and Fortuin found a better fit to the experimental data when they used a slightly different specific heat for the inlet stream, $(C_p)_{in}$.

Figure 8.5 shows a comparison between experimental results and the model. Numerical values used in the model are summarized in Problem 8.12. The startup

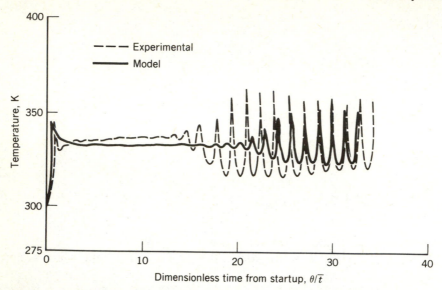

Figure 8.5 Experimental and model results on the acid-catalyzed hydration of 2,3-epoxy-1-propanol to glycerol. [D. P. Vermeulen and J. M. H. Fortuin, Paper presented at ISCRE, Philadelphia, 1986.]

transient shows an initial overshoot followed by an apparent approach to steady state. But oscillation begins after a long delay, $\theta > 10\bar{t}$, and the system goes into a limit cycle. The long delay before the occurrence of the oscillations is remarkable. So is the good agreement between model and experiment. Two facts are apparent: quite complex behavior is possible with a simple model, and one should wait a long time before reaching firm conclusions regarding stability.

8.2 Unsteady Tubular Reactors

8.2.1 Isothermal Piston Flow

Dynamic analysis of piston flow reactors is fairly straightforward and, frankly, rather unexciting. Piston flow causes the dynamic response of the system to be especially simple. The form of response is a limiting case of that found in real systems. We have seen that piston flow is usually a desirable regime from the viewpoint of reaction yields and selectivities. It turns out to be somewhat undesirable from a control viewpoint since there is no natural dampening of disturbances.

Unlike stirred tanks, piston flow reactors are distributed systems with a one-dimensional gradient in composition and physical properties. Their steady state performance is governed by a set of ordinary differential equations. Their dynamic performance is governed by a set of partial differential equations, albeit

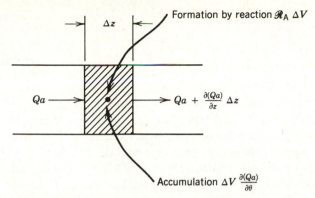

Δz

Formation by reaction $\mathscr{R}_A \, \Delta V$

$Qa \longrightarrow$

$Qa + \dfrac{\partial (Qa)}{\partial z} \Delta z$

Accumulation $\Delta V \dfrac{\partial (Qa)}{\partial \theta}$

Figure 8.6 Component balance for a differential volume element in an unsteady state piston flow reactor.

simple, first order ones. Figure 8.6 illustrates a component balance for a differential volume element:

$$\underbrace{\mathscr{R}_A \, \Delta V + aQ}_{\text{Input}} - \underbrace{\left[aQ + \frac{\partial (aQ)}{\partial z} \Delta z \right]}_{\text{Output}} = \underbrace{\frac{\partial a}{\partial \theta} \Delta V}_{\text{Accumulation}} \qquad (8.46)$$

or

$$\frac{\partial a}{\partial \theta} + \frac{1}{A_c} \frac{\partial (aQ)}{\partial z} = \mathscr{R}_A \qquad (8.47)$$

where $A_c = \Delta V / \Delta z$ is the cross-sectional area of the tube. Note that we have implicitly assumed a tube with inelastic walls, $\partial V / \partial \theta = 0$. An overall mass balance gives a form of the continuity equation

$$\frac{\partial \rho}{\partial \theta} + \frac{1}{A_c} \frac{\partial (\rho Q)}{\partial z} = 0 \qquad (8.48)$$

If ρ is constant, Equation 8.48 shows Q to be constant as well. Then Equation 8.47 becomes

$$\frac{\partial a}{\partial \theta} + \bar{u} \frac{\partial a}{\partial z} = \mathscr{R}_A \qquad (8.49)$$

where $\bar{u} = Q/A_c$ is constant if A_c is constant. Equation 8.49 has the following solution:

$$\frac{z}{\bar{u}} = \int_{a_{in}(\theta - z/\bar{u})}^{a(\theta, z)} \frac{da'}{\mathscr{R}_A} \tag{8.50}$$

Formal verification that this result actually satisfies Equation 8.49 is an exercise in differentiation. However, a physical interpretation will readily confirm its validity. Material in the reactor at time θ and position z entered the reactor at time $\theta' = \theta - (z/\bar{u})$. Its composition at the time of entrance was $a(\theta', 0) = a_{in}(\theta') = a_{in}(\theta - z/\bar{u})$, and the composition since that time has evolved according to batch reaction kinetics as indicated by the right-hand side of Equation 8.50.

The above result may be generalized to variable density, variable cross section reactors. Simply replace the lower limit of Equation 8.50 with $a_{in}(\theta - \alpha)$, where α represents the age of material at position z. The exit composition is obtained by using $a_{in}(\theta - \bar{t})$ as the lower limit:

$$\bar{t} = \int_{a_{in}(\theta - \bar{t})}^{a_{out}(\theta)} \frac{da}{\mathscr{R}_A} \tag{8.51}$$

Molecules entering the reactor at time θ will leave at time $\theta + \bar{t}$ without mixing or interacting with any other molecules. In piston flow, there is no mixing between molecules having different ages.

Concentration disturbances entering a piston flow reactor are not dampened by mixing within the system. An inert substance with input signal $C_{in}(\theta)$ will emerge with output signal $C_{out}(\theta) = C_{in}(\theta - \bar{t})$. This pure time delay is known as **deadtime**. Systems with substantial amounts of deadtime oscillate when feedback control is attempted. This is caused by the controller responding to an output signal that may be completely different than that corresponding to the current input.

Example 8.9   Suppose a fast, acid–base neutralization is occurring in a piston flow reactor:

$$A + B \overset{k}{\to} Product \qquad k \gg 0 \tag{8.52}$$

and suppose it is desired to maintain constant pH in the outlet stream by feeding controlled amounts of the base B. Show the consequences of using the signal from a pH meter installed at the reactor outlet to regulate the input base concentration $b_{in}(\theta)$.

Solution   Imagine the inlet acid concentration has been constant at a_0 for $\theta < 0$. The controller has (somehow) achieved the correct base concentration b_0

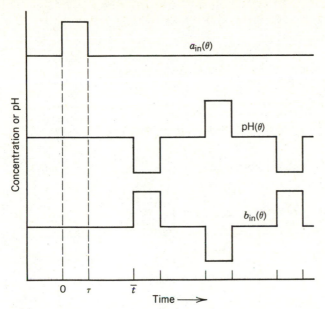

Figure 8.7 Sustained oscillation due to feedback control of a system with deadtime.

to give the desired pH. Suppose now that the inlet acid concentration undergoes a short-term fluctuation:

$$a_{in}(\theta) = a_0 \qquad \theta < 0$$

$$a_{in}(\theta) = a_1 \qquad 0 < \theta < \theta_1$$

$$a_{in}(\theta) = a_0 \qquad \theta > \theta_1 \tag{8.53}$$

How does the pH at the reactor outlet respond to this disturbance? For concreteness, suppose $a_1 > a_0$ and $\theta_1 < \bar{t}$. There will be no outlet response until time $\theta = \bar{t}$, when the pH suddenly drops, reflecting the additional acid supply. The controller will begin feeding more base, but this has no immediate effect on the outlet pH. At time $\theta = \bar{t} + \theta_1$, the pH will return to its initial, undisturbed value and will remain there until time $\theta = 2\bar{t}$, when the pH suddenly rises, reflecting the additional base feed at the inlet during the interval $\bar{t} < \theta < \bar{t} + \theta_1$. Continuing this logic will show that the original, short-term upset in acid concentration will lead to sustained oscillations. See Figure 8.7

Feedforward control represents a theoretically sound approach to controlling systems with appreciable deadtime. Sensors are installed at the inlet to the system to measure fluctuating inputs. The appropriate responses to these inputs

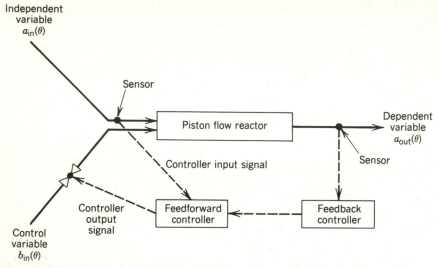

Figure 8.8 Adaptive control scheme for a system with appreciable deadtime.

are then calculated from a model of the system. In the previous example, $a_{in}(\theta)$ would be measured and $b_{in}(\theta)$ would be calculated. The model used for the calculations may be imperfect but can be improved using feedback of actual responses. In **adaptive** control, this feedback of results is done automatically using a special error signal to correct the model:

$$\varepsilon = \text{Predicted response} - \text{Measured response}$$

Thus there is feedback correction of the model and feedforward control of the system. See Figure 8.8.

8.2.2   Unsteady, Nonisothermal Piston Flow

The unsteady counterpart to Equation 4.27 is

$$\frac{\partial T}{\partial \theta} + \bar{u}\frac{\partial T}{\partial z} = \frac{-\Delta H_{\mathscr{R}}\mathscr{R}}{\rho C_p} - \frac{2U(T - T_{\text{ext}})}{R\rho C_p} \tag{8.54}$$

For the special case of an adiabatic, piston flow reactor, $U = 0$; and a formal solution to Equation 8.54 can be written in direct analogy to Equation 8.50:

$$\frac{z}{\bar{u}} = \int_{T_{\text{in}}(\theta - z/\bar{u})}^{T(\theta, z)} \frac{\rho C_p}{-\Delta H_{\mathscr{R}}\mathscr{R}} dT' \tag{8.55}$$

This formal solution is not useful for finding $T(\theta, z)$ since $\mathscr{R}$ will depend on unknown compositions. It does, however, show that the temperature at time θ and position z is determined by inlet conditions at time $\theta - (z/\bar{u})$, or more generally at time $\theta - \alpha$. Temperatures, like compositions, progress in a batchlike trajectory from their entering values to their final ones without regard for conditions occurring in other portions of the reacting mass. Including heat exchange to the environment, $U > 0$, does not change this fact. A solution for $T_{out}(\theta)$ can be found by solving the ordinary differential equation

$$\bar{u}\frac{dT}{dz} = \frac{-\Delta H_{\mathscr{R}}\mathscr{R}}{\rho C_p} - \frac{2U(T - T_{ext})}{R\rho C_p} \tag{8.56}$$

subject to the initial condition that

$$T = T_{in}(\theta - \bar{t}) \quad \text{at} \quad z = 0 \tag{8.57}$$

The solution of Equations 8.56 and 8.57, when evaluated at $z = L$, gives $T_{out}(\theta)$. This solution is done simultaneously with a version of Equation 8.49:

$$\bar{u}\frac{da}{dz} = \mathscr{R}_A \tag{8.58}$$

where

$$a = a_{in}(\theta - \bar{t}) \quad \text{at} \quad z = 0 \tag{8.59}$$

The solution of Equations 8.58 and 8.59 is evaluated at $z = L$ to give $a_{out}(\theta)$.

Piston flow reactors lack any internal mechanisms for memory. There is no axial dispersion of heat or mass. What has happened previously has no effect on what is happening now. This fact prevents **steady state multiplicity** in piston flow reactors (unless there is some form of external feedback). Given a set of operating conditions $(a_{in}, T_{in}, T_{ext})$, only one output (a_{out}, T_{out}) is possible. Piston flow reactors can, however, exhibit **parametric sensitivity**, where a small change in inlet conditions can lead to a large change in outlet conditions.

Example 8.10 Suppose the styrene polymerization of Section 4.2.1 is conducted in a large-diameter tubular reactor in which motionless mixers have been installed. Assume $\bar{t} = 2$ hr and pure monomer feed. Explore the sensitivity of a_{out}/a_{in} and T_{out} to changes in T_{in}.

Solution Large-diameter reactors tend to be adiabatic, and motionless mixers alleviate radial gradients without causing mixing in the axial direction. Thus an assumption of adiabatic piston flow is reasonable, at least from the viewpoint of investigating parametric sensitivity. Equation 8.56 is solved simultaneously with Equation 8.58. Substituting for the constants (see Section 4.2.1) gives

$$\frac{d(a/a_{\text{in}})}{d(z/L)} = -\bar{t}k\left(\frac{a}{a_{\text{in}}}\right) = -2 \times 10^{10}\left(\frac{a}{a_{\text{in}}}\right)\exp\left(-\frac{10{,}000}{T}\right)$$

$$\frac{dT}{d(z/L)} = \frac{-a_{\text{in}}\,\Delta H_{\mathscr{R}}\,\bar{t}k(a/a_{\text{in}})}{\rho C_p} = 8 \times 10^{12}\frac{a}{a_{\text{in}}}\exp\left(-\frac{10{,}000}{T}\right) \qquad (8.60)$$

Some solutions are:

T_{in}	T_{out}	ΔT	$a_{\text{out}}/a_{\text{in}}$
350	354	4	0.990
360	370	10	0.975
370	422	52	0.870
370.5	441.3	70.8	0.823
371	771	400	0

Parametric sensitivity is indicated by the dramatic changes in T_{out} and a_{out} that occur for T_{in} in the range 370 to 371 K.

The reader will observe that the results in the above example were determined by steady state calculations. Indeed, parametric sensitivity is a steady state phenomenon. If the operating conditions are constant, the time derivatives in Equations 8.49 and 8.54 (these being $\partial a/\partial\theta$ and $\partial T/\partial\theta$) are zero. Thus there are an infinite number of possible steady states. However, most of the steady states show either very low conversions or very high ones. Operation with intermediate conversions and intermediate temperatures is possible only for a very restricted range of operating parameters. For the above example with adiabatic operation, this range is almost impossibly narrow.

True, steady state multiplicity can arise in piston flow reactors when there is an external mechanism for feedback. For example, the effluent from the reactor may be used to heat the incoming feed as in Figure 8.9. See Figure 5.3 for another possibility. In either of these situations, a given inlet condition can give rise to a runaway or an extinction depending on previous operating history.

8.2.3 Unsteady Convective Diffusion

The unsteady version of the convective diffusion equation is obtained just by adding a time derivative to the steady version. Equation 6.13 for the convective diffusion of mass becomes

$$\frac{\partial a}{\partial \theta} + v_z(r)\frac{\partial a}{\partial z} = \mathscr{D}_A\left(\frac{1}{r}\frac{\partial a}{\partial r} + \frac{\partial^2 a}{\partial r^2} + \frac{\partial^2 a}{\partial z^2}\right) + \mathscr{R}_A \qquad (8.61)$$

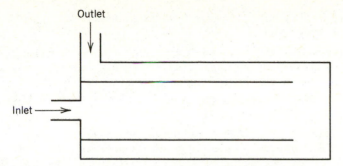

Figure 8.9 Piston flow reactor with integral heat exchange.

For the diffusion of heat,

$$\frac{\partial T}{\partial \theta} + v_z(r)\frac{\partial T}{\partial z} = \alpha_T\left(\frac{1}{r}\frac{\partial T}{\partial r} + \frac{\partial^2 T}{\partial r^2} + \frac{\partial^2 T}{\partial z^2}\right) - \frac{\Delta H_{\mathscr{R}}\mathscr{R}}{\rho C_p} \tag{8.62}$$

Simultaneous solution of these equations is appropriate to the design of control systems for reactors in which diffusion (or dispersion) is important. Diffusion provides an internal mechanism for the feedback of previous operating conditions, possibly giving rise to steady state multiplicity. Material entering the reactor at time θ will have its reaction environment influenced by material that entered previously.

In principle, Equations 8.61 and 8.62 can be solved numerically using the simple methods of Chapters 6 and 7. The two-dimensional problem in r and z is solved for a fixed value of θ. A step forward in θ is taken, the two-dimensional problem is resolved at the new θ, and so on. In practice, computer time can become a problem; and even with the current generation of computers, more sophisticated numerical techniques may be warrented. See Davis (1984)[3] or Lapidus and Pinder (1982)[4] for a general description of numerical techniques.

Suggestions for Further Reading

This chapter has presented time-domain solutions of unsteady material and energy balances. The more usual undergraduate treatment of dynamic systems is given in a course on control and relies heavily on Laplace transform techniques. One suitable reference to this approach is:

G. Stephanopoulos, *Chemical Process Control: An Introduction to Theory and Practice*, Prentice-Hall, Englewood Cliffs, NJ, 1984.

Another standard text with a good introduction to Laplace transforms is:

[3] M. E. Davis, *Numerical Methods and Modeling for Chemical Engineers*, Wiley, New York, 1984.
[4] L. Lapidus and G. F. Pinder, *Numerical Solution of Partial Differential Equations*, Wiley, New York, 1982.

D. R. Coughanour and L. B. Koppel, *Process Systems Analysis and Control*, McGraw-Hill, New York, 1965.

A rather different method of looking at reactor dynamics uses residence time theory in the unsteady state. Chapter 9 introduces the steady state version of residence time theory while an introduction to the unsteady version is provided in:

E. B. Nauman, and B. A. Buffham, *Mixing in Continuous Flow Systems*, Wiley, New York, 1983.

Unsteady reaction data are often an excellent means for estimating physical parameters which would be difficult or impossible to elucidate from steady state measurements. However, the associated problems in nonlinear optimization can be formidable. A recent review and comparison of methods is given by:

L. T. Biegler, J. J. Damiano, and G. E. Blau, "Nonlinear Parameter Estimation: A Case Study Comparison," *AIChE J.*, **32**, 29–45 (1986).

Problems

8.1   Determine the fraction unreacted at time $\theta = \bar{t}$ for the startup trajectory represented by Equation 8.17. Use $k = 1 \text{ hr}^{-1}$ and $\bar{t} = 3$ hr.

8.2   Determine the fractional filling rate Q_{fill}/Q that will fill an isothermal, constant-density, stirred tank reactor while simultaneously achieving the steady state conversion corresponding to flow rate Q. Assume a first order reaction with $k = 1 \text{ hr}^{-1}$ and $\bar{t} = V/Q = 3$ hr. Note that Equation 8.17 applies to this situation for θ in the range $0 < \theta < \theta_{\max} = V/Q_{\text{fill}}$.

8.3   Repeat Problem 8.2 for a second order reaction with $ka_{\text{in}} = 1 \text{ hr}^{-1}$ and $\bar{t} = 3$ hr.

8.4   Suppose the consecutive elementary reactions

$$2\text{A} \xrightarrow{k_{\text{I}}} \text{B} \xrightarrow{k_{\text{II}}} \text{C}$$

occur in an isothermal perfect mixer. Use the transient technique to determine the steady state outlet composition and to explore system stability. Suppose $k_{\text{I}} a_{\text{in}} \bar{t} = 2$, $k_{\text{II}} \bar{t} = 1$, $b_{\text{in}} = c_{\text{in}} = 0$.

8.5   Suppose limitations in the product recovery system of a stirred tank reactor prevent effluent flow rates above $1.05Q$, where Q is the steady state flow rate. How long will it take to shut down the reactor using constant RTD control?

8.6   The autorefrigerated styrene polymerizer of Section 4.2.1 is to be started up. Cold monomer is charged to the vessel and is heated using an external heat exchanger with $UA_{\text{ext}}/V\rho C_p = 0.5$ and $T_{\text{ext}} = 413$ K. The reactor pressure is set for reflux at 404 K. Steady throughput with $Q_{\text{in}} = Q_{\text{out}} = Q$ begins when this steady operating temperature is achieved. Plot T and a/a_{in} versus time for the startup period (batch reaction period) and for the first 2 hr of continuous operation.

8.7   Suppose the following reactions are occurring in an isothermal perfect mixer:

$$A + 2B \rightarrow 3B \qquad R_I = k_I ab^2$$

$$B \rightarrow C \qquad R_{II} = k_{II} b$$

Since the autocatalytic reaction is third order, a steady state material balance gives a cubic in b_{out}. This means that either one or three real, steady state solutions for b_{out} are possible. Suppose $b_{in}/a_{in} = 1/15$ and explore the stability of the single or middle steady state for each of the following cases:

(a) $k_I \bar{t} a_{in}^2 = 190$, $k_{II} \bar{t} = 4.750$ (a small disturbance from the steady state gives damped oscillations)

(b) $k_I \bar{t} a_{in}^2 = 225$, $k_{II} \bar{t} = 5.625$ (a small disturbance from the steady state gives sustained oscillations)

(c) $k_I \bar{t} a_{in}^2 = 315$, $k_{II} \bar{t} = 7.875$ (a small disturbance from the steady state gives undamped oscillations and divergence to a new steady state)

See P. Gray and S. K. Scott, *Chem. Eng. Sci.*, **39**, 1087–1097 (1984), for a detailed analysis of this reaction system.

8.8   Suppose the following reactions are occurring in an isothermal perfect mixer:

$$A + B \rightarrow 2B \qquad R_I = k_I ab$$

$$B \rightarrow C \qquad R_{II} = k_{II} b$$

Suppose there is no B in the feed but that some B is charged to the reactor at startup. Can this form of startup lead to stable operation with $b_{in} = 0$ but $b_{out} > 0$?

8.9   Entering engineering freshmen tend to choose their curriculum based on job demand for the current graduating class. It is easy to change curricula in the freshman year but it becomes difficult in subsequent years. Thus we might model the engineering education process as a stirred tank with $t_s = 1$ year followed by a piston flow reactor with $t_p = 3$ years. Does this model predict a good balance of supply and demand? What strategy would you suggest to a freshman whose sole concern was being in high demand upon graduation?

8.10   Standard thermodynamic texts give a more general version of Equation 8.6. See Equation 7.4 of J. M. Smith and H. C. Van Ness, *Introduction to Chemical Engineering Thermodynamics*, 4th ed., McGraw-Hill, New York, 1986. This more general version can be written as

$$\frac{d}{dt}[\rho(H - PV)V] = \left(H_{in} + \frac{\bar{u}_{in}^2}{2} + Z_{in} g_{in} \right) Q_{in} \rho_{in}$$

$$- \left(H_{out} + \frac{\bar{u}_{out}^2}{2} + Z_{out} g_{out} \right) Q_{out} \rho_{out}$$

$$- V \Delta H_{\mathscr{R}} \mathscr{R} + UA_{ext}(T_{ext} - T_{out})$$

where u is the velocity of the inlet or outlet streams and Zg represents the pressure head due to gravity. Show how this more general version relates to Equation 8.6. The neglected terms are

$$\frac{d}{dt}(-\rho P V^2) \qquad \Delta\left(\frac{\bar{u}^2 Q \rho}{2}\right) \qquad \Delta[ZgQ\rho]$$

Are these apt to be important compared to the included terms? Show by a specific numerical example.

8.11 Blood vessels have elastic walls which expand or contract due to changes in pressure or the passage of corpuscles. How should Equations 8.47 and 8.48 be modified to reflect this behavior?

8.12 Refering to Example 8.8, Vermeulen and Fortuin estimated all the parameters in their model from physical data. They then compared model predictions to experimental results and from this they made improved estimates using nonlinear regression. Their results were as follows:

Parameter	Estimate from Physical Data	Estimate from Regression Analysis	Units
ρQ	1.9×10^{-3}	1.881615×10^{-3}	kg/s
ρV	0.30	0.2998885	kg
$m_R C_R$	392	405.5976	J/K
e_{in}	8.55	8.532488	mol/kg
$(C_p)_{in}$	2.65×10^3	2.785279×10^3	J kg^{-1} K^{-1}
C_p	$2.65 \times 10_3$	2.517017×10^3	J kg^{-1} K^{-1}
s_{in}	0.15	0.1530875	mol/kg
k_0	8.5×10^{10}	8.534612×10^{10}	kg mol^{-1} s^{-1}
UA_{ext}	30	32.93344	J s^{-1} K^{-1}
$-\Delta H_{\mathscr{R}}$	8.82×10^4	8.792731×10^4	J/mol
q	30	32.62476	J/s
T_{in}	273.91	273.9100	K
T_{ext}	298.34	298.3410	K
E/R_g	8827	8815.440	K
K_1	10^3	10^3	mol/kg
K_2	0.012023	0.012023	mol/kg
T_0	300.605	300.605	K
e_0	0	0	mol/kg
s_0	0.894	0.894	mol/kg

(a) Show that Equation 8.42 is consistent with the dissociation equilibria of Equations 8.41 and that, to a good approximation,

$$h_{out} = 0.5\left(s_{out} - k_2 + \sqrt{s_{out}^2 + 6 s_{out} k_2 + k_2^2}\right)$$

(b) Using the parameter estimates obtained by regression analysis, confirm the qualitative behavior shown in Figure 8.5. *Hint*: Use a sophisticated integration routine. Do not attempt to match Vermeulen and Fortuin exactly.

(c) The parameter fitting procedure used experimental data from a single run. Determine the sensitivity of the model by replacing the regression estimates with the physical estimates. What does this suggest about the reproducibility of the experiment?

(d) Devise a means for achieving steady operation at high conversion to glycerol. Undesirable side reactions may become significant at 423 K. At atmospheric pressure and complete conversion, the mixture boils at 378 K.

9

Mixing in Continuous-Flow Systems

Reactor *design* usually begins in the laboratory with a kinetic study. Data are taken in small-scale, specially designed equipment which hopefully (but not inevitably) approximates an ideal, isothermal reactor: batch, perfect mixing, or piston flow. The laboratory data are fit to a kinetic model using the methods of Section 4.3. The kinetic model is then combined with a transport model to give the overall design.

Suppose now that a pilot-plant or full-scale reactor has been built and operated. How can its performance be used to confirm the kinetic and transport models and to improve future designs? Reactor *analysis* begins with an operating reactor and seeks to understand several interrelated aspects of actual performance: kinetics, flow patterns, mixing, mass transfer, and heat transfer. The present chapter is concerned with the analysis of flow and mixing processes and their interactions with kinetics. It uses *residence time theory* as the major tool for the analysis.

In a batch reactor, all molecules enter and leave together. If the system is isothermal, reaction yields depend only on the elapsed time and on the initial composition. The situation in flow systems is more complicated but not impossibly so. The counterpart of the batch reaction time is the age of a molecule. Aging begins when a molecule enters the reactor and ceases when it leaves. The total time spent within the boundaries of a system is known as the exit age, or *residence time* t. Except in batch and piston flow reactors, molecules leaving the system will have a variety of residence times. The distribution of residence times provides considerable information about homogeneous, isothermal reactions. For single, first order reactions, knowledge of the residence time distribution allows the yield to be calculated exactly, even in flow systems of arbitrary complexity. For other reaction orders, it is usually possible to calculate tight limits, within which the yield must lie. Even if the system is nonisothermal and heterogeneous, knowledge of the *residence time distribution* provides substantial insight regarding the flow processes occurring within it.

9.1   Residence Time Theory

9.1.1   Inert Tracer Experiments

Suppose that a nonreactive tracer has been fed to a perfect mixer for an extended period of time, giving $C_{in} = C_{out} = C_0$ for $\theta < 0$. At time $\theta = 0$, the tracer supply is suddenly stopped so that $C_{in} = 0$ for $\theta > 0$. The transient response of the system is given by Equation 8.4:

$$V \frac{dC}{d\theta} = Q_{out}C \tag{9.1}$$

where constant-volume operation with $\mathcal{R}_C = 0$ has been assumed. The solution is

$$\frac{C(t)}{C_0} = \exp\left(-\frac{Q_{out}\theta}{V}\right) = \exp\left(-\frac{\theta}{\bar{t}}\right) \tag{9.2}$$

Equation 9.2 represents the washout of tracer molecules that were originally in the system. Consider some time $t > 0$ when the fraction of molecules remaining in the system was $W(t) = C(t)/C_0$. These molecules must necessarily have entered the reactor before time $\theta = 0$ since no tracer was fed after this. Thus these molecules have residence times of t or longer:

$$\boxed{\begin{array}{l} W(t) = \text{Fraction of molecules leaving the system that} \\ \qquad\qquad \text{had a residence time of } t \text{ or longer} \end{array}} \tag{9.3}$$

It is apparent that $W(0) = 1$ since all molecules must have a residence time of zero or longer and that $W(\infty) = 0$ since all molecules will eventually leave the system. Also, the function $W(t)$ must be nonincreasing.

Equation 9.3 defines a basic function of residence time theory known as the *washout function*. Properties of $W(t)$ will be discussed shortly. For the moment note that

$$\boxed{W(t) = \exp\left(-\frac{t}{\bar{t}}\right)} \tag{9.4}$$

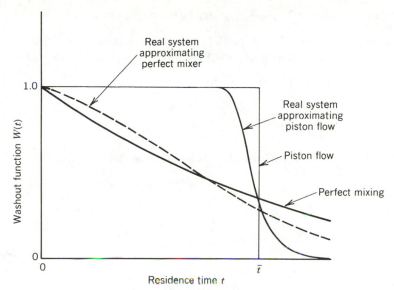

Figure 9.1   Residence time washout functions.

for a perfect mixer. This residence time distribution is called an *exponential distribution*. For a piston flow reactor

$$
\begin{aligned}
W(t) &= 1 \qquad 0 < t < \bar{t} \\
W(t) &= 0 \qquad t > \bar{t}
\end{aligned}
\qquad (9.5)
$$

Figure 9.1 shows these functions together with some washout functions that might be expected in real systems.

Washout experiments can be used to measure the residence time distribution in continuous-flow systems. A good step change must be made at the reactor inlet. The concentration of tracer molecules leaving the system must be accurately measured at the outlet. The flow properties of the tracer molecules must be similar to those of the reactant molecules.

It is usually possible to meet these requirements in practice. The major theoretical requirement is that the inlet and outlet streams have unidirectional flows so that molecules that once enter the system stay in until they exit, never to return. Systems with unidirectional inlet and outlet streams are said to be *closed*, and most systems of chemical engineering importance are closed to a reasonable approximation. Transient experiments with inert tracers can be used to determine residence time distributions in closed systems. In real systems, these experiments will indeed be experiments. For mathematical analysis, the experiments are mathematical ones applied to a dynamic model of the system. Equations 9.4 and 9.5 were in fact derived from mathematical models of a perfect mixer and a piston flow reactor.

9.1.2   Residence Time Functions and Properties

This book regards the washout function $W(t)$ as the fundamental function of residence time theory. It is defined over the range $0 \leq t < \infty$ and satisfies three conditions:

(i) $W(0) = 1$
(ii) $W(\infty) = 0$ (9.6)
(iii) $dW/dt \leq 0$ (nonincreasing)

Closely related to $W(t)$ is the *cumulative distribution function* $F(t)$, defined as

$$F(t) = 1 - W(t) \tag{9.7}$$

$F(t)$ can be directly measured using a step change of the turn-on variety: $C_{in} = 0$, $\theta < 0$; $C_{in} = 1$, $\theta > 0$. A word definition is

$$F(t) = \text{Fraction of molecules leaving the system which} \atop \qquad\qquad \text{had a residence time of } t \text{ or less} \tag{9.8}$$

and the following conditions apply:

(i) $F(0) = 0$
(ii) $F(\infty) = 1$ (9.9)
(iii) $dF/dt \geq 0$ (nondecreasing)

A final function of importance is the residence *density function* $f(t)$, also called the differential distribution function or frequency function, and defined as

$$f(t) = \frac{dF}{dt} = -\frac{dW}{dt} \tag{9.10}$$

The density function has range $0 \leq t < \infty$ and satisfies two conditions:

(i) $f(t) \geq 0$ (nonnegativity)   (9.11a)

(ii) $\int_0^\infty f(t)\, dt = 1$ (unit integral)   (9.11b)

Experimental determination of the density function requires a very rapid injection of tracer molecules at the inlet to the system. Ideally, a finite number of molecules will be injected in an infinitesimal period of time. Mathematically, $f(t)$

can be determined by differentiation according to Equation 9.10. It can also be determined as the response of a dynamical model to a unit impulse or Dirac *delta function*. The delta function is a convenient mathematical artifact which is usually defined as

$$\delta(t) = 0 \qquad t \neq 0 \tag{9.12a}$$

$$\int_{-\infty}^{\infty} \delta(t)\, dt = 1 \tag{9.12b}$$

The delta function is everywhere zero except at the origin, where it has an infinite discontinuity, a discontinuity so large that the integral under it is unity. The limits of integration need only include the origin itself; Equation 9.12b can equally well be written as

$$\int_{0-}^{0+} \delta(t)\, dt = 1 \tag{9.13}$$

The delta function has another integral of substantial use:

$$\int_{-\infty}^{\infty} \phi(t)\, \delta(t - t_0)\, dt = \phi(t_0) \tag{9.14}$$

where $\phi(t)$ is any "ordinary" function. We suggest by this that $\delta(t)$ is not an ordinary function. Instead, it is sometimes best to regard $\delta(t)$ as a special kind of limit, an aspect of $\delta(t)$ that will be illustrated in Example 9.3.

Example 9.1   Apply a positive step change in tracer concentration to a perfect mixer to determine the cumulative distribution function $F(t)$.

   Solution   The positive step change can be expressed as $C_{in} = 0$, $\theta < 0$, $C_{in} = 1$, $\theta > 0$. The system model for $\theta > 0$ is

$$V \frac{dC}{d\theta} = Q(1 - C) \tag{9.15}$$

Subject to the initial condition that $C = 1$ at $\theta = 0$, the transient response is

$$C(\theta) = 1 - \exp\left(-\frac{\theta Q}{V}\right) \tag{9.16}$$

and thus

$$F(t) = 1 - \exp\left(-\frac{t}{\bar{t}}\right) \tag{9.17}$$

Example 9.2   Apply a delta function input to a piston flow reactor to determine $f(t)$.

Solution For a nonreactive tracer, the dynamical model for a piston flow reactor is given by Equation 8.49 with $\mathscr{R}_C = 0$:

$$\frac{\partial C}{\partial \theta} = \bar{u}\,\frac{\partial C}{\partial z} = 0 \tag{9.18}$$

It can be verified (see Section 8.2.1) that

$$C(\theta, z) = C_{in}\left(\theta - \frac{z}{\bar{u}}\right) \tag{9.19}$$

thus

$$C_{out}(\theta) = C(\theta, L) = C_{in}(\theta - \bar{t}) \tag{9.20}$$

Since $C_{in}(\theta) = \delta(\theta)$, the outlet response is $\delta(\theta - \bar{t})$. Thus

$$f(t) = \delta(t - \bar{t}) \tag{9.21}$$

for a piston flow reactor. In light of this result, the residence time distribution for piston flow is called a **delta distribution**.

Example 9.3 Apply a delta function input to a perfectly mixed reactor to determine $f(t)$.

Solution This solution illustrates a possible definition of the delta function as the limit of an ordinary function. Suppose the system is disturbed by a rectangular pulse of duration τ and height A/τ so that A units of tracer are injected. The input signal is $C_{in} = 0$, $\theta < 0$; $C_{in} = A/\tau$, $0 < \theta < \tau$; $C_{in} = 0$, $\theta > \tau$. The outlet response is found from the dynamical model of a perfect mixer. The result is

$$C_{out}(\theta) = 0 \qquad\qquad\qquad\qquad\qquad \theta < 0$$

$$C_{out}(\theta) = \left(\frac{A}{\tau}\right)\left[1 - \exp\left(-\frac{\theta}{\bar{t}}\right)\right] \qquad\qquad 0 < \theta < \tau \tag{9.22}$$

$$C_{out}(\theta) = \left(\frac{A}{\tau}\right)\left[1 - \exp\left(-\frac{\tau}{\bar{t}}\right)\right]\exp\left(-\frac{\theta}{\bar{t}}\right) \qquad \theta > \tau$$

Now consider the limit as τ approaches zero. L'Hospital's rule may be used to show that

$$\lim_{\tau \to 0} \frac{A}{\tau}\left[1 - \exp\left(-\frac{\tau}{\bar{t}}\right)\right] = \frac{A}{\bar{t}} \tag{9.23}$$

The transient response to this limiting pulse of infinitesimal duration is

$$C_{out}(\theta) = 0 \qquad\qquad\qquad \theta < 0$$

$$C_{out} = \frac{A}{\bar{t}} \qquad\qquad\qquad \theta = 0 \qquad\qquad\qquad (9.24)$$

$$C_{out} = \left(\frac{A}{\bar{t}}\right) \exp\left(-\frac{\theta}{\bar{t}}\right) \qquad \theta > 0$$

A unit impulse of tracer gives the density function. Therefore, set $A = 1$ to obtain

$$f(t) = \left(\frac{1}{\bar{t}}\right) \exp\left(-\frac{t}{\bar{t}}\right) \qquad t \geq 0 \qquad\qquad\qquad (9.25)$$

as the residence time density function for a perfect mixer.

Equation 9.25 obviously can be obtained from Equation 9.17 by differentiation. It could also be obtained by the transient response technique using pulse shapes other than rectangular. Triangular or Gaussian pulses could be used, for example. In these cases, however, the limit would have to be taken as the pulse duration becomes infinitesimally short while the amount of injected tracer remains finite. Any of these limits will correspond to a delta function input.

A good experimental approximation to a delta function input requires that the pulse duration τ be short compared to the mean residence time; $\tau < 0.01\bar{t}$ is usually sufficient. It is not necessary that the amount of injected tracer be known. Instead, the density function is found by normalizing the outlet response

$$f(t) = \frac{C_{out}(t)}{\int_{-\infty}^{\infty} C_{out}(\theta)\, d\theta} \qquad\qquad\qquad (9.26)$$

where the lower limit of the integral is zero if it is understood that injection begins at time $\theta = 0$. Normalization ensures that Equation 9.11b, the unit integral property, is satisfied. Setting $A = 1$ in Example 9.3 represents normalization. Other forms of normalization are necessary when the input function is a step change. For a negative step change, the appropriate normalization is

$$W(t) = \frac{C_{out}(t) - C_{\infty}}{C_0 - C_{\infty}} \qquad\qquad\qquad (9.27)$$

where C_0 is the initial tracer concentration and C_{∞} is the final or background concentration. For a positive step change,

$$F(t) = \frac{C_{out}(t) - C_0}{C_\infty - C_0} \tag{9.28}$$

These normalizations for step changes show that any background tracer concentration should be subtracted from the measured response. Although not explicitly shown in Equation 9.26, subtraction of the background concentration is also needed for the impulse response technique.

9.1.3 Means and Moments

Residence time distributions can be described by any of the functions $W(t)$, $F(t)$, or $f(t)$. They can also be described using an infinite set of parameters known as **moments**:

$$\mu_n = \int_0^\infty t^n f(t)\, dt = n \int_0^\infty t^{n-1} W(t)\, dt \tag{9.29}$$

where $n = 1, 2, \ldots$. These moments are also called **moments about the origin**. The first moment is the mean of the distribution or the mean residence time:

$$\boxed{\bar{t} = \int_0^\infty t f(t)\, dt = \int_0^\infty W(t)\, dt} \tag{9.30}$$

Thus $\bar{t}$ can be found from inert tracer experiments. It can also be found from measurements of the system volume and throughput since

$$\bar{t} = \frac{V}{Q} \tag{9.31}$$

Agreement of the $\bar{t}$'s calculated by these two methods provides a good check on experimental accuracy. Occasionally, Equation 9.31 is used to calculate an unknown volume from inert tracer data.

Equation 9.31 is limited to single-phase, constant-density systems. It is part of a more general result that can be stated as

$$\bar{t} = \frac{\text{Inventory of molecules of a given type}}{\text{Rate at which these molecules leave the system}} \tag{9.32}$$

This result is valid for any conserved (nonreactive) species. The system must be at steady state. It may be multiphase. By "inventory" we mean the total number of molecules within the system that are free to leave, permanently bound molecules being excluded from the total.

Roughly speaking, the first moment $\bar{t}$ measures the size of a residence time distribution, while higher moments measure its shape. The ability to characterize shape is enhanced by using *moments about the mean*:

$$\mu'_n = \int_0^\infty (t - \bar{t})^n f(t) \, dt \tag{9.33}$$

Of these, the second is the most interesting and has a special name, the *variance*:

$$\sigma_t^2 = \mu'_2 = \int_0^\infty (t - \bar{t})^2 f(t) \, dt \tag{9.34}$$

Expanding the parenthetical term and integrating (see Example 9.4) gives

$$\sigma_t^2 = \mu_2 - \bar{t}^2 \tag{9.35}$$

It is this equation that is normally used to calculate the variance from experimental data, μ_2 being calculated from Equation 9.29 with $n = 2$. Note that either $f(t)$ or $W(t)$ can be used to calculate μ_2. Use the one that was obtained directly from the experimental measurements. If moments of the highest possible accuracy are desired, use a negative step change to get $W(t)$ directly. However, accurate moments beyond the second are difficult to obtain under the best of circumstances. The weightings of t^n or t^{n-1} in Equation 9.29 place too much emphasis on the tail of the residence time distribution to allow accurate numerical results.

After σ_t^2 has been calculated, it is usually normalized by $(\bar{t})^2$ to give the *dimensionless variance*:

$$\sigma^2 = \frac{\sigma_t^2}{(\bar{t})^2} = \frac{\mu_2}{(\bar{t})^2} - 1 \tag{9.36}$$

The dimensionless variance has been used extensively, perhaps excessively, to characterize mixing. For piston flow, $\sigma^2 = 0$, and for perfect mixing, $\sigma^2 = 1$. Most turbulent flow systems have dimensionless variances that lie between zero and 1, and σ^2 can then be used to fit a variety of residence time models as will be discussed in Section 9.1.4. The dimensionless variance is generally unsatisfactory for characterizing laminar flows where $\sigma^2 > 1$ is normal in liquid phase systems.

Example 9.4   Determine the first three moments about the origin and about the mean for piston flow and perfect mixing.

Solution   Begin with piston flow for which $f(t) = \delta(t - \bar{t})$. The moments about the origin are

$$\mu_n = \int_0^\infty t^n \delta(t - \bar{t}) = (\bar{t})^n \tag{9.37}$$

where Equation 9.14 was used to evaluate the integral. The moments about the mean are

$$\mu'_n = \int_0^\infty (t - \bar{t})^n \, \delta(t - \bar{t}) = (\bar{t} - \bar{t})^n = 0 \tag{9.38}$$

For perfect mixing,

$$\mu_n = \int_0^\infty t^n \frac{1}{\bar{t}} \, e^{-t/\bar{t}} \, dt$$

$$= (t)^n \int_0^\infty x^n e^{-x} \, dx = (\bar{t})^n \, \Gamma(n+1) = n!(\bar{t})^n \tag{9.39}$$

where $\Gamma(n+1) = n!$ is the gamma function. To find the moments about the mean, the parenthetical term in Equation 9.33 is expanded. For $n = 2$,

$$\mu'_2 = \int_0^\infty (t - \bar{t})^2 \, f(t) \, dt = \int_0^\infty \left[t^2 - 2\bar{t}t + (\bar{t})^2 \right] f(t) \, dt$$

$$= \int_0^\infty t^2 f(t) \, dt - 2 \int_0^\infty \bar{t} f(t) \, dt + (\bar{t})^2 \int_0^\infty f(t) \, dt$$

$$= \mu_2 - 2(\bar{t})^2 + (\bar{t})^2 = \mu_2 - (\bar{t})^2 \tag{9.40}$$

and for a perfect mixer $\mu_n = (\bar{t})^n n!$. Thus

$$\mu'_2 = 2(\bar{t})^2 - (\bar{t})^2 = (\bar{t})^2 \tag{9.41}$$

For $n = 3$,

$$\mu'_3 = \int_0^\infty (t - \bar{t})^3 \, f(t) \, dt = \int_0^\infty \left[t^3 - 3t^2\bar{t} + 3t(\bar{t})^2 - (\bar{t})^3 \right] f(t) \, dt$$

$$= \mu_3 - 3\mu_2\bar{t} + 3(\bar{t})^3 - (\bar{t})^3 = \mu_3 - 3\mu_2\bar{t} + 2(\bar{t})^3 \tag{9.42}$$

and for a perfect mixer,

$$\mu'_3 = (2 \cdot 3)(\bar{t})^3 - (2)(\bar{t})^2 t + 2(\bar{t})^3 = 2(\bar{t})^3 \tag{9.43}$$

Note that Equations 9.40 and 9.42 apply to any residence time distribution, while Equations 9.41 and 9.43 are specialized to the exponential distribution of a stirred tank reactor.

9.1.4 Residence Time Models

Experimental residence time data are often fit to **empirical models**. These models typically consist of stirred tanks and piston flow reactors in series and parallel combinations. Configurations can always be devised to fit the data. In fact, piston flow elements alone can match an arbitrary residence time distribution. Figure 9.2 shows how a parallel arrangement of piston flow reactors can be fit to an experimental washout function, and it is apparent that any desired quality of fit can be achieved just by increasing the number of piston flow elements. Generally, one desires to use a model with as few components as possible. Hopefully, the model will even provide some insight into the physical processes occurring within the system; but one should not expect too much of empirical models.

One of the simplest empirical models consists of a stirred tank in series with a piston flow reactor as indicated in Figure 9.3a. Other than the mean residence time itself, the model contains only one adjustable parameter. This parameter is called the **fractional tubularity** τ_p and is the fraction of the system volume that is occupied by the piston flow element. Figure 9.3b shows the washout function for the fractional tubularity model. The equation for the washout function is

$$W(t) = 1 \qquad\qquad t < \tau_p \bar{t}$$

$$W(t) = \exp\left[\frac{-(t/\bar{t} - \tau_p)}{1 - \tau_p}\right] \qquad t > \tau_p \bar{t} \qquad\qquad \textbf{(9.44)}$$

This equation can be fit to experimental data in several ways. The model

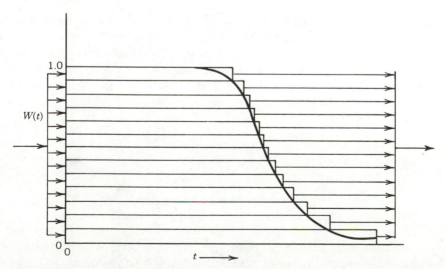

Figure 9.2   Modeling of a residence time distribution using piston flow reactors in parallel.

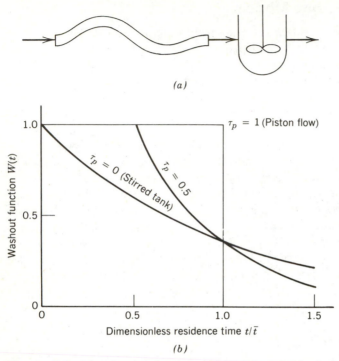

Figure 9.3 The fractional tubularity model: (a) physical representation; (b) normalized washout function.

indicates a sharp *first appearance time* $t_{min} = \tau_p \bar{t}$, which corresponds to the fastest material moving through the system. The mean residence time is found using Equation 9.30 or Equation 9.32, and $\tau_p = t_{min}/\bar{t}$ is found by observing the time when the experimental washout function first drops below 1.0. Probably better is to plot $\log W$ versus t. This should give a straight line (for $t > t_{min}$) with slope $= (\bar{t} - \tau_p \bar{t})^{-1}$. Another approach is to calculate the dimensionless variance and then to obtain τ_p from

$$\tau_p = 1 - \sqrt{\sigma^2} = 1 - \sigma \tag{9.45}$$

where $0 < \sigma^2 < 1$ is necessary if the model is to fit the data.

The fractional tubularity model has been used to fit residence time data in fluidized-bed reactors. It is also appropriate for modeling real stirred tank reactors which have small amounts of dead time, as would perhaps be caused by the inlet and outlet piping. It is not well suited to modeling systems that are nearly in piston flow since such systems rarely have sharp first appearance times.

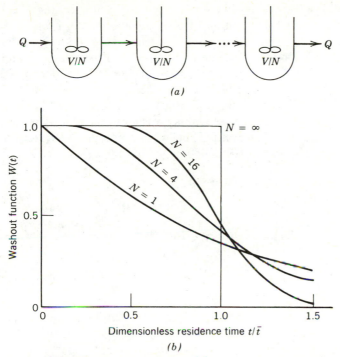

Figure 9.4 The tanks-in-series model: (a) physical representation; (b) normalized washout function.

A simple model having fuzzy first appearance times is the tanks-in-series model illustrated in Figure 9.4. The washout function is

$$W(t) = e^{-Nt/\bar{t}} \sum_{i=0}^{N-1} \frac{N^i t^i}{i!(\bar{t})^i} \tag{9.46}$$

and the density function is

$$f(t) = \frac{N^N t^{N-1} e^{-Nt/\bar{t}}}{(\bar{t})^{N-1}(N-1)!} \tag{9.47}$$

where N (an integer) is the number of tanks in series. Each tank, individually, has volume V/N and mean residence time $\bar{t}/N$. The model naturally gives the exponential distribution of a single stirred tank for $N = 1$. It approaches the delta distribution of piston flow as $N \to \infty$. The model is well suited to modeling small deviations from piston flow. Physical systems that consist of N tanks (or compartments, or cells) in series are fairly common, and the model has obvious utility for these situations. The model is poorly suited for characterizing small deviations from the exponential distribution of a single stirred tank because N takes only integer values. However, extensions to the basic tanks-in-series model exist which allow N to take nonintegral values and even $N < 1$ (see Nauman and Buffham, 1983). The case of $N < 1$ corresponds to **bypassing** where a portion of the fluid has a residence time much smaller than the mean.

Figure 9.5a shows a physical situation which can be expected to give bypassing, and Figure 9.5b shows the corresponding washout function. Bypassing is said to exist when the washout function falls below that of a perfect mixer for low values of time. Specifically, we might define bypassing to exist if

$$W_{\text{system}} < \exp\left(-\frac{t}{\bar{t}}\right) \tag{9.48}$$

for some $t < \bar{t}$. On the other hand, **stagnancy** exists when

$$W_{\text{system}} > \exp\left(-\frac{t}{\bar{t}}\right) \tag{9.49}$$

for some $t > \bar{t}$. According to these rather imprecise definitions, the system illustrated in Figure 9.5 simultaneously shows bypassing and stagnancy. The reason for this is that the experimental washout function is compared to the exponential distribution of a perfect mixing having the same $\bar{t}$. The integral under the experimental $W(t)$ curve, Equation 9.30, must be the same as that under the exponential distribution. Thus material that travels through the system "too slowly" will compensate for material that travels "too quickly." Which of these flow streams is the problem depends on the design intent of the system. Did the engineer place the inlet and outlet too close together, thus causing bypassing, or did he build too large a tank, thus causing stagnancy?

Systems with bypassing and stagnancy can often be modeled as subsystems in parallel. A washout experiment applied to a parallel system gives a flow-averaged result:

$$W(t) = qW_1(t) + (1 - q)W_2 \tag{9.50}$$

where q is the fraction of the total flow sent to subsystem 1, which has washout function W_1. The mean residence time for the composite system is found by applying Equation 9.30 to Equation 9.50:

$$\bar{t} = q\bar{t}_1 + (1 - q)\bar{t}_2 \tag{9.51}$$

where $\bar{t}_1 = V_1/Q_1 = V_1/qQ$ is the mean residence time for subsystem 1. If $q \gg 1 - q$, we might infer that the original design intent was to channel the flow through a region like subsystem 1 and that the flow through subsystem 2 is more or less accidental. Then the inadvertent flow represents bypassing if $\bar{t}_2 \ll \bar{t}$ and

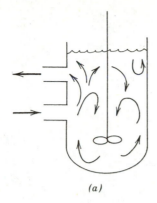

(a)

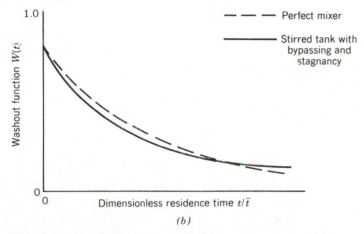

(b)

Figure 9.5 Pathological residence time behavior in a poorly designed stirred tank: (*a*) physical representation; (*b*) washout function.

stagnancy if $\bar{t}_2 \gg \bar{t}$. Bypassing and stagnancy are easier to detect and distinguish in systems that are nearly in piston flow than in systems close to perfect mixing. See Figure 9.6. They may be harder to eliminate than in the mechanically misdesigned stirred tank of Figure 9.5*a*. Most instances of bypassing, stagnancy, or other flow maldistributions stem from hydrodynamic problems such as the viscosity variations discussed in Chapter 6. Residence time measurements are a diagnostic aid in such cases, but the actual solution will probably require a hydrodynamic analysis.

When the hydrodynamics of a system are known, the residence time distribution can be calculated, at least when molecular diffusion is negligible. A hydrodynamic model is considered a ***theoretical model*** as opposed to the em-

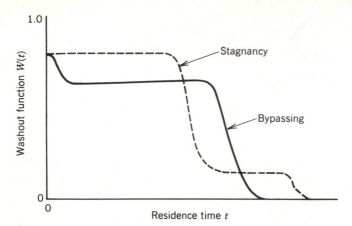

Figure 9.6 Bypassing and stagnancy in systems near piston flow.

pirical models discussed previously in this chapter. A simple but useful example is the parabolic velocity profile for laminar, Newtonian flow in a tube:

$$u(r) = 2\bar{u}\left(1 - \frac{r^2}{R^2}\right) \tag{9.52}$$

Suppose the system has length L. Then the residence time of a molecule passing through the system at radial position r is

$$t = t(r) = \frac{L}{u(r)} = \frac{\bar{t}}{2[1 - (r^2/R^2)]} \tag{9.53}$$

To find the distribution function of residence times, note that u decreases as r increases. Thus all the flow in the region from 0 to r has a residence time of $t(r)$ or less. The fraction of the flow having a residence time of t or less is

$$F(t) = F(r) = \frac{1}{Q}\int_0^r 2\pi u r'\, dr'$$

$$= \frac{1}{\pi R^2 \bar{u}}\int_0^r 2\pi(2\bar{u})\left[1 - \left(\frac{r'}{R}\right)^2\right]r'\, dr' \tag{9.54}$$

$$= \frac{2r^2 R^2 - r^4}{R^4}$$

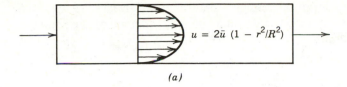

(a)

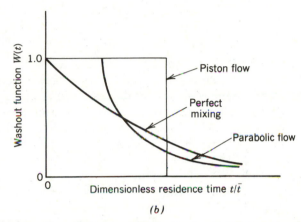

(b)

Figure 9.7 Residence time distribution for laminar flow in a circular tube: (a) physical representation; (b) washout function.

Elimination of r between Equations 9.53 and 9.54 gives

$$F(t) = 1 - \frac{(\bar{t})^2}{4t^2} \qquad t > \frac{\bar{t}}{2} \tag{9.55}$$

The washout function is

$$\boxed{\begin{aligned} W(t) &= 0 & t &< \frac{\bar{t}}{2} \\ W(t) &= \frac{(\bar{t})^2}{4t^2} & t &> \frac{\bar{t}}{2} \end{aligned}} \tag{9.56}$$

This function is shown in Figure 9.7. It has a sharp first appearance time, $t_{min} = \bar{t}/2$, and a slowly decreasing tail. For $t < 4.3\bar{t}$, the washout function for parabolic flow is higher than that for the exponential distribution. High residence times are associated with material near the tube wall ($r/R = 0.94$ for $t = 4.3\bar{t}$), and this material is stagnant in the sense of condition 9.49. In fact, the system is so stagnant that the second moment μ_2, and thus the variance of the residence time distribution, is infinite:

$$\mu_2 = 2 \int_0^\infty t W(t)\, dt = 2 \int_0^{\bar{t}/2} t\, dt + 2 \int_0^\infty t \left(\frac{\bar{t}^2}{4t^2} \right) dt$$

(9.57)

$$= \frac{\bar{t}^2}{2} + 4\bar{t}^2 \ln t \, \bigg|_{\bar{t}/2}^\infty = \infty$$

In the absence of diffusion, all hydrodynamic models show infinite variances. This is a consequence of the "zero slip" condition of hydrodynamics which forces $u = 0$ at the walls of a vessel. In real systems, molecular diffusion will ultimately remove molecules from the stagnant regions near walls. For real systems, $W(t)$ will asymptotically approach an exponential distribution and will have finite moments of all orders. However, molecular diffusivities are low in liquid phase systems, and σ^2 may be large indeed. This fact suggests the general inappropriateness of σ^2 for characterizing residence time distributions.

Rigorous models for residence time distributions require use of the convective diffusion equation, Equation 6.15 or Equation 6.79. Such solutions, either analytical or numerical, are rather difficult. We solved the simplest possible version, Equation 9.18, to determine the residence time distribution of a piston flow reactor. The preceding derivation of $W(t)$ for parabolic flow was actually equivalent to solving

$$\frac{\partial C}{\partial \theta} + 2\bar{u} \left(1 - \frac{r^2}{R^2} \right) \frac{\partial C}{\partial z} = 0$$

(9.58)

subject to a negative step change of the inert tracer. This model corresponds to the diffusion-free laminar flow reactor discussed in Chapter 6. We now go to the simplest version of the convective diffusion equation, which actually involves diffusion or a diffusionlike term:

$$\frac{\partial C}{\partial \theta} + \bar{u} \frac{\partial C}{\partial z} = D \frac{\partial^2 C}{\partial z^2}$$

(9.59)

This, of course, is the axial dispersion model discussed in Chapter 7. Its solution using closed boundary conditions, Equations 7.8 and 7.9, and subject to a negative step change is fairly complex. See Brenner[1]. For Peclet numbers, $\mathrm{Pe} = \bar{u}L/D$, above 16, the following result provides an excellent approximation

---

[1] H. Brenner, *Chem. Eng. Sci.*, **17**, 229 (1962).

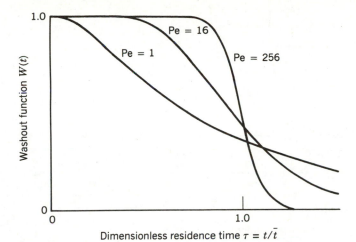

Figure 9.8 Residence time distributions for the axial dispersion model.

to the exact solution:

$$W(\tau) = 1 - \int_0^\tau \frac{\mathrm{Pe}}{4\pi\theta^3} \exp\left[\frac{-\mathrm{Pe}(1-\theta)^2}{4\theta}\right] d\theta \qquad (9.60)$$

where $\tau = t/\bar{t}$ is the dimensionless residence time. Figure 9.8 shows the washout function for the axial dispersion model, including Brenner's exact solution for $\mathrm{Pe} = 1$. The model is defined for $0 \leq \mathrm{Pe} \leq \infty$, and the extreme values correspond to perfect mixing and piston flow, respectively. The axial dispersion model shows a fuzzy first appearance time. It is competitive with and generally preferable to the tanks-in-series model for modeling small deviations from piston flow. It should be used with caution for large deviations. Detailed predictions of the model at small Pe are likely to fail under close scrutiny. See Nauman and Buffham[2] for a more comprehensive discussion.

Correlations for Pe versus Re for flow in circular tubes and packed beds were given in Section 7.1. The experimental data for these correlations were obtained from transient response experiments using inert tracers. Experimental values for the dimensionless variance are used to calculate the Peclet number:

$$\sigma^2 = \frac{2}{\mathrm{Pe}} - \frac{2}{\mathrm{Pe}^2}(1 - e^{-\mathrm{Pe}}) \qquad (9.61)$$

This result is theoretically valid for $0 < \sigma^2 < 1$ but should be used with caution for $\sigma^2 > 0.25$. Again, the real system is unlikely to follow the detailed predictions of the axial dispersion model at low Pe and high σ^2.

[2] E. B. Nauman and B. A. Buffham, *Mixing in Continuous Flow Systems*, Wiley, New York, 1983.

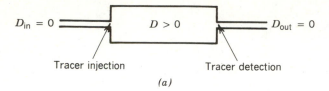

Tracer injection   Tracer detection

(a)

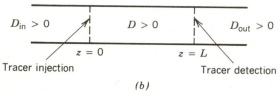

Tracer injection   Tracer detection

(b)

Figure 9.9 Transient response measurements based on the axial dispersion model: (*a*) closed system with probes for tracer injection and detection; (*b*) open system with probes for tracer injection and detection.

As mentioned previously, the unambiguous interpretation of transient response experiments requires that the system be closed. The axial dispersion model provides a good means for understanding the problems caused by open systems. See Figure 9.9. The closed system has no dispersion in the inlet and outlet streams and satisfies the Danckwerts boundary conditions. Transient response experiments, depending on the nature of the injection, give the residence time functions $W(t)$, $F(t)$, or $f(t)$. In particular, Equation 9.30 gives the correct value for $\bar{t}$ using the experimental results for $W(t)$ or $f(t)$.

Figure 9.9*b* illustrates an open version of the axial dispersion model. Measurement problems arise whenever $D_{in} > 0$ or $D_{out} > 0$. Suppose an impulse is injected into the system at $z = 0$. Unlike the closed case, not all the tracer enters the system immediately. Some will diffuse backward, up the inlet stream, and its entry into the system will be delayed. At the system outlet, some exiting material will diffuse backward and will reenter the system. These molecules will be counted more than once by the tracer detection probe. Not surprisingly, the measured response function is not $f(t)$ but another function, $g(t)$, which has a larger mean:

$$\mu_1 = \frac{L}{\bar{u}} + \frac{D_{in} + D_{out}}{\bar{u}^2} = \left[1 + \frac{(D_{in} + D_{out})}{\bar{u}L}\right]\bar{t} \qquad (9.62)$$

If erroneously intepreted as $\bar{t}$, Equation 9.62 is seen to give significant errors for inlet and outlet Peclet numbers less than about 100. If transient response measurements must be made on a open system, the recommended approach is to rescale $g(t)$ so that it has the correct mean:

$$[g(t)]_{\text{rescaled}} = \frac{\bar{t}\, g(t)}{\int_0^\infty \theta\, g(\theta)\, d\theta} \qquad (9.63)$$

See Nauman[3] for details. However, even this approach does not give $f(t)$, since the second and higher moments remain in error.

9.2 Reaction Yields

We suppose now that the residence time distribution is known and seek to determine the yield for an isothermal, homogeneous reaction occurring in the system having that residence time distribution. The case of a first order reaction gives the simplest and most general result.

9.2.1 First Order Reactions

For an isothermal, first order reaction, the probability that a particular molecule reacts depends only on the time it has spent in the system:

$$P_R = 1 - e^{-kt} \qquad (9.64)$$

To find the conversion, we need the average reaction probability for a great many molecules which have flowed through the system. The averaging is done with respect to residence time since residence time is what determines the individual reaction probabilities:

$$1 - \frac{a_{\text{out}}}{a_{\text{in}}} = \bar{P}_R = \int_0^\infty (1 - e^{-kt}) f(t)\, dt = 1 - \int_0^\infty e^{-kt} f(t)\, dt \qquad (9.65)$$

The fraction unreacted is given by

$$\boxed{\frac{a_{\text{out}}}{a_{\text{in}}} = \int_0^\infty e^{-kt} f(t)\, dt = 1 - k \int_0^\infty e^{-kt} W(t)\, dt} \qquad (9.66)$$

In performing these intergrations numerically, use whichever of $f(t)$ or $W(t)$ was determined directly. If a positive step change was used to find $F(t)$, convert to $W(t)$ using Equation 9.7

[3] E. B. Nauman, *Instr. Chem. Eng. Symp. Ser.* **87**, 569 (1984).

Example 9.5 Use residence time theory to predict the yield of a first order reaction occurring in an isothermal stirred tank. Repeat for a piston flow reactor.

Solution For the stirred tank, $W(t) = \exp(-t/\bar{t})$. Substitution into Equation 9.66 gives

$$\frac{a_{\text{out}}}{a_{\text{in}}} = 1 - k\int_0^\infty e^{-(k+1/\bar{t})t}\, dt = \frac{1}{1 + k\bar{t}} \qquad (9.67)$$

For the piston flow reactor, use $f(t) = \delta(t - \bar{t})$. Then

$$\frac{a_{\text{out}}}{a_{\text{in}}} = \int_0^\infty e^{-kt}\,\delta(t - \bar{t}) = e^{-k\bar{t}} \qquad (9.68)$$

Equation 9.66 can be generalized to include operation with unsteady inlet concentrations, $a_{\text{in}} = a_{\text{in}}(\theta)$:

$$a_{\text{out}}(\theta) = \int_0^\infty a_{\text{in}}(\theta - t)\, e^{-kt} f(t)\, dt = \int_{-\infty}^\theta a_{\text{in}}(t)\, e^{-k(\theta - t)} f(\theta - t)\, dt \quad (9.69)$$

This result allows the unsteady output to be calculated when the input concentration $a_{\text{in}}(\theta)$ is of a compound that reacts with first order kinetics. The case $k = 0$, corresponding to an inert tracer, is of special interest:

$$C_{\text{out}}(\theta) = \int_0^\infty C_{\text{in}}(\theta - t) f(t)\, dt = \int_{-\infty}^\theta C_{\text{in}}(t) f(\theta - t)\, dt \qquad (9.70)$$

Example 9.6 Determine the outlet response of a continuous flow mixing system when subjected to inlet disturbances of a negative step change, a positive step change, and a delta input of inert tracer.

Solution We suppose that all concentrations are normalized so that the negative step change has the form

$$C_{\text{in}}(\theta) = 1 \qquad \theta < 0 \qquad (9.71)$$

$$C_{\text{in}}(\theta) = 0 \qquad \theta > 0$$

The outlet response is found from Equation 9.69:

$$C_{out}(\theta) = \int_0^\infty C_{in}(\theta - t)f(t)\,dt$$

$$= \int_0^\theta C_{in}(\theta - t)f(t)\,dt + \int_\theta^\infty C_{in}(\theta - t)f(t)\,dt \qquad (9.72)$$

$$= \int_0^\theta 0 \cdot f(t)\,dt + \int_\theta^\infty 1 \cdot f(t)\,dt = \int_\theta^\infty f(t)\,dt$$

The final integral in Equation 9.72 may be evaluated using Equation 9.10:

$$C_{out}(\theta) = \int_\theta^\infty \left(\frac{-dW}{dt}\right)dt = \int_\infty^\theta dW = W(\theta) \qquad (9.73)$$

This confirms what we already know: The washout function is measured using a negative step change.

For the positive step change,

$$C_{in}(\theta) = 0 \qquad \theta < 0$$
$$\qquad\qquad\qquad\qquad\qquad\qquad\qquad (9.74)$$
$$C_{in}(\theta) = 1 \qquad \theta > 0$$

and the reader can show that

$$C_{out}(\theta) = F(\theta) \qquad (9.75)$$

For the delta input,

$$C_{in}(\theta) = \delta(\theta) \qquad (9.76)$$

and

$$C_{out}(\theta) = \int_0^\infty \delta(\theta - t)f(t)\,dt = f(\theta) \qquad (9.77)$$

The property that $\delta(x) = \delta(-x)$ was used to evaluate the integral in Equation 9.77.

Example 9.7   Suppose the input of an inert tracer varies sinusoidally:

$$C_{in}(\theta) = C_0(1 + \beta \sin \omega\theta) \qquad \beta \le 1 \qquad (9.78)$$

Find the outlet response of a stirred tank reactor. What is the maximum deviation in C_{out} from its midpoint C_0?

Solution   For a stirred tank, $f(t) = (1/\bar{t})\exp(-t/\bar{t})$. The second integral in Equation 9.70 becomes

$$
C_{\text{out}}(\theta) = \int_{-\infty}^{\theta} \frac{C_0(1 + \beta \sin \omega t)}{\bar{t}} \exp\left[\frac{-(\theta - t)}{\bar{t}}\right] dt
$$

$$
= \frac{C_0 e^{-\theta/\bar{t}}}{\bar{t}} \int_{-\infty}^{\theta} (1 + \beta \sin \omega t) e^{+t/\bar{t}} dt \tag{9.79}
$$

$$
= C_0 \frac{1 + \beta(\sin \omega\theta - \omega\bar{t} \cos \omega\theta)}{1 + \omega^2 \bar{t}^2}
$$

which is the outlet response. To find the maximum deviation, note that $\sin \omega\theta - \omega\bar{t} \cos \omega\theta$ achieves extreme values when $\cot \omega\theta = -\omega\bar{t}$. Some trigonometry then shows that the maximum deviation is

$$
|C_{\text{out}} - C_0|_{\max} = \frac{\pm\beta}{\sqrt{1 + \omega^2 \bar{t}^2}} \tag{9.80}
$$

This result is useful in designing stirred tanks to damp out errors in feed concentration.

9.2.2   Micromixing

A perfect mixer has an exponential distribution of residence times: $W(t) = \exp(-t/\bar{t})$. Can any other continuous flow system have this distribution? Perhaps surprisingly, the answer to this question is a firm yes. To construct an example, suppose the feed to a reactor is encapsulated. The size of the capsules is not critical. They must be large enough to contain many molecules but must remain small compared to the dimensions of the reactor. Imagine them as small ping-pong balls as in Figure 9.10a. The balls are agitated, gently enough not to break them but well enough to randomize them within the vessel. In the limit of high agitation, the vessel can approach perfect mixing with respect to the ping-pong balls. A sample of balls collected from the outlet stream will have an exponential distribution of residence times:

$$
W_b(t_b) = \exp\left(-\frac{t_b}{\bar{t}}\right) \tag{9.81}
$$

Since the balls are small, the molecules within them will also participate in the exponential distribution, but they are far from perfectly mixed. Molecules that entered together stayed together. While in the reactor, they mixed only with molecules having the same age and initial composition. The composition within each ball evolved with time spent in the system as though the balls were small

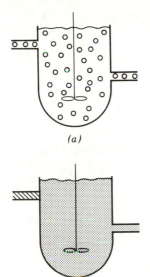

(a)

(b)

Figure 9.10 Extremes of micromixing in a stirred tank reactor: (a) Ping-Pong balls in an agitated vessel—the segregated stirred tank; (b) molecular homogeneity—the perfect mixer.

batch reactors. The exit concentration for a ball, $a(t_b)$, is the same as that in a batch reactor after reaction time t_b.

We have just described a ***completely segregated stirred tank***. It has an exponential distribution of residence times but a reaction environment that is very different from that within a perfect mixer. It can be modeled as a set of piston flow reactors in parallel, with the lengths of the individual piston flow elements being distributed exponentially. We saw in Figure 9.2 that any residence time distribution can be modeled as piston flow elements in parallel. Thus the encapsulation or ping-pong ball concept can be applied to any residence time distribution. A reactor modeled in this way is said to be ***completely segregated***. Its outlet concentration is found by averaging the concentrations of the individual batch or piston flow reactors:

$$a_{\text{out}} = \int_0^\infty a_{\text{batch}}(t) f(t) \, dt \qquad\qquad (9.82)$$

where $a_{\text{batch}}(t) = a_t$ is the concentration of a batch or piston flow reactor after reaction time t. The inlet concentration is the same for each batch and provides the initial condition for finding $a_{\text{batch}}(t)$.

Example 9.8   Find the outlet concentration from a completely segregated stirred tank for a first order reaction. Repeat for a second order reaction with $\mathscr{R}_A = -ka^2$.

Solution The residence time distribution is exponential, $f(t) = (1/\bar{t})\exp(-t/\bar{t})$. For first order kinetics, $a_{batch}(t) = a_{in}\exp(-kt)$. Equation 9.82 becomes

$$a_{out} = \int_0^\infty a_{in}e^{-kt}\frac{e^{-t/\bar{t}}}{\bar{t}}\,dt = \frac{1}{1+k\bar{t}} \tag{9.83}$$

which is identical to the outlet concentration of a perfect mixer. This result confirms the observation made in Section 9.2.1: *The yield of a first order reaction is uniquely determined by the residence time distribution*.

For a second order reaction, $a_{batch}(t) = a_{in}/(1 + a_{in}kt)$. Equation 9.82 becomes

$$\frac{a_{out}}{a_{in}} = \int_0^\infty \frac{\exp(-t/\bar{t})\,dt}{(1+a_{in}kt)\bar{t}} = \frac{\exp(1/a_{in}k\bar{t})}{a_{in}k\bar{t}}\int_{(a_{in}k\bar{t})^{-1}}^\infty \frac{e^{-x}\,dx}{x} \tag{9.84}$$

$$= xe^x E_1(x)\Big|_{x=(a_{in}k\bar{t})^{-1}}^\infty$$

where $E_1(x)$ is the exponential integral function. This is a tabulated function which can be found in many handbooks or mathematical tables. Note, however, that Equation 9.82 can be integrated numerically as it would have to be for more complicated batch kinetics.

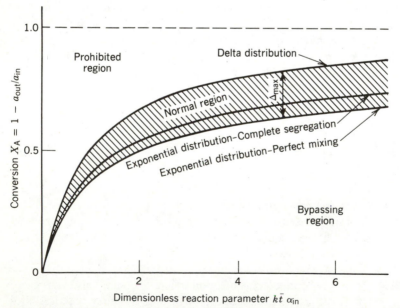

Figure 9.11 Conversion limits for second order reactions.

Equation 9.84 is different from the equivalent result for a perfect mixer:

$$\frac{a_{\text{out}}}{a_{\text{in}}} = \frac{\sqrt{1 + 4a_{\text{in}}k\bar{t}} - 1}{2a_{\text{in}}k\bar{t}} \tag{9.85}$$

The difference is not very large, however. See Figure 9.11. The maximum difference in $a_{\text{out}}/a_{\text{in}}$ between Equations 9.84 and 9.85 is 0.07, which occurs at $a_{\text{in}}k\bar{t} = 16$, complete segregation giving the higher conversions. By comparison, the biggest difference between piston flow and perfect mixing is 0.192, which occurs at $a_{\text{in}}k\bar{t} = 4.9$.

For reactions other than first order, the conversion depends not only on the residence time distribution but also on mixing between molecules having differing ages. This is called **micromixing**. Completely segregated reactors have no mixing between molecules of differing ages. This zero level of micromixing is possible with any residence time distribution. At the opposite extreme, perfect mixers have complete mixing between molecules, but perfect mixing implies an exponential distribution of residence times. Complete micromixing is impossible except with the exponential distribution. Other residence time distributions have some maximum possible level of micromixing, which is known as **maximum mixedness**. Less micromixing than this is always possible. More would force a change in the residence time distribution.

A qualitative picture of micromixing is given in Figure 9.12. The x-axis, labeled **macromixing**, measures the breadth of the residence time distribution. It is zero for piston flow, fairly broad for the exponential distribution of stirred tanks, and broader yet for situations involving bypassing or stagnancy. The y-axis is micromixing, which varies from none to complete. The y-axis also

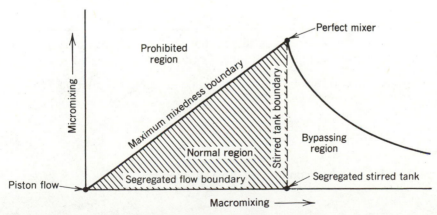

Figure 9.12 Schematic representation of mixing space.

measures how important micromixing effects can be. They are unimportant for piston flow and have maximum importance for stirred tank reactors. Well-designed reactors will usually fall in the *normal region* bounded by the three apexes, which correspond to piston flow, perfect mixing, and complete segregation with an exponential distribution. Without even measuring the residence time distribution, we can determine limits on the performance of most real reactors just by calculating the performance at the three apexes of the normal region. The calculations require knowledge only of the rate constants and $\bar{t} = V/Q$. Thus far in this book, we have dealt with two ideal reactors: the piston flow reactor and the perfect mixer. We now add a third, the *segregated stirred tank*. In terms of micromixing, this third ideal reactor differs as much from perfect mixing as piston flow differs from perfect mixing.

When the residence time distribution is known, the uncertainty about reactor performance is greatly reduced. A real system must lie somewhere along a vertical line in Figure 9.12. Performance will usually be bounded by the upper and lower points on this line: maximum mixedness and complete segregation. The complete segregation limit can be calculated from Equation 9.82. The maximum mixedness limit is found by solving *Zwietering's differential equation*:

$$\frac{da}{d\lambda} + \frac{f(\lambda)}{W(\lambda)}\left[a_{\mathrm{in}} - a(\lambda)\right] + \mathscr{R}_{\mathrm{A}} = 0 \qquad (9.86)$$

The solution does not use an initial value of a as a boundary condition. Instead, the usual boundary condition associated with Equation 9.86 is

$$\lim_{\lambda \to \infty} \frac{da}{d\lambda} = 0 \qquad (9.87)$$

The outlet concentration from a maximum mixedness reactor is found by evaluating the solution to Equation 9.86 at $\lambda = 0$: $a_{\mathrm{out}} = a(0)$.

Example 9.9 Solve Zwietering's differential equation for arbitrary reaction kinetics and an exponential residence time distribution.

Solution Since $f(\lambda)/W(\lambda) = 1/\bar{t}$ for this situation, Equation 9.86 becomes

$$\frac{da}{d\lambda} + \frac{\left[a_{\mathrm{in}} - a(\lambda)\right]}{\bar{t}} + \mathscr{R}_{\mathrm{A}} = 0 \qquad (9.88)$$

Observe that the boundary condition will be satisfied if

$$\frac{a_{\mathrm{in}} - a(\lambda)}{\bar{t}} + \mathscr{R}_{\mathrm{A}} = 0 \qquad (9.89)$$

for all λ since this gives $da/d\lambda = 0$ for all λ. Set $\lambda = 0$ to obtain

$$a_{out} = a(0) = a_{in} + \bar{t}\mathscr{R}_A \tag{9.90}$$

This is the expected result since a stirred tank at maximum mixedness is a perfect mixer.

Analytical solutions to Equation 9.86 are difficult since the $\mathscr{R}_A$ term is usually nonlinear. For numerical solutions, Equation 9.86 can be treated as though it were an initial value problem. Guess a value for $a_{out} = a(0)$. Integrate Equation 9.86. If $a(\lambda)$ achieves a steady state value, the correct $a(0)$ has been guessed. For other $a(0)$, $a(\lambda)$ will tend toward $\pm\infty$ as $\lambda \to \infty$. This numerical approach is similar to the shooting methods of Chapter 7. The computed results are very sensitive to the guessed values for $a(0)$, and small changes in $a(0)$ will cause $a(\infty)$ to range from $-\infty$ to $+\infty$. Unlike the forward shooting example of Section 7.1.1, the sensitivity is now beneficial since it allows $a(0) = a_{out}$ to be calculated with high precision.

Example 9.10 Solve Zwietering's differential equation for the residence time distribution corresponding to two stirred tanks in series. Use second order kinetics with $a_{in}k\bar{t} = 5$.

Solution For $N = 2$, the tanks-in-series model (see Equation 9.47) gives $f(\tau) = 4\tau\exp(-2\tau)$ and $W(\tau) = (1 + 2\tau)\exp(-2\tau)$, where $\tau = t/\bar{t}$. Equation 9.86 becomes

$$\frac{da}{d\tau} + \frac{4\tau}{1 + 2\tau}(a_{in} - a) - 5a^2 = 0 \tag{9.91}$$

This equation may be solved as an initial value problem using any convenient technique. Marching ahead with $\Delta\tau = 0.0625$ gives the following results:

$a(0)$	$a(\infty)$
0	$-\infty$
0.1	$-\infty$
0.2	$-\infty$
0.3	$+\infty$
0.25	$-\infty$
$\vdots$	$\vdots$
0.276	$-\infty$
0.277	$+\infty$

Obviously, $a_{out} = a(0)$ can be calculated with high precision. It happens that the estimate for a_{out} is not very accurate because of the large step size, but this can be overcome using a smaller $\Delta\tau$ or a more sophisticated integration technique. The correct value is $a_{out} = 0.287$.

We have now outlined the major points of residence time theory. The next example shows how the theory can be used to systematically analyze the performance of a nonideal reactor. The example is contrived to simplify computations and avoid disclosing proprietary information. However, the analytical approach and results are representative of actual experience while still illustrating general principles.

Example 9.11   You have been assigned the problem of improving the performance of an existing polymerization reactor. Initially, you know only that it operates at an input flow rate of 10,000 lbs/hr and gives a conversion of 62 $\pm$ 1 percent at a nominal operating temperature of 140°C. The reactor drawings show a complicated arrangement of stirring paddles and cooling coils. The design intent was to approximate piston flow, but a detailed hydrodynamic analysis would be difficult to impossible. The drawings do show the working volume of the reactor and you calculate that the fluid inventory should be about 12,500 lbs. Thus you estimate $\bar{t} = 1.25$ hr.

The company library contains the original kinetics study, which seems to have been of excellent quality. The major reaction is a linear self-condensation (see Chapter 12), the kinetics of which can be closely approximated by

$$\mathcal{R}_A = -ka^2$$

where $a_{in}k = 4$ hr^{-1} at 140°C. Thus you expect $a_{in}k\bar{t} = 5$ and calculate the following results:
Piston flow (Equation 1.42):

$$\frac{a_{out}}{a_{in}} = 0.167$$

Perfect mixing (Equation 1.78):

$$\frac{a_{out}}{a_{in}} = 0.358$$

Segregated stirred tank (Equation 9.84):

$$\frac{a_{out}}{a_{in}} = 0.299$$

The actual result, $a_{out}/a_{in} = 0.38 \pm 0.01$, is worse than any of the ideal reactors!

There are two main possibilities:

1. The residence time distribution lies outside the normal region.
2. The calculated value for $a_{in}k\bar{t}$ is too high. This in turn leads to several subpossibilities:

    (i) The laboratory kinetics are wrong.

    (ii) The actual temperature is lower than the measured one.

    (iii) The value for $\bar{t}$ is too high.

The good engineer will consider all these possibilities and perhaps a few more. Temperature errors are very common, particularly in viscous, low-thermal-conductivity systems typical of polymers; and they lead to sizable errors in concentration. However, measured temperatures are usuallly lower than actual rather than higher. Suppose you decide that the original kinetic study was sound, that there are no apparent changes in the process chemistry, and that the analytical techniques are accurate. This makes flow distribution or mixing the likely culprit. Besides, you would like to see just how that strange agitation/cooling system performs from a flow viewpoint.

    A residence time study requires an inert tracer, a means of injecting it near the reactor inlet, and a means of detecting it at the reactor outlet. The experiment can usually be done without disrupting normal production. Your management will often insist that it be done this way. Suppose you find an inert hydrocarbon which is not normally present in the system, which is easily detected by gas chromatography, and which can be tolerated in the product stream. You arrange for the tracer injection port and the product sampling ports to be installed during a maintenance shutdown. It is important that the tracer be well mixed in the inlet stream. Otherwise it might channel though the system and give nonrepresentative results. You accomplish this by injecting the tracer at the suction side of the transfer pump that is feeding the reactor. You also dissolve a little polymer in the tracer stream to match its viscosity more closely to that of the reactor feed stream.

    Having carefully prepared as indicated above, you perform a tracer washout experiment and obtain the following results:

Time, min	C_{out}/C_{in}
0	1.00
15	0.79
30	0.60
45	0.45
60	0.38
75	0.27
90	0.21
105	0.18
120	0.12
150	0.07
180	0.05

These data indicate a mean residence time much less than the calculated 1.25 hr. Were it not for your careful preparations, you might doubt the experiment. Instead, you arrange for the reactor to be opened and inspected during the next shutdown. You find it partially filled with crosslinked polymer. When this is removed, the conversion increases to 74 percent, $a_{out}/a_{in} = 0.26$.

Suppose that a residence time experiment performed on the clean reactor gives $\bar{t} = 1.25$ hr as expected. Suppose also that the washout curve closely matches that for two stirred tanks in series so that

$$ f(t) = \frac{4te^{-2t/\bar{t}}}{\bar{t}} \qquad \bar{t} = 1.25 $$

Equation 9.82 is integrated numerically using

$$ a_{batch}(t) = \frac{1}{1 + a_{in}kt} = \frac{1}{1 + 4t} $$

Equation 9.86 is integrated numerically as in Example 9.10. The results are: Complete segregation (Equation 9.82):

$$ \frac{a_{out}}{a_{in}} = 0.290 $$

Maximum mixedness (Equation 9.86):

$$ \frac{a_{out}}{a_{in}} = 0.287 $$

so that the micromixing limits provide tight bounds on the conversion. Since the actual result is outside these limits, something else is wrong. Quite likely it is the measured temperature

---

9.2.3 The Bounding Theorem

The states of complete segregation and maximum mixedness represent limits on the extent of micromixing that is possible with a given residence time distribution. In complete segregation, molecules that enter together stay together. They mix with other molecules only in the exit stream. While in the reactor, they are surrounded by molecules having the same age and mix with molecules having different ages only when they leave the reactor. This mixing situation can be represented by a parallel collection of piston flow elements as shown in Figure 9.2. It can also be represented as a single piston flow reactor with a large number of side exits. See Figure 9.13a. The size and spacing of the side exits can be

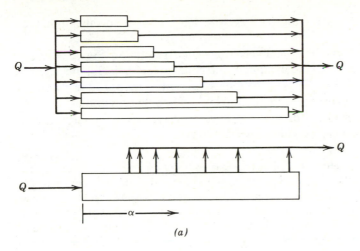

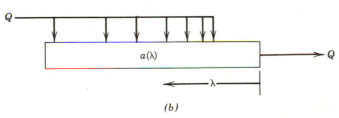

Figure 9.13 Extreme mixing models for an arbitrary residence time distribution: (*a*) complete segregation; (*b*) maximum mixedness.

adjusted to model any residence time distribution. Thus piston flow with side exits is capable of modeling any residence time distribution. It is a completely segregated model since molecules in the reactor mix only with other molecules having exactly the same age.

Another way of modeling an arbitrary residence time distribution is to use a single piston flow reactor with a large number of side entrances. See Figure 9.13*b*. The size and spacing of the entrances are adjusted like the exits were adjusted in Figure 9.13*a*. Thus Figure 9.13 shows two ways of representing the same residence time distribution. The second way is quite different than the first. Molecules flow in through the side entrances and immediately mix with molecules that are already in the system. This is a maximum mixedness reactor, and there is substantial mixing between molecules having different ages. Since there is only one exit, molecules that are mixed together will leave together, but they may have entered at different times. By way of contrast, there is only one entrance to the completely segregated reactor. Molecules that are mixed together in a completely segregated reactor must necessarily have entered together but they may leave separately.

The performance of the completely segregated reactor in Figure 9.13a is governed by Equation 9.82. The maximum mixedness reactor in Figure 9.13b is governed by Equation 9.86. These reactions represent extremes in the kind of mixing that can occur between molecules having different ages. Do these reactors also represent extremes of performance as measured by conversions or selectivities? The **bounding theorem** provides a partial answer:

Suppose $\mathscr{R}_A$ is a function of a alone and that both $d\mathscr{R}_A/da$ and $d^2\mathscr{R}_A/da^2$ do not change sign over the range of concentrations encountered in the reactor. Then, for a system having a fixed residence time distribution, Equations 9.82 and 9.86 provide absolute bounds on the conversion of component A, the conversion in a real system necessarily falling within the bounds. If $d^2\mathscr{R}_A/da^2 > 0$, conversion is maximized by maximum mixedness and minimized by complete segregation. If $d^2\mathscr{R}_A/da^2 < 0$, the converse is true. If $d^2\mathscr{R}_A/da^2 = 0$, micromixing has no effect on conversion.

Example 9.11   Apply the bounding theory to an nth order reaction, $\mathscr{R}_A = -ka^n$.

Solution

$$\frac{d\mathscr{R}_A}{da} = -nka^{n-1} \tag{9.92}$$

and

$$\frac{d^2\mathscr{R}_A}{da^2} = -n(n-1)ka^{n-2} \tag{9.93}$$

The first derivative is always negative. The second derivative is negative if $n > 1$, is zero if $n = 1$, and is positive if $n < 1$. Thus it does not change sign for a fixed n and the bounding theory applies. For $n > 1$ (e.g., second order reactions), $d^2\mathscr{R}_A/da^2 < 0$ and conversion is highest in a completely segregated reactor. For $n = 1$, the reaction is first order, and micromixing does not affect conversion. For $n < 1$ (e.g., half-order), $d^2\mathscr{R}_A/da^2 > 0$ and maximum mixedness gives the highest possible conversion.

The bounding theory gives **sufficient** conditions for reactor performance to be bounded by complete segregation and maximum mixedness. These conditions may not be **necessary**. In particular, the requirement that $d\mathscr{R}_A/da$ keep the same sign for $0 < a < a_{in}$ is not necessary. Some reactions show maximum rates ($d\mathscr{R}_A/da = 0$ for some $a = a_{crit}$), yet the bounding theory still applies provided $d^2\mathscr{R}_A/da^2$ does not change sign for $0 < a_{crit} < a_{in}$. If $d^2\mathscr{R}_A/da^2$ does change sign, examples have been found which give a maximum conversion at an intermediate level of micromixing.

9.2.4 Micromixing Models

Micromixing effects tend to be fairly subtle in systems with premixed feed. There is no effect at all when the reaction is first order or when the residence time distribution corresponds to piston flow. The effects are largest when the distribution is exponential and when the reaction is highly complex as in a polymerization. Even here, however, the effects are usually so small that they cannot be measured with accuracy. Uncertainties in the kinetic model are often larger than the predicted differences between complete segregation and maximum mixedness. For these reasons, few data and correlations exist for micromixing in systems with premixed feed.

In systems with separate feed streams for the reactants, molecular mixing must occur for reaction to occur. The mechanisms for this mixing are the same as those responsible for micromixing in a premixed feed system, but experimental measurements are much easier. For a liquid phase, stirred tank reactor, the *packet diffusion* model is believed qualitatively correct and has been experimentally verified to a reasonable extent. According to this model, hydrodynamic forces rapidly disperse the incoming fluid into small packets. These packets are approximately the same size as the Kolomogoroff scale of turbulence:

$$\eta = \left(\frac{\mu^3}{\rho^3 \nu} \right)^{1/4} \tag{9.94}$$

where ν is the power dissipation per unit mass of fluid. Following this rapid initial dispersion, the packets continue to evolve in size and shape but at a relatively slow rate. Molecular level mixing occurs by diffusion between packets, and the rates of diffusion and of the consequent chemical reaction can be calculated. Early versions of the model assumed spherical packets of constant and uniform size. Variants now exist that allow the packet size and shape to evolve with time.

In laminar flow stirred tanks, the packet diffusion model can be replaced by a slab-diffusion model. The diffusion and reaction calculations are similar to those for the turbulent flow case. Unfortunately, no good methods have emerged for measuring or predicting slab thicknesses.

Suggestions for Further Reading

The ideas explored in this chapter are discussed at length in:

E. B. Nauman, and B. A. Buffham, *Mixing in Continuous Flow System*, Wiley, New York, 1983.

Problems

9.1   How much gel was removed from the reactor in Example 9.11?

9.2   A step change experiment of the turnoff variety gave the following results:

θ	$C_{out}(\theta)/C_{out}(0)$
0	1.00
5	1.00
10	0.98
15	0.94
20	0.80
30	0.59
45	0.39
60	0.23
90	0.08
120	0.04

where θ is in seconds.

(a) Estimate $\bar{t}$ for the system.

(b) Estimate the conversion for an isothermal, first order reaction with $k = 0.093$ s^{-1}.

9.3   Determine the dimensionless variance of the residence time distribution in Problem 9.2. Then use Equation 9.61 to fit the axial dispersion model to this system. Is axial dispersion a reasonable model for this situation?

9.4   Apply the bounding theorem to the reversible, second order reaction

$$A + B \rightleftharpoons C + D \qquad \mathcal{R} = k_f ab - k_r cd$$

Assume A, B, C, and D have similar diffusivities so that local stoichiometry is preserved. Under what circumstances will conversion be maximized by complete segregation? By maximum mixedness?

9.5   Suppose a piston pump operating at 100 strokes per minute is used to meter one component into a reactant stream. The concentration of this component should not vary by more than 0.1%. Devise a method for achieving this.

9.6   Mixing experts often use a simple rule of thumb for the scaleup of agitated vessels: maintain constant power per unit volume. Is this a reasonable approach for scaling a stirred tank reactor? Assume the reaction is

$$A + B \rightarrow C + D$$

with A and B entering through separate, unmixed streams? Does it make any difference if the reactor is batch, semibatch, or continuous?

9.7   A typical power input for vigorous agitation is 10 hp per 1000 gallons in systems with "waterlike" physical properties.

(a) Calculate the Kolomogoroff scale of turbulence.

(b) Assume that a spherical droplet having a diameter equal to the Kolomogoroff size is placed in large, homogeneous mass of fluid. How long will it take for concentrations inside the drop to closely approach those in the homogeneous fluid? Use $D = 2 \times 10^{-9}$ m^2/s and require a 95% response to the homogeneous phase concentration?

(c) Suppose a second order reaction with unmixed feed streams is occurring in the agitated vessel. How large can the rate constant ka_0 be if mixing and diffusion times are to remain an order of magnitude smaller than reaction times?

9.8 Experimental conditions prevented the application of a good step change at the inlet to the reactor, but it was possible to monitor both $C_{in}(\theta)$ and $C_{out}(\theta)$:

Time, s (seconds)	$C_{in}(\theta)$ (% Tracer)	C_{out}, (% Tracer)
< 0	0	0
3	0.072	0
6	0.078	0
9	0.081	0.008
15	0.080	0.017
20	0.075	0.020
30	0.065	0.027
40	0.057	0.035
60	0.062	0.043
80	0.068	0.051
100	0.068	0.057
120	0.068	0.062

The reactor happens to be a gas-fluidized bed for which the fractional tubularity model is usually appropriate.

(a) Write the model as

$$f(t) = \alpha e^{-\alpha(t-\tau)} \qquad t > \tau$$

and estimate the parameters α and τ.

(b) Use this estimate and the convolution integral, Equation 9.70, to predict $C_{out}(\theta)$ given the experimental values for $C_{in}(\theta)$. Can your estimates for α and τ be improved by this approach? *Hint:* A reasonable approximation to the input signal might be

θ	C_{in}
0 to 20	0.078
20 to 40	0.066
40 to 60	0.060
60 to 80	0.065
80 to 100	0.068

9.9 Solid-catalyzed gas reactions are often modeled as if they were homogeneous. A frequently encountered form for the rate equation is

$$R_A = \frac{ka}{1 + K_A a}$$

Suppose $k = 2 \text{ s}^{-1}$ and $K_A = 0.8 \text{ m}^3/\text{mol}$. Determine bounds on the yield for a reactor having $\bar{t} = 3$ s and an inlet feed concentration of $2 \text{ mol}/\text{m}^3$.

9.10 Suppose the reactor in Problem 9.9 obeys the fractional tubularity model with $\tau_p = 0.5$. Calculate bounds on the yield.

9.11 Suppose the activation energy for the self-condensation reaction in Example 9.11 is 44 kJ mol^{-1}.

(a) What size error in temperature would account for the observed result of $a_{\text{out}}/a_{\text{in}} = 0.26$ compared to the limits calculated from residence time theory?

(b) Is the temperature error in part (a) consistent with the yield and residence time measurements made before the gel was removed?

10

Heterogeneous Catalysis

Heterogeneous reactions involve more than one phase. Catalytic reactions involve a component which may participate in several elementary reaction steps but does not appear in the overall reaction, the consumption and generation rates being equal or nearly so. By convention, *heterogeneous catalysis* refers to the common situation where the transport of the noncatalytic reactants occurs in one phase and the catalytic activity occurs in another phase. The transport phase is a fluid and the catalytic phase is typically a solid. There are myriad examples of heterogeneous catalysis. A few important ones are:

- Oxidation of gaseous hydrocarbons on a solid catalyst. Oxygen and the hydrocarbons are supplied from the gas phase and the oxidation products are removed by it. The catalyst is usually stationary (fixed bed) but sometimes moves with the gas stream (fluid bed).

- Gas phase polymerization of ethylene using a coordination metal catalyst. The monomer is transported in the gas phase. The reaction occurs on a small piece of catalyst suspended in the gas phase. The polymer molecules form another solid phase around the catalyst particle and are transported out of the system as a dry, free-flowing powder.

- Many biochemical reactions involve heterogeneous catalysis. Reactant and product transport typically occur in an aqueous phase. The catalyst, an enzyme, is a protein molecule. It may be immobilized on a solid substrate, it may be freely suspended, or it may be in a second liquid phase (a lipid phase).

- Hydrocracking of heavy hydrocarbons on a solid catalyst. This involves three phases: a gas phase to supply the hydrogen, a liquid phase to supply the hydrocarbons, and a solid phase for the catalyst. In a trickle bed reactor, the solid is stationary. The liquid and gas are both fed at the top and removed at the bottom. The product is usually a mixture of gaseous and liquid components.

All these examples involve interphase mass transfer and chemical reaction. Mathematical analysis requires that the transport equations be solved in conjunction with the chemical kinetics. The exact nature of the solution depends on the specific system being considered. The next section treats a case which is common in the petrochemical industry: transport of reactants and products in a gas phase and reaction on a stationary solid phase. This corresponds to a catalytic, fixed-bed reactor such as that used in the exhaust system of most cars. Understanding solid-catalyzed gas reactions is very important to the chemical industry and also provides insight into the type of analysis needed to describe other examples of heterogeneous catalysis.

10.1 Solid-Catalyzed Gas Reactions

The catalytic oxidation or hydrogenation of light hydrocarbons is typical of solid-catalyzed reactions. Most solid catalysts are supplied as cylindrical pellets with lengths and diameters in the range of 2 to 10 mm. The pellets are actually porous, with micropores ranging in diameter from a few angstroms to a few microns. The internal surface area, accessible through the pores, is enormous, ranging from tens to hundreds of square meters per gram. It is this internal area that accounts for most of the catalytic activity. The actual catalytic sites are atoms or molecules on the surface. The structural material of the catalytic particle is often a metal oxide (e.g., alumina, silica, vanadium pentoxide). This structural material may act as a catalyst directly or it may *support* a more expensive substance such as finely divided platinum. When the reaction exotherm is large, the catalyst pellets are randomly packed in small-diameter (10 to 50 mm) tubes which are often quite long (2 to 10 m). If the adiabatic temperature change is small, large-diameter vessels can be used. Annular flow reactors (See Problem 3.7) are employed when it is important to minimize the gas phase pressure drop. Another approach is to flow the gas through the internal labyrinth of a monolithic catalyst as in automobile exhaust systems. Regardless of the specific geometry used to contact the gas and the solid, all these schemes require a complex set of mass transfer and reaction steps:

1. Bulk transport of reactants to the vicinity of a catalyst particle.
2. Interphase mass transfer, typically by film diffusion, from the bulk gas phase to the external surface of the porous catalyst.
3. Transport of reactants to the internal surface of the catalyst particle by diffusion through the pores.
4. Adsorption of reactant molecules onto the catalytic surface.
5. Reaction between adsorbed components on the solid surface.

6. Desorption of product molecules into the catalyst pore.

7. Pore diffusion of product molecules to the external surface of the pellet.

8. Interphase mass transfer into the bulk gas phase.

9. Bulk transport of products to the reaction outlet.

All these steps can influence the overall reaction rate. It is generally desirable to eliminate the mass transfer resistances, Steps 1 to 3 and 7 to 9. When this is done, the concentrations of reactants immediately adjacent to the catalytic surface are equal to their concentrations in the bulk gas phase. The resulting kinetics are known as **intrinsic kinetics** since they are intrinsic to the catalytic surface and not to the design of the pores, or the pellets, or the reactor. Much of this chapter and most current research in heterogeneous catalysis centers on the measurement, understanding, and modification of intrinsic kinetics. However, it is first useful to consider all of Steps 1 to 9 to gain an overall appreciation of heterogeneous reactor design. We consider an overall reaction of the form

$$A + B + \cdots \rightleftharpoons P + Q + \cdots \tag{10.1}$$

occurring in a packed-bed reactor.

Step 1   Refer to Figure 10.1. The entering gas has composition $a_{in}, b_{in}, \ldots$. The exiting gas has composition $a_{out}, b_{out}, \ldots$. These are related through a

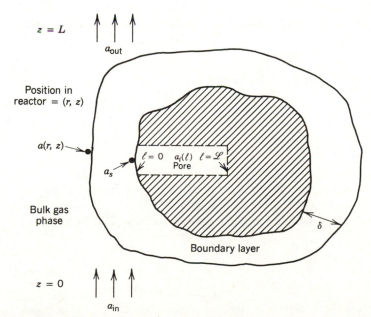

Figure 10.1   Gas and surface concentration in a gas–solid catalytic reactor.

system model that applies to the bulk fluid phase. For a packed bed, a *pseudo-homogeneous* model, Equation 7.43, is usually used to describe bulk mass transport. For component A,

$$\bar{u}_s \frac{\partial a}{\partial z} = D_r \left(\frac{1}{r} \frac{\partial a}{\partial r} + \frac{\partial^2 a}{\partial r^2} \right) + \varepsilon \mathscr{R}_A(a, b, \dots) \tag{10.2}$$

where $\mathscr{R}_A$ is a pseudohomogeneous rate expression based on the gas phase volume and on the bulk, gas phase concentrations $a, b, \dots$. When $\mathscr{R}_A$ is known, Equation 10.2 can be integrated as in Chapter 7. Steps 2 to 8 are used to find $\mathscr{R}_A(a, b, \dots)$. Step 9 is the equivalent of Step 1 but for product species.

Step 2 The gas at position (r, z) has a bulk concentration $a(r, z)$. It has a concentration $a_s(r, z)$ at the external surface of a catalyst pellet that is located at position (r, z). The bulk and surface concentrations are related through a mass transfer coefficient, and the steady-state flux across the interface must be equal to the reaction rate. Thus

$$\mathscr{R}_A = k_s(a_s - a) \tag{10.3}$$

where k_s is a mass transfer coefficient which presumably depends on the boundary layer thickness and the molecular diffusivity of component A. We will treat it as an empirical constant. Note that k_s has dimensions of reciprocal time. It is actually the product of a mass transfer coefficient and a mass transfer area per unit volume of catalytic bed.

Step 3 Pore diffusion is usually modeled by a one-dimensional diffusion equation:

$$\mathscr{D}_A \frac{d^2 a_i}{d\ell^2} + \mathscr{R}_A = 0 \tag{10.4}$$

The solution to this equation, which will be detailed in Section 10.3.1, gives the concentration at position ℓ in a pore of length $\mathscr{L}$, $0 < \ell < \mathscr{L}$. Denote this concentration as $a_i(z, \ell)$. One boundary condition associated with Equation 10.4 is

$$a_i(z, 0) = a_s(z) \tag{10.5}$$

where the appearance of z serves to remind us that all these concentrations depend on location within the overall reactor. (In the more general case of a nonisothermal packed bed, the concentrations will depend on both z and r.)

Step 4 This is the first step of what we have called the intrinsic reaction kinetics. Component A is adsorbed on the surface at a catalytically active site S:

$$A + S \rightleftharpoons AS \tag{10.6}$$

This kinetic relationship provides the necessary link between the gas phase concentration a_i and the concentration of A in its adsorbed form, which is denoted as [AS]. The units for *surface concentration* are moles per unit area of catalyst surface.

Step 5 A surface reaction occurs between adsorbed species.[1] A possible reaction might be

$$AS + BS \rightleftharpoons PS + QS \tag{10.7}$$

where the products, P and Q, are formed as chemisorbed species. This surface reaction mechanism provides the first link between reactant concentrations and product concentrations.

Step 6 The products are desorbed to give the gas phase concentrations $p_i(z, \ell)$. The desorption mechanism is written as

$$PS \rightleftharpoons P + S \tag{10.8}$$

Thus the catalytic sites consumed in Step 4 are released in Step 6.

Step 7 Product species diffuse outward through the pores, the governing equation being

$$D_P \frac{d^2 p_i}{d\ell^2} = \mathscr{R}_P(a_i, b_i, \dots) \tag{10.9}$$

Step 8 Product species diffuse across the particle boundary layer:

$$\mathscr{R}_P = k_s(p_s - p) \tag{10.10}$$

The products enter the gas phase at axial position z with concentration $p(z)$. For solid-catalyzed reactions, the overall rate $\mathscr{R}(a, b, \dots)$ normally depends on product concentrations as well as on reactant concentrations. Thus $\mathscr{R}(a, b, \dots) = \mathscr{R}(a, b, \dots, p, q, \dots)$. Step 8 provides the final link needed to calculate the reaction rate for reactant A back in Step 1.

Step 9 Product species are transported to the reactor outlet, with concentrations $p(r, z)$ governed by convective diffusion or another suitable model.

---

[1] The Eley–Rideal mechanism envisions reaction of an adsorbed molecule with one still in the fluid phase. See Problem 10.13.

10.2 Intrinsic Kinetics

10.2.1 Recycle Reactors and Contact Time

The intrinsic kinetics are the preferred starting point for designing a heterogeneous reactor. They can be measured experimentally by using finely ground catalyst in a high-velocity gas stream. Doing this eliminates pore diffusion and film transfer as significant resistances to mass transfer. Thus $a_i = a_s = a$. Heat transfer resistances are also minimized so that isothermality is achieved within a catalyst particle. The recycle reactor of Figure 3.2 is an excellent means for measuring intrinsic kinetics. At high recycle rates, the system behaves as a perfect mixer. It is sometimes called a *gradientless reactor* since composition and temperature are everywhere uniform within the catalyst bed. The recycle reactor has a low net throughput, which ensures a reasonable difference between a_{in} and a_{out} so that the reaction rate can be accurately measured:

$$\mathscr{R}_A = \frac{a_{\text{out}} - a_{\text{in}}}{\bar{t}} \tag{10.11}$$

The recycle reactor is preferred to a once-through reactor. If the once-through reactor operates at the same net throughput as the recycle reactor, it will have a relatively high conversion. This high conversion achieved in a single pass through the catalyst bed gives rise to spatial gradients in composition and perhaps in temperature that confound the rate data. A once-through reactor operated at low throughputs and high conversions is called an *integral reactor*. If a once-through reactor is operated at high throughputs and low conversions, it is called a *differential reactor*. The concentration gradients vanish but the difference between a_{in} and a_{out} becomes too small for accurate measurement.

    The mean residence time has a primary influence on the yield of a homogeneous reaction. What is the corresponding variable for heterogeneous, solid-catalyzed reactions? Homogeneous reactions occur throughout the volume of the fluid phase, and the mean residence time $\bar{t} = \varepsilon V/Q$ is a ratio of the volume available for reaction divided by the net throughput. This formulation recognizes that $\bar{t} = \varepsilon V/Q$ for gas phase (nonadsorbed) molecules in a packed bed. We should no longer use the total reactor volume to calculate the mean residence time in the fluid phase since some of the volume is now filled with solid. Note, however, that pore volume should be included with the interparticle volume in calculating ε.

    Solid-catalyzed reactions occur on the catalyst surface, and the appropriate ratio compares the surface area to the net throughput:

$$\bar{t}_c = \frac{V\rho_c a_c}{Q} \tag{10.12}$$

where $\bar{t}_c$ is called the mean contact time. It has units of time per length. The catalyst density ρ_c is the catalyst mass divided by the reactor volume, and a_c is the specific surface area of the catalyst, area per unit mass. The mean contact time can also be expressed as

$$\bar{t}_c = \bar{t}\frac{\rho_c a_c}{\varepsilon} \qquad (10.13)$$

where ε is the void fraction in the bed.

The performance of isothermal, solid-catalyzed reactors should be compared on the basis of constant contact time just as isothermal homogeneous reactors are compared on the basis of constant residence time. A theory of contact time distributions—including measurement techniques using adsorbable tracers—has been developed which is the exact heterogeneous analog of residence time theory. See Nauman and Buffham.[2] This theory is needed to understand the yield performance of fluidized beds and other moving particle reactors where the catalyst density varies from point to point in the system. The theory is less important for fixed-bed reactors since $\rho_c a_c/\varepsilon$ is generally constant from point to point provided the scale of scrutiny is larger than a few particles in size. Even for fixed beds, however, it provides a useful theoretical link between true, heterogeneous rates (moles reacted per time per area of catalytic surface) and pseudohomogeneous rates (mole reacted per time per volume of fluid phase):

$$[\mathscr{R}]_{\text{pseudohomogeneous}} = \frac{\rho_c a_c}{\varepsilon}[\mathscr{R}]_{\text{heterogeneous}} \qquad (10.14)$$

In practice, the kinetics of heterogeneous reactions are usually expressed in terms of a pseudohomogeneous rate. The proper volume to use in defining such a rate is the fluid phase volume εV, and the proper $\bar{t}$ to use in Equation 10.11 is the fluid phase residence time $\bar{t} = \varepsilon V/Q$. This is a subtlety, but failure to heed it could cause a systematic error in going from intrinsic rate studies using finely ground catalyst to a scaled-up reactor design using normal pellets.

Example 10.1 A recycle reactor containing 101 gm of catalyst is used in an experimental study. The catalyst is packed into a segment of the reactor having a volume of 125 cm³. The recycle lines and pump have an additional volume of 150 cm³. The particle density of the catalyst is 1.12 gm/cm³, its internal void fraction is 0.505, and its surface area is 400 m²/gm. A gas mixture is fed to the system at 150 cm³/s. The inlet reactant concentration is 0.0016 mol/l. The outlet concentration is 0.0004 mol/l. Determine the pseudohomogeneous reaction rate, the rate per unit mass of catalyst, and the rate per unit surface area of catalyst.

Solution To find the pseudohomogeneous rate, we use Equation 10.11 with $\bar{t}$ defined as $\varepsilon V/Q$. The total free volume εV is the sum of the intra- and

[2] E. B. Nauman and B. A. Buffham, *Mixing in Continuous Flow Systems*, Wiley, New York, 1983.

interparticle void volumes plus the recycle line and pump volumes:

$$\varepsilon V = \frac{101}{1.12}(0.505) + \left(125 - \frac{101}{1.12}\right) + 150 = 230 \text{ cm}^3$$

$$\varepsilon = \frac{230}{275} = 0.836$$

$$\bar{t} = \frac{230 \text{ cm}^3}{150 \text{ cm}^3/\text{s}} = 1.53 \text{ s}$$

$$[\mathscr{R}_A]_{\text{pseudohomogeneous}} = \frac{0.0004 - 0.0016}{1.53} = -7.83 \times 10^{-4} \text{ mol l}^{-1} \text{ s}^{-1}$$

The catalyst density is defined using the total volume V. Thus

$$\rho_c = \frac{101}{125 + 150} = 0.367 \text{ gm/cm}^3 = 367 \text{ gm/l}$$

and

$$[\mathscr{R}_A]_{\text{catalyst mass}} = \frac{\varepsilon [\mathscr{R}_A]_{\text{pseudohomogeneous}}}{\rho_c} = -1.78 \times 10^{-6} \text{ mol gm}^{-1} \text{ s}^{-1}$$

$$[\mathscr{R}_A]_{\text{surface area}} = \frac{\varepsilon [\mathscr{R}_A]_{\text{pseudohomogeneous}}}{\rho_c a_c} = -4.45 \times 10^{-9} \text{ mol m}^{-2} \text{ s}^{-1}$$

Example 10.2   Suppose the reaction in the previous example is first order. Determine the pseudohomogeneous rate constant, the rate constant based on catalyst mass, and the rate constant based on catalyst surface area.

Solution   Since $\mathscr{R}_A = -ka_{\text{out}}$ for a CSTR, the rates in the previous example are just divided by the appropriate exit concentrations to obtain k. The ordinary, gas phase concentration is used for the pseudohomogeneous rate:

$$[k]_{\text{pseudohomogeneous}} = \frac{-[\mathscr{R}_A]_{\text{pseudohomogeneous}}}{a_{\text{out}}} = 1.96 \text{ s}^{-1}$$

The reactant concentration per unit mass is used for the rate based on catalyst mass:

$$[a_{\text{out}}]_{\text{catalyst mass}} = \frac{a_{\text{out}}\varepsilon}{\rho_c} = 9.11 \times 10^{-7} \text{ mol/gm}$$

and

$$[k]_{\text{catalyst mass}} = \frac{-[\mathscr{R}_A]_{\text{catalyst mass}}}{a_{\text{out}}\varepsilon/\rho_c} = 1.96 \text{ s}^{-1}$$

Similarly,

$$[k]_{\text{surface area}} = \frac{-[\mathscr{R}_A]_{\text{surface area}}}{a_{\text{out}}\varepsilon/\rho_c a_c} = \frac{-4.45 \times 10^{-9}}{2.28 \times 10^{-9}} = 1.96 \text{ s}^{-1}$$

Example 10.3 Suppose your technician "improves" the recycle reactor by reducing the recycle line and pump volume to 100 cm³. What effect will this have on a_{out} and $\mathscr{R}_A$ if all other conditions are held constant?

Solution If the reaction is truly heterogeneous, the rate per surface area of catalyst should be unaffected. Since the surface area is held constant, we expect $Q(a_{\text{out}} - a_{\text{in}})$ to be constant and thus a_{out} to be constant. (If a_{out} decreases, there is probably a homogeneous reaction occurring in parallel with the heterogeneous one. If a_{out} increases, the wall of the recycle line may have a catalytic effect.) The change does not affect the specific surface area a_c. Thus the rate based on catalyst mass will be unaffected since

$$[\mathscr{R}_A]_{\text{catalyst mass}} = a_c[\mathscr{R}_A]_{\text{surface area}}$$

The technician's change will affect the bulk catalyst density, the void fraction, and the pseudohomogeneous rate:

$$\rho_c = \frac{101}{125 + 100} = 0.449 \text{ gm/cm}^3 = 449 \text{ gm/}\ell$$

$$\varepsilon = \frac{180}{225} = 0.800$$

$$[\mathscr{R}_A]_{\text{pseudohomogeneous}} = \frac{\rho_c[\mathscr{R}_A]_{\text{catalyst mass}}}{\varepsilon} = -9.99 \times 10^{-4} \text{ mol } l^{-1} \text{ s}^{-1}$$

The fluid phase residence time also changes:

$$\bar{t} = \frac{180}{150} = 1.20 \text{ s}$$

10.2.2 Surface Mechanisms and Models

As discussed in Section 10.1, solid-catalyzed reactions have three steps that are subject to kinetic limitations (as opposed to mass transfer limitations): adsorption of reactants, surface reaction, and desorption of products. These three steps constitute the surface reaction mechanism and determine the intrinsic kinetics. The approach we take is to write down an assumed reaction mechanism and from this deduce the functional form of the rate expression. Adsorption, possibly reversible, is truly heterogeneous:

$$A(gas) + S(solid) \underset{k_a^-}{\overset{k_a^+}{\rightleftharpoons}} AS(solid) \tag{10.15}$$

Treating this reaction as an elementary step, the rate should be

$$\mathscr{R}_I = k_a^+ a[S] - k_a^-[AS] \tag{10.16}$$

where a is the usual gas phase concentration and $[S]$ and $[AS]$ are surface concentrations, in moles per unit area of catalyst surface. There are as many versions of Equation 10.16 as there are adsorbable species.

Surface reaction steps have the form

$$AS + BS \underset{k_{\mathscr{R}}^-}{\overset{k_{\mathscr{R}}^+}{\rightleftharpoons}} PS + QS \tag{10.17}$$

with rates

$$\mathscr{R}_{II} = k_{\mathscr{R}}^+[AS][BS] - k_{\mathscr{R}}^-[PS][QS] \tag{10.18}$$

Typically, solid-catalyzed reactions can be explained using a single surface reaction step although nothing prevents several versions of Equation 10.17 being written for complex reactions.

The desorption steps have the form

$$PS(solid) \underset{k_d^-}{\overset{k_d^+}{\rightleftharpoons}} P(gas) + S(solid) \tag{10.19}$$

with rate

$$\mathscr{R}_{III} = k_d^+[PS] - k_d^- p[S] \tag{10.20}$$

In a continuous-flow system at steady state, all the gas and surface concentrations will have steady state values. This requires

$$\mathscr{R}_I = \mathscr{R}_{II} = \mathscr{R}_{III} = \mathscr{R} \tag{10.21}$$

A final assumption is needed: the concentration of active catalyst sites is constant:

$$[S] + [AS] + [BS] + [PS] + [QS]\ldots = S_0 \tag{10.22}$$

where S_0 is a constant. It is now a matter of algebra to deduce the overall rate expression.

Example 10.4   Consider a solid-catalyzed reaction having the overall form

$$A(\text{gas}) \rightleftharpoons P(\text{gas}) \tag{10.23}$$

Suppose the mechanism is

$$A + S \underset{k_a^-}{\overset{k_a^+}{\rightleftharpoons}} AS$$

$$AS \underset{k_{\mathscr{R}}^-}{\overset{k_{\mathscr{R}}^+}{\rightleftharpoons}} PS \tag{10.24}$$

$$PS \underset{k_d^-}{\overset{k_d^+}{\rightleftharpoons}} P + S$$

Determine the functional form for the pseudohomogeneous, intrinsic kinetics.

Solution   Equation 10.21 gives

$$k_a^+ a[S] - k_a^-[AS] = k_{\mathscr{R}}^+[AS] - k_{\mathscr{R}}^-[PS] = k_d^+[PS] - k_d^- p[S] = \mathscr{R} \tag{10.25}$$

and Equation 10.22 gives

$$S_0 = [S] + [AS] + [PS] \tag{10.26}$$

There are four equations in this set. This allows us to solve for $\mathscr{R}$ while eliminating the unknown surface concentrations [S], [AS], and [PS]. After much algebra,

$$\mathscr{R} = \frac{(S_0 k_a^+ k_{\mathscr{R}}^+ k_d^+)a - (S_0 k_a^- k_{\mathscr{R}}^- k_d^-)p}{(k_a^- k_{\mathscr{R}}^- + k_a^- k_d^+ + k_{\mathscr{R}}^+ k_d^+) + (k_{\mathscr{R}}^+ + k_{\mathscr{R}}^- + k_d^+)k_a^+ a + (k_a^- + k_{\mathscr{R}}^+ + k_{\mathscr{R}}^-)k_d^- p} \tag{10.27}$$

Redefining constants gives the functional form

$$\mathcal{R} = \frac{ka - k'p}{1 + k_A a + k_P p} \tag{10.28}$$

The numerator of Equation 10.28 is what you would expect for a reversible first order reaction. The denominator reveals the influence of adsorption on the reaction rate. Interestingly, both A and P retard the reaction compared to simple, first order kinetics. This is typical of heterogeneous kinetics. Reaction rates are retarded by all adsorbable species, whether or not they participate in the surface reaction.

The above approach has avoided the concept of **rate-determining step**, the idea that a single step, say, adsorption, is so much slower than the other two that it determines the overall reaction rate. The concept of rate-determining step has been widely employed in the literature starting with Hougen and Wason[3]. It has the advantage of giving kinetic models with less algebra than the equal rate approach employed above. It has the disadvantage of giving less general models. It may also mislead the unwary experimentalist into thinking that surface mechanisms can be unambiguously determined from steady state experiments. This is rarely possible.

Example 10.5   Suppose that adsorption is much slower than surface reaction or desorption for the overall reaction A $\rightleftharpoons$ P. Deduce the functional form for the pseudohomogeneous, intrinsic kinetics.

Solution   The adsorption step is reversible and rate determining:

$$\mathcal{R} = k_a^+ a[S] - k_a^-[AS] \tag{10.29}$$

The reaction and equilibrium steps are assumed to be so fast compared with adsorption that they achieve equilibrium:

$$\frac{[PS]}{[AS]} = \frac{k_{\mathcal{R}}^+}{k_{\mathcal{R}}^-} = K_{\mathcal{R}} \tag{10.30}$$

$$\frac{p[S]}{[PS]} = \frac{k_d^+}{k_d^-} = K_d \tag{10.31}$$

As in the previous example,

$$S_0 = [S] + [AS] + [PS] \tag{10.32}$$

[3] O. A. Hougen and K. M. Watson, *Chemical Process Principles*, Vol. III, Wiley, New York, 1947.

Somewhat simpler algebra gives

$$\mathcal{R} = \frac{(S_0 k_a^+ K_{\mathcal{R}} K_d) a - (S_0 k_a^-) p}{K_{\mathcal{R}} K_d + (1 + K_{\mathcal{R}}) p} = \frac{ka - k'p}{1 + k_P p} \tag{10.33}$$

When the adsorption step is rate determining, A no longer retards the reaction.

Example 10.6 Repeat the previous example assuming now that the surface reaction is rate determining.

Solution Appropriate equations for the adsorption, reaction, and desorption steps are

$$\frac{[AS]}{a[S]} = K_a$$

$$\mathcal{R} = k_{\mathcal{R}}^+ [AS] - k_{\mathcal{R}}^- [PS] \tag{10.34}$$

$$\frac{p[S]}{[PS]} = K_d$$

Equation 10.32 holds as before. Elimination of [S], [AS], and [PS] gives

$$\mathcal{R} = \frac{(S_0 k_{\mathcal{R}}^+ K_a K_d) a - (S_0 k_{\mathcal{R}}^-) p}{K_d + K_a K_d a + p} = \frac{ka - k'p}{1 + k_A a + k_P p} \tag{10.35}$$

which is indistinguishable from the general form.

Example 10.7 Derive a Hougen and Watson-type model for the overall reaction 2A → P assuming that the surface reaction is the rate-determining step.

Solution A plausible mechanism for the observed reaction is

$$A + S \rightleftharpoons AS$$
$$2AS \rightarrow PS \tag{10.36}$$
$$PS \rightleftharpoons P + S$$

The corresponding "rate" equations are

$$\frac{[AS]}{a[S]} = K_a$$

$$\mathcal{R} = k_{\mathcal{R}} [AS]^2 \tag{10.37}$$

$$\frac{p[S]}{[PS]} = K_d$$

Equation 10.32 holds true as usual. Elimination of the surface concentrations gives

$$\mathscr{R} = \frac{S_0^2 K_a^2 k_{\mathscr{R}} a^2}{\left[1 + K_a a + (p/K_d)\right]^2} = \frac{ka^2}{(1 + k_A a + k_P p)^2} \tag{10.38}$$

The retardation due to adsorption appears as a square because two catalytic sites are involved in the surface reaction step.

Examples of Hougen and Watson kinetic models, which are also called Langmuir–Hinshelwood models, can be continued endlessly. See Butt (1980)[4] for a good summary. The models usually have numerators that are the same as if the surface reactions were homogeneous. The denominators reveal the heterogeneous nature of the reactions. They typically consists of factors like $(1 + k_A a + k_B b + \cdots)$. Thus consider an overall reaction of the form

$$A + B \rightleftharpoons P + Q \tag{10.39}$$

A plausible function to fit to the data would be

$$\mathscr{R} = \frac{kab - k'pq}{\left(1 + k_A a + k_B b + k_P p + k_Q q\right)^2} \tag{10.40}$$

The square in the denominator is more or less arbitrary. It was chosen because both the forward and reverse reactions *could be* second order. For typical data, a function without the square but with larger numerical values for k_A, k_B, k_P, k_Q would fit just as well. For irreversible reactions, $k' = 0$, and a power law model

$$\mathscr{R} = k a^\alpha b^\beta p^\gamma q^\sigma \tag{10.41}$$

would fit the data equally well.[5] This fact shows the futility of trying to obtain true surface mechanisms from steady state kinetics data. For aesthetic and historic reasons, most of us still prefer the Hougen and Watson forms to power law models, but mechanisms suggested by the models should not be taken too seriously.

10.3   Effectiveness Factors

Few industrial reactors operate in a region where the intrinsic kinetics are applicable. The particles are usually made so large to minimize pressure drop in

---

[4] J. B. Butt, *Reaction Kinetics and Reactor Design*, Prentice-Hall, Englewood Cliffs, NJ, 1980.
[5] See Section 4.3 for a simple approach to fitting this model to experimental data. If not too close to equilibrium, this model can even fit reversible reactions using negative exponents.

the packed bed that intraparticle resistances to heat and mass transfer become important. The superficial fluid velocity may be below the point where external film resistances can be ignored. An approach is needed for estimating actual reaction rates given the intrinsic kinetics and operating conditions within the reactor. The usual approach is to use an **effectiveness factor** η, defined as

$$\boxed{\eta = \frac{\text{Actual reaction rate}}{\text{Rate as predicted by intrinsic kinetics}}} \qquad (10.42)$$

The global design equation, Equation 10.2, for example, is modified:

$$\bar{u}_s \frac{da}{dz} = D_r \left[\frac{1}{r} \frac{\partial a}{\partial r} + \frac{\partial^2 a}{\partial r^2} \right] + \eta \varepsilon \mathscr{R}_A \qquad (10.43)$$

where $\mathscr{R}_A$ now represents the intrinsic kinetics. What is now needed is a correlation or other means for estimating η at every point in the reactor. This may be done empirically, for example, by running a single tube of what ultimately will be a multitubular reactor. However, some progress has been made in determining η from first principles. We now outline the salient results achieved to date.

10.3.1 Pore Diffusion

For isothermal reactions, the most important mass transfer limitation is pore diffusion. As a simplified model of pore diffusion, suppose the pores are long, narrow cylinders. The narrowness allows us to neglect radial gradients so that concentrations depend only on the distance ℓ from the mouth of the pore. The governing equation is

$$\mathscr{D}_A \frac{d^2 a_i}{d\ell^2} + \mathscr{R}_A = 0 \qquad (10.44)$$

with a similar equation for all other reactant and product species. The boundary conditions are

$$a_i = a_s \quad \text{at } \ell = 0 \qquad (10.45)$$

which applies to the pore mouth and

$$\frac{da_i}{d\ell} = 0 \quad \text{at} \quad \ell = \mathscr{L} \qquad (10.46)$$

which supposes either that the pore is dead ended at $\ell = \mathscr{L}$ or else has a second opening to the catalyst surface at $\ell = 2\mathscr{L}$. Given the intrinsic kinetics, $\mathscr{R}(a_i, b_i, \ldots)$, solution of Equations 10.44 (one equation for each component) is relatively straightforward. A general case requires numerical solution, but the method of approach can be illustrated by considering a single, first order reaction. Specifically, suppose $\mathscr{R}_A = -ka_i$. Then the solution to Equation 10.44, which satisfies Equations 10.45 and 10.46, is

$$\frac{a_i}{a_s} = \frac{e^{-2\mathscr{L}\sqrt{k/\mathscr{D}_A}}e^{\ell\sqrt{k/\mathscr{D}_A}} + e^{-\ell\sqrt{k/\mathscr{D}_A}}}{1 + e^{-2\mathscr{L}\sqrt{k/\mathscr{D}_A}}} \tag{10.47}$$

This gives the concentration profile in the pore, $a_i(\ell)$. The total rate of reaction within a pore must equal the net flux of reactant across the pore mouth. Suppose the pore has radius R_{pore}. Then

$$\text{Actual rate} = \pi R_{\text{pore}}^2 \left(-\mathscr{D}_A \frac{da_i}{d\ell}\bigg|_{\ell=0} \right)$$

$$= \pi R_{\text{pore}}^2 a_s \mathscr{D}_A \sqrt{\frac{k}{\mathscr{D}_A}} \frac{1 - e^{-2\mathscr{L}\sqrt{k/\mathscr{D}_A}}}{1 + e^{-2\mathscr{L}\sqrt{k/\mathscr{D}_A}}} \tag{10.48}$$

is the actual rate as affected by pore diffusion. If there were no diffusional limitation, the entire volume of the pore would have concentration a_s and the intrinsic rate would apply:

$$\text{Intrinsic rate} = \pi R_{\text{pore}}^2 \mathscr{L}(ka_s) \tag{10.49}$$

Equation 10.42 gives

$$\eta = \frac{\mathscr{D}_A}{k\mathscr{L}} \sqrt{\frac{k}{\mathscr{D}_A}} \frac{1 - e^{-2\mathscr{L}\sqrt{k/\mathscr{D}_A}}}{1 + e^{-2\mathscr{L}\sqrt{k/\mathscr{D}_A}}} \tag{10.50}$$

or

$$\boxed{\eta = \frac{\tanh\left(\mathscr{L}\sqrt{k/\mathscr{D}_A}\right)}{\mathscr{L}\sqrt{k/\mathscr{D}_A}}} \tag{10.51}$$

where the dimensionless group $\mathscr{L}\sqrt{k/\mathscr{D}_A}$ is known as the **Thiele modulus**. The Thiele modulus can be measured experimentally by comparing actual rates to intrinsic rates for a first order, isothermal, solid-catalyzed reaction. It can also be predicted from first principles given an estimate of the pore length $\mathscr{L}$. Note that the pore radius does not enter the calculations (although the effective diffusivity will be affected by pore radius when R_{pore} is very small, say, 10^{-8} m).

Example 10.8 The dehydrogenation of ethylbenzene considered in Example 3.1 uses a 3-mm spherical catalyst particle. The rate constant is 3.752 s^{-1} and the diffusivity of ethylbenzene in steam is 5×10^{-5} m^2/s under reaction condisitions. Assume that the pores are large enough in diameter so that this bulk diffusivity is applicable. What is the lowest isothermal effectiveness factor that you would expect for this catalyst?

Solution The lowest η corresponds to the largest value for $\mathscr{L}$. The average value of $\mathscr{L}$ cannot realistically exceed 3 mm, and 1.5 mm is probably a better estimate. Thus

$$\left(\mathscr{L} \sqrt{\frac{k}{\mathscr{D}_A}} \right)_{max} = 0.41$$

and

$$\eta_{min} = \frac{\tanh 0.41}{0.41} = 0.95$$

Many theoretical embellishments have been made to the basic model of pore diffusion as presented here. Effectiveness factors have been derived for reaction orders other than first and for Hougen and Watson kinetics. Shape and tortuosity factors have been introduced to treat pores other than the idealized cylinders considered here. The Knudsen diffusivity or a combination of Knudsen and bulk diffusivities has been used for very small pores. While these studies have theoretical importance and may help explain some observations, they are not yet well enough developed for predictive use. Our knowledge of the internal structure of a porous catalyst is still rather rudimentary and imposes a basic limitation on theoretical predictions. We give only a brief account of Knudsen diffusion.

In bulk diffusion, the predominant interaction of molecules is with other molecules in the fluid phase. This is the ordinary kind of diffusion, and we have denoted the corresponding diffusivity as $\mathscr{D}_A$. At low gas densities and in small-diameter pores, the mean free path of molecules may become comparable to the pore diameter. Then the predominant interaction is with the walls of the pore, and diffusion within a pore is governed by the **Knudsen diffusivity** $\mathscr{D}_K$. This diffusivity is predicted by the kinetic theory of gases to be

$$\mathscr{D}_K = \frac{2 R_{pore}}{3} \sqrt{\frac{8 R_g T}{\pi M_A}} \tag{10.52}$$

where M_A is the molecular weight of the diffusing species.

Example 10.9 Repeat Example 10.8 assuming a pore diameter of 200 Å = 2 × 10⁻⁸ m. The reaction temperature is 700°C.

Solution

$$\mathscr{D}_K = \frac{2 \times 10^8 \text{ m}}{3} \sqrt{\frac{8}{\pi} \frac{8.717 \text{ J}}{\text{mol K}} \frac{973 \text{ K}}{0.108 \text{ kg/mol}}} = 3 \times 10^{-6} \text{ m}^2/\text{s}$$

This is a factor of 17 less than the bulk diffusivity. Thus

$$\left(\mathscr{L} \sqrt{\frac{k}{\mathscr{D}_K}} \right)_{\text{max}} = 1.68$$

and

$$\eta_{\text{min}} = \frac{\tanh 1.68}{1.68} = 0.56$$

Example 10.10 How fine would you have to grind the ethylbenzene catalyst for laboratory kinetic studies to give the intrinsic kinetics?

Solution Assume $\eta = 0.95$ represents an adequate approach to the intrinsic kinetics. Thus, by Example 10.8, $\mathscr{L}\sqrt{k/\mathscr{D}_K} = 0.41$. Since k and $\mathscr{D}_K$ are known, we obtain

$$\mathscr{L} = \frac{0.41}{\sqrt{3.752/3 \times 10^{-6}}} = 3.7 \times 10^{-4} \text{ m} = 370 \ \mu\text{m}$$

This is a perfectly feasible size for use in a laboratory reactor. Due to pressure drop limitations, it is infeasibly small for a full-scale packed bed. However, even smaller catalyst particles, $d_p \approx 60 \ \mu\text{m}$, are used in fluidized-bed reactors. See Chapter 11. For such small particles we can assume $\eta = 1$ even for the 25-Å pore diameters found in some cracking catalysts.

When the Knudsen and bulk diffusivities are significantly different, η is determined by the smaller of the two. The pore diameters for most commercial catalysts are in the range 10 to 1000 Å. At typical operating temperature of about 700 K, this gives Knudsen diffusivities in the range of 10^{-6} to 10^{-8} m²/s. Bulk diffusivities at atmospheric pressure will usually be in the range of 10^{-4} to 10^{-6} m²/s. The Knudsen diffusivity is independent of pressure but the bulk diffusivity varies approximately as P^{-1}. Thus Knudsen diffusion will determine η at low to moderate pressures, but the bulk diffusivity can be limiting at high pressures. When the two diffusivities are commensurate, the combined effect is actually worse than either acting alone. The following equation is adequate for most purposes:

$$\frac{1}{\mathscr{D}_{net}} = \frac{1}{\mathscr{D}_A} + \frac{1}{\mathscr{D}_K} \tag{10.53}$$

A more rigorous result together with theoretical justification is available in the literature.[6]

10.3.2 Interphase Mass Transfer

The pore diffusion model provides the necessary link between concentrations at the surface of the catalyst and those within the pores. For a first order reaction,

$$\mathscr{R}_A = \text{Actual rate} = -\left[\frac{\tanh\left(\mathscr{L}\sqrt{k/\mathscr{D}_A}\right)}{\mathscr{L}\sqrt{k/\mathscr{D}_A}}\right] ka_s \tag{10.54}$$

We now link a_s to the bulk concentration a. Using Equation 10.3, the actual rate must be equal to the mass transfer rate from the bulk fluid to the external surface of the catalyst particle:

$$\mathscr{R}_A = k_s(a_s - a) \tag{10.55}$$

Equations 10.54 and 10.55 allow a_s to be eliminated. The result is

$$\mathscr{R}_A = -\eta ka \tag{10.56}$$

where

$$\eta = \frac{k_s \tanh\left(\mathscr{L}\sqrt{k/\mathscr{D}_A}\right)}{k_s \mathscr{L}\sqrt{k/\mathscr{D}_A} + k \tanh\left(\mathscr{L}\sqrt{k/\mathscr{D}_A}\right)} \tag{10.57}$$

Equation 10.56 is the appropriate pseudohomogeneous rate to use in global models such as Equation 10.2. The term $-ka$ is what the kinetics would be if there were no mass transfer limitations. In this example, the effectiveness η takes two forms of mass transfer resistance into account.

More complicated situations involving external mass transfer can be treated theoretically. Unfortunately, a priori predictions are not yet possible even for the simple cases considered here. In practice, effectiveness factors are determined experimentally.

[6]L. B. Rothfield, *AIChE J.*, **9**, 19 (1963).

10.3.3 Nonisothermal Effectiveness

Catalyst pellets often operate at temperatures that are substantially different from the bulk gas temperature. Large heats of reaction and the low thermal conductivities typical of catalyst supports make temperature gradients possible in all but the finely ground powders used for intrinsic kinetic studies. There may also be a film resistance to heat transfer at the external surface of the catalyst. When the reaction is exothermic, internal temperatures will exceed the bulk gas temperature. For endothermic reactions, internal temperatures will be less than the bulk gas temperature.

Equation 10.42 continues to define the effectiveness factor; but, with exothermic reactions, reaction rates inside the pellet will be higher than would be predicted using the bulk gas temperature. Thus $\eta > 1$ is expected for exothermic reactions in the absence of mass diffusion limitations. (The case $\eta > 1$ is also possible for isothermal reactions given adequately weird kinetics.) With systems having low thermal conductivities but high molecular diffusivities and high heats of reaction, the actual rate can be an order of magnitude higher than the intrinsic rate. Thus $\eta \gg 1$ is theoretically possible for exothermic reactions. When mass transfer limitations emerge, concentrations will be lower inside the pellet than outside. The decreased concentration may have a larger effect on the rate than the increased temperature. Thus $\eta < 1$ is also possible for an exothermic reaction. For an endothermic reaction, $\eta < 1$ (except perhaps for some esoteric kinetic schemes).

The theory of nonisothermal effectiveness is sufficiently well advanced to allow order-of-magnitude estimates for η. The analysis requires simultaneous solutions for the concentration and temperature profiles within a pellet. The solutions are necessarily numerical. Solutions are feasible for actual pellet shapes (such as cylinders) but are significantly easier for spherical pellets since this allows a one-dimensional form for the energy equation:

$$\lambda_{\text{eff}} \left(\frac{d^2T}{dr_p^2} + \frac{2}{r_p} \frac{dT}{dr_p} \right) = -\Delta H_{\mathscr{R}} \mathscr{R} \tag{10.58}$$

where r_p is the radial coordinate within a pellet and λ_{eff} is an effective thermal conductivity. The boundary conditions associated with Equation 10.58 are $T = T_s$ at the external surface and $dT/dr_p = 0$ at the center of the pellet.

Equation 10.58 must be solved simultaneously with material balance equations. The pore diffusion model of Section 10.3.1 is inappropriate for this purpose since the pores may not originate at the surface of the catalyst pellet. Instead, the diffusion-controlling pores of Equation 10.44 might originate at the walls of macropores that are deep inside the pellet. Figure 10.2 depicts a possible pore structure having a bimodal pore size distribution. The large macropores pipe reactants into the interior with little change in concentration compared to a_s. The micropores represent the diffusional resistance in the pellet, and it is their length that is measured by the Thiele modulus $\mathscr{L}\sqrt{k/\mathscr{D}_A}$. To deal with simultaneous heat

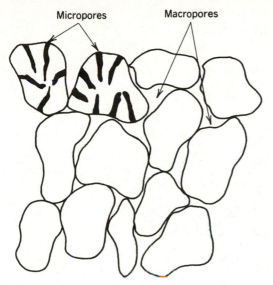

Micropores Macropores

Figure 10.2 Possible bimodal pore structure for a catalyst pellet.

and mass transfer limitations, it is necessary to use a different model of the diffusion process. This model is

$$\mathscr{D}_{\text{eff}}\left(\frac{d^2a}{dr_p^2} + \frac{2}{r_p}\frac{da}{dr_p}\right) = \mathscr{R}_A \qquad (10.59)$$

subject to the boundary conditions that $a = a_s$ at the external surface and that $da/dr_p = 0$ at the center. Equation 10.59 is obviously compatible with Equation 10.58. However, the reader should recognize that it will predict concentration profiles $a(r_p)$ that are quite different than one would expect for a bimodal pore structure. This is a price paid for computational simplicity.

Numerical solutions to Equations 10.58 and 10.59 have been given by Weisz and Hicks[7] for the case of a first order, irreversible reaction. The solution for η depends on three dimensionless groups:

An Arrhenius number:

$$\frac{E}{R_g T_s}$$

[7]P. B. Weisz and J. S. Hicks, *Chem. Eng. Sci.*, **17**, 265 (1962).

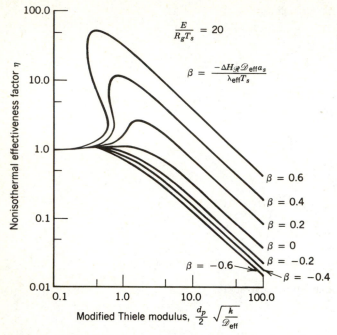

Figure 10.3 Effectiveness factors for nonisothermal, first order reactions in spherical pellets. [Adopted from P. B. Weisz and J. S. Hicks, *Chem. Eng. Sci.*, **17**, 265 (1962).]

A heat generation number:

$$\frac{(-\Delta H_{\mathscr{R}})\mathscr{D}_{eff}a_s}{\lambda_{eff}T_s}$$

A modified Thiele modulus:

$$\frac{d_p}{2}\sqrt{\frac{k}{\mathscr{D}_{eff}}}$$

Figure 10.3 gives their results for an Arrhenius number of 20. Given plausible estimates for λ_{eff} and $\mathscr{D}_{eff}$, the magnitude of η can be calculated. For the special case of $\Delta H_{\mathscr{R}} = 0$, Equation 10.59 is an alternate to the pore diffusion model for isothermal effectiveness. It predicts rather different results. For example, suppose $(d_p/2)\sqrt{k/\mathscr{D}_{eff}} = \mathscr{L}\sqrt{k/\mathscr{D}_A} = 1$. Then Equation 10.51 gives $\eta = 0.76$ while Equation 10.59 gives $\eta = 0.99$. The lesson from this is that $\mathscr{D}_{eff}$ and $\mathscr{D}_A$ are fundamentally different quantities and assume different values when applied to the same physical system.

10.3.4 Deactivation

The definition of effectiveness factor, Equation 10.42, can be expanded to account for deactivation processes, which decrease the activity of a catalyst as a function of time. In this context, the intrinsic kinetics in Equation 10.42 should be determined using a new, freshly prepared catalytic surface. The activity of the surface and thus the actual rate of reaction will change with time so that $\eta = \eta(\theta)$, where θ is the time the catalyst has been "on stream." It is necessary to consider deactivation processes in reactor designs since they can have a marked effect on process economics and even operability. The catalyst in a fluidized-bed cracker, for example, deactivates in seconds; and process feasibility requires continuous regeneration.

Deactivation is usually classified as being either *physical* or *chemical*. Physical deactivation includes such factors as blocking of pores by entrained solids, loss of active surface by unfavorable migration of catalytic metal atoms on or into the surface of the support, loss of surface by sintering, and loss of surface by physical adsorption of impurities. Chemical deactivation includes loss of active sites through chemisorption of impurities and either loss of sites or pore blockage due to coking. Many of these deactivation processes decrease the number of active sites S_0. Others add new, diffusional resistances to the transport of reactants to and from the internal surface. In either case, they cause a multiplicative reduction in observed reaction rates:

$$[\text{Actual rate}] = \eta(\theta)[\text{Intrinsic rate of fresh catalyst}] \tag{10.60}$$

Significant progress has been made in developing theoretical expressions for $\eta(\theta)$ for deactivation processes such as coking. However, practical results still require experimental determination. A life test should be included in any catalyst development effort. The data from this test will allow η to be calculated as a function of time on stream.

It should be noted that many of the deactivation processes are potentially reversible. Physical adsorption occurs whenever there is a gas phase impurity that is below its critical point, and it is immediately reversible just by eliminating the impurity from the feed stream. Reversal of deactivation due to chemisorption requires a chemical treatment to remove the chemically bonded poison from the catalytic surface. Decoking is usually accomplished by a high-temperature oxidation. Except for physical adsorption, however, it is rare that catalytic activity can ever be restored completely. Even with periodic or continuous reactivation, the catalyst will have a finite life that must be considered in the overall process economics.

10.4 Biochemical Reactions

Most biochemical reactions involve catalysts called enzymes. Mass transport to and from the catalyst is usually done in the liquid phase. Reaction rates, in moles per volume per time, are several orders of magnitude lower than rates typical of solid–catalyzed gas reactions. These lower rates are a consequence of slower diffusion in the liquid phase and of the generally lower operating temperatures. The great majority of biochemical reactions have temperature optima in the range 4 to 40°C. Also, many of the reactions require very specific molecular orientations before they can proceed. As compensation for the lower rates, biochemical reactions have wonderful selectivities. They usually give pure compounds with specific structural isomerization.

10.4.1 Enzymes and Michaelis – Menten Kinetics

Enzymes are protein molecules that catalyze biochemical reactions. The enzyme contains one or more active sites. These are often associated with metal atoms that are incorporated in the structure of the protein. The enzyme usually forms a colloidal suspension in a liquid medium, but it may also be bound to a surface (an *immobilized enzyme*). The reactant molecule will be absorbed onto the active site; it reacts on the surface of the protein, and the product is desorbed. These steps give rise to kinetic rate expressions rather similar to those encountered in gas–solid heterogeneous systems.

Suppose the reaction

$$A \rightarrow P$$

occurs using an enzyme as the catalyst. We postulate the following reaction mechanism:

$$A + E \rightleftharpoons AE \qquad \frac{[AE]}{a[E]} = K \tag{10.61}$$

$$AE \rightarrow P + E \qquad \mathscr{R} = k[AE] \tag{10.62}$$

where E denotes the active site and AE represents the adsorbed complex. This mechanism is somewhat different than those used for gas–solid catalysis since there is no explicit desorption step. One can imagine that the reaction

$$PS \rightarrow P + E \tag{10.63}$$

exists but is fast and irreversible so that desorption is essentially instantaneous. As for gas–solid reactions, we suppose the total concentration of sites is constant:

$$[AE] + [E] = E_0 \tag{10.64}$$

Substitution for [AE] and [E] gives

$$\frac{\mathscr{R}}{k} + \frac{[AE]}{aK} = \frac{\mathscr{R}}{k} + \frac{\mathscr{R}}{kKa} = E_0 \tag{10.65}$$

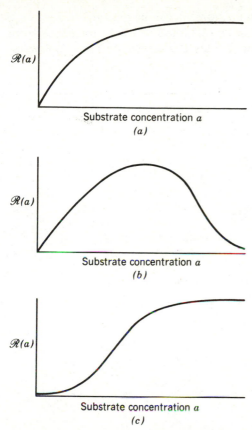

Figure 10.4 Effects of substrate (reactant) concentration on rate for enzymatic reactions: (a) simple Michaelis-Menten kinetics; (b) substrate inhibition; (c) activation.

and

$$\mathcal{R} = \frac{E_0 k a}{1/K + a} = \frac{k'a}{1 + k_A a} \qquad \textbf{(10.66)}$$

This form is equivalent to Hougan and Watson (or Langmuir–Hinshelwood) kinetics for the reaction A → P with negligible adsorption of P. In the biochemical field, Equation 10.66 is called ***Michaelis–Menten kinetics*** and was in fact derived considerably earlier than the equivalent example for gas–solid catalysis.

The reactant A is commonly called the ***substrate***. Figure 10.4a shows the effect of substrate concentration a on the reaction rate according to simple Michaelis–Menten kinetics. Enzymatic reactions frequently exhibit a phenomenon known as ***substrate inhibition*** where the reaction rate reaches a maximum and subsequently declines as illustrated in Figure 10.4b. Enzymatic reactions can also exhibit ***substrate activation*** as illustrated at the sigmoidal rate dependence in

Figure 10.4c. Inhibition and activation phenomena can be modeled with variants of Michaelis–Menten kinetics involving two or more substrate molecules being adsorbed at a single site or enzyme molecules having two or more interacting sites.

Example 10.11 Suppose an enzymatic reaction has the following mechanism:

$$A + E \rightleftharpoons AE \qquad \frac{[AE]}{a[E]} = K_I \tag{10.67}$$

$$A + AE \rightleftharpoons A^2E \qquad \frac{[A^2E]}{a[AE]} = K_{II} \tag{10.68}$$

$$AE \rightarrow P + E \qquad \mathscr{R} = k[AE] \tag{10.69}$$

Determine the functional form of the rate equation.

Solution The total concentration of active sites is

$$[A^2E] + [AE] + [E] = E_0 \tag{10.70}$$

Eliminating the unknown surface concentration gives

$$\mathscr{R} = \frac{kE_0 a}{(1/K_I) + a + K_{II}a^2} \tag{10.71}$$

which has $\mathscr{R}(0) = 0$, $\mathscr{R}' = 0$, a maximum at $a = (2K_{II})^{-1/2}$, and $\mathscr{R}(\infty) = 0$. The assumed mechanism involves a first order surface reaction with inhibition of the reaction if a second substrate molecule is adsorbed. A similar functional form for $\mathscr{R}(a)$ can be obtained by assuming a second order, dual site model. As in the case of gas–solid heterogeneous catalysis, it is not possible to verify reaction mechanisms simply by steady state rate measurements.

Example 10.12 Suppose the reaction mechanism is now

$$A + E \rightleftharpoons AE \qquad \frac{[AE]}{a[E]} = K_I \tag{10.72}$$

$$A + AE \rightleftharpoons A^2E \qquad \frac{[A^2E]}{a[AE]} = K_{II} \tag{10.73}$$

$$A^2E \rightarrow AE + P + E \qquad \mathscr{R}_{II} = k_{II}[A^2E] \tag{10.74}$$

$$AE \rightarrow P + E \qquad \mathscr{R}_I = k_I[AE] \tag{10.75}$$

Determine the functional form for $\mathscr{R}_P(a)$.

Solution This case is somewhat different than any we considered in gas–solid catalysis since product can be formed by two distinct reactions. The overall rate is

$$\mathscr{R}_P = \mathscr{R}_{\mathrm{I}} + \mathscr{R}_{\mathrm{II}} = k_{\mathrm{II}}[A^2E] + k_{\mathrm{I}}[AE] \tag{10.76}$$

Equation 10.70 holds as in the previous example. Eliminating the unknown surface concentrations gives

$$\mathscr{R}_P = \frac{E_0\left(k_{\mathrm{I}}a + k_{\mathrm{II}}K_{\mathrm{II}}a^2\right)}{(1/K_{\mathrm{I}}) + a + K_{\mathrm{II}}a^2} \tag{10.77}$$

As written, this rate equation exhibits neither inhibition nor activation. However, the inhibition of the previous example is found if $k_{\mathrm{II}} = 0$, and activation is found if $k_{\mathrm{I}} = 0$. In the latter case,

$$\mathscr{R}_P = \frac{E_0 k_{\mathrm{II}} K_{\mathrm{II}} a^2}{(1/K_{\mathrm{I}}) + a + K_{\mathrm{II}}a^2} \tag{10.78}$$

which gives $\mathscr{R}_P(0) = 0$, $\mathscr{R}_P'(0) = 0$, and $\mathscr{R}_P(\infty) = \mathscr{R}_{\max} = E_0 k_{\mathrm{II}}$.

More complicated situations involving multiple substrates or inhibitors which can occupy enzyme sites but do not react can be treated in the framework of Michaelis–Menten kinetics. Mass transfer limitations often arise in biochemical reactions, particularly with intercellular reactions where a substrate has to cross cell walls. Effectiveness factors can be defined but have little utility since there is no standard geometry, such as pore, which causes the mass transfer limitations. The temperature dependence of enzymatic reactions is usually modeled with an Arrhenius form although the operating window is often so narrow that an exponential dependence, $k = k_0 \exp(\alpha T)$, would fit the data equally well. Most enzymes deactivate rapidly at temperatures of 50°C although some can function at temperatures above 100°C. Typically, deactivation is irreversible.

10.4.2 Fermentations

Fermentations represent a form of heterogeneous catalysis where the catalyst is a living organism. Three types of microbes are commonly used for fermentations: bacteria, yeasts, and molds. Bacteria and yeasts are single celled. Molds are multicelled but individual plants are still microscopic in size. Cells derived from higher order plants and animals can also be fermented. The product of a fermentation may be the cells themselves as when yeast is grown for baking, or it may be a metabolic product such as alcohol. It is the latter situation that conforms to the traditional definition of catalysis, particularly when the fermentation is continuous with recycle of live cells.

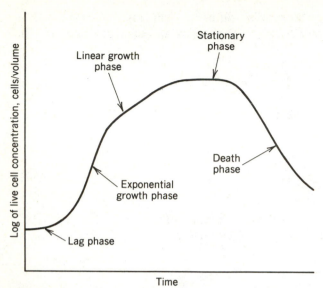

Figure 10.5   Typical growth phases in a batch fermentation.

Most fermentations are batch, and the time required to grow the necessary cell mass occupies a substantial portion of the batch cycle time. The chemistry of cell growth is extremely complex and, even in the simplest living cells, is only partially understood. The nutrient mixture must include a carbon source, which is typically a sugar such as glucose; sources for nitrogen, phosphorus, potassium; and a large variety of trace elements. Oxygen may be required directly (*aerobic* fermentations) or may be obtained from a carbohydrate which also serves as the carbon source (*anaerobic* fermentations). Some bacteria, called *facultative anaerobes*, utilize molecular oxygen when available but can survive and grow without it. For others, called *strict anaerobes*, oxygen is a poison.

A batch[8] fermentation begins with an initial charge of cells called an *inoculum*. Growth follows the typical curve illustrated in Figure 10.5. During the initial lag phase, cells are adjusting themselves to the new environment. Most microbes, and particularly bacteria, are extremely adaptable and can utilize a great variety of carbon sources. However, one or more unique enzymes are generally required for each source and these must be manufactured by the cell in response to the new environment. These are called *induced enzymes.* The induction period will be short if the fermentation medium is similar to that used in culturing the inoculum. If the chemistry is dramatically different, appreciable cell death may occur before the surviving cells have retooled for the new environment. The manufacture of desired products is negligible during the lag phase.

[8]Aerobic fermentations are semibatch with respect to oxygen that is supplied continuously.

Exponential growth occurs after cell metabolisms have adjusted and before a key nutrient becomes limiting or toxic products accumulate. In the exponential growth phase, the total cell mass will increase by a fixed percentage during each time interval, typically doubling every few hours. Ultimately, however, the growth rate must slow and stop.

A linear growth phase can occur in semibatch fermentations when some key reactant is supplied at a fixed rate. Typically, the limiting reactant is oxygen, and the supply rate is dictated by mass transfer considerations. In *fed-batch* fermentations, the limiting reactant will be the substrate (e.g., glucose) which is fed to the system at a fixed rate. Ultimately, cell mass achieves a maximum and the culture enters a stationary or maintenance phase. A stationary population can be sustained using continuous culture techniques as described below, but the stationary phase is typically rather brief in batch fermentations. The stationary phase is followed by a phase where cells die or sporulate. The number of viable cells usually follows an exponential decay curve during this period. The cells can reduce their nutritional requirements in time of stress, and surviving cells will cannibalize the bodies of cells that *lyse* (rupture).

Models for batch growth curves can be constructed by assuming mechanisms for cell births and deaths. These must be reasonably complicated if the model is to account for a lag phase and for a prolonged stationary phase.

Example 10.13   A simple model for births and deaths is

$$\frac{dc}{dt} = k_b ac - k_d c \tag{10.79}$$

where c is the concentration of live cells. This equation assumes that the birth rate is proportional to the existing cell population and to the substrate concentration a. Spontaneous deaths occur at a rate proportional to the cell concentration. Assume the nutrient is consumed at a constant rate per living cell until the supply is exhausted:

$$a = a_0 \left[1 - \beta \int_0^t c(\alpha) \, d\alpha \right] \quad \text{for} \quad \beta \int_0^t c(\alpha) \, d\alpha < 1$$

$$= 0 \quad \text{for} \quad \beta \int_0^t c(\alpha) \, d\alpha \geq 1 \tag{10.80}$$

Solve this set of equations to obtain the batch growth curve $c(t)$.

Solution   Equation 10.79 is an integrodifferential equation due to the manner in which the substrate concentration is obtained. Growth models frequently contain integrals as a means of incorporating the history of the batch into the model. The integral can be eliminated by differentiation, but the resulting set of ODEs will not be analytically solvable. A numerical solution is

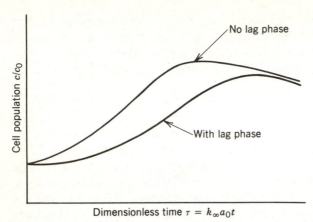

Figure 10.6 Simulation of cell growth in a batch fermentor.

most easily achieved by retaining the integrodifferential form. First order difference approximations give

$$c_{j+1} = c_j + \Delta t \left(k_b c_j a_j - k_d c_j \right) \tag{10.81}$$

where

$$a_j = a_0 \left(1 - \beta \sum_{i=1}^{j} c_i \, \Delta t_i \right) \tag{10.82}$$

Results for the case $k_d/(k_b a_0) = 0.1$, $\beta c_0 k_b a_0 = 0.05$ are illustrated in Figure 10.6. Note that $k_b a_0$ has dimensions of reciprocal time so that $\tau = k_b a_0 t$ is dimensionless.

Example 10.14 Modify the above model to simulate a lag phase.

Solution A simple way of introducing a lag phase is to allow k_b to vary with time. Suppose

$$k_b = k_\infty (1 - e^{-\alpha t}) \tag{10.83}$$

so that the birth rate proportionality constant gradually evolves to a final value as enzymes are induced in response to the new environment. Figure 10.6 illustrates this model for $k_d/(k_\infty a_0) = 0.1$, $\beta c_0 k_\infty a_0 = 0.05$, $\alpha/(k_\infty a_0) = 1$.

The major products of a fermentation result from the microbe's energy metabolism. These products typically include carbon dioxide; water in the case of aerobic ferementations; and small organic molecules such as ethanol, acetic acid, lactic acid, and 2,3-butanediol. These products are produced when the cells are growing and, possibly at a reduced rate, in the stationary phase. The organic products are excreted through the cell walls and to some extent are toxic to the organisms that produce them. Thus a batch fermentation can be limited by accumulation of products as well as by depletion of the substrate. The specific mix of organic products, even for a fixed substrate and organism, depends on factors such as the phase of growth in the batch cycle, the pH of the medium, and whether or not molecular oxygen is available. For example, the fermentation of glucose by *Bacillus polymxya* gives a mixture of ethanol, acetic acid, lactic acid, and butanediol; but butanediol can be obtained in almost stoichiometric yield;

$$C_6H_6(OH)_6 + \tfrac{1}{2}O_2 \rightarrow C_4H_8(OH)_2 + 2CO_2 + H_2O \qquad \text{(10.84)}$$

late in the batch cycle with an acidic medium and aeration.

Occasionally, optimal growing conditions are suggested just by the overall stoichiometry and energetics of the metabolic reactions. For example, ethanol can be produced anaerobically:

$$C_6H_6(OH)_6 \rightarrow 2C_2H_5OH + 2CO_2 \qquad \text{(10.85)}$$

This reaction has a negative change in free energy so it can support microbial growth. An anaerobic fermentation for butanediol releases free hydrogen and has less favorable energetics. Thus one might expect more alcohol than butanediol under strictly anaerobic conditions. With oxygen available, the butanediol route produces more energy and is favored although, with enough oxygen, the reaction may go all the way to CO_2 and water. More detailed predictions require kinetic models of the actual metabolic pathways in a cell. The energy-producing pathways are now known to a large extent, and some models are becoming possible.

Secondary products of fermentation include complex organic molecules such as vitamins and antibiotics. For the most part these are obtained in aerobic fermentations, the products are usually excreted, and production occurs late in the growth phase or in the stationary phase. Such products are not directly associated with growth of the cells and are called *secondary metabolites*.

For products of the basic energy metabolism and for secondary metabolic products, continuous fermentation is an attractive possibility. Figure 10.7 gives a possible flow sheet for a fermentor with recycle of live cells. The fermentation is started in a batch mode but is converted to continuous operation when the desired cell mass is achieved. The feed stream to the fermentor in continuous operation lacks some key nutrient so that further cell growth is restricted, but it does contain the carbon source needed for the basic energy metabolism and whatever additional components are necessary for the desired product. The cells go into a special form of stationary phase where there is no net growth but metabolism continues. It is often possible to maintain this state for weeks or months and to achieve high volumetric productivities. Yields also tend to be good

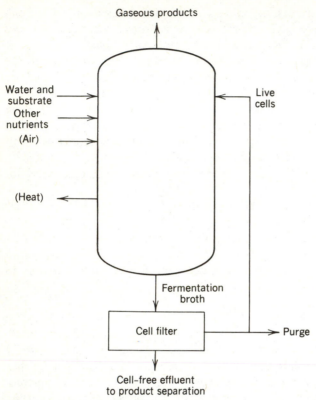

Figure 10.7 Continuous fermentor with recycle of live
cells.

since there is no consumption of substrate to increase cell mass once continuous
operation begins. This form of fermentation is obviously limited to excreted
products that are not growth associated.

The *activated sludge* process used for wastewater treatment is a varient of
the cell recycle reactor shown in Figure 10.7. In activated sludge processes, the
feedstream contains all necessary nutrients so that there is continuous growth of
biomass. However, the growth rate is insufficient to maintain the desired cell
density in the reactor. A portion of the exiting cells is separated and returned. If
the activated sludge achieves steady state, the net outflow of cells exactly matches
the growth rate, but the cell density is much higher than would be possible
without recycle of cells.

Cell harvesting is needed for nonexcreted products or when the cell itself is
the desired product. Continuous fermentation is still a possibility but the feed
stream must now contain all necessary nutrients for growth. The growth rate of
cells adjusts so that it exactly matches their washout rate.

Example 10.15 Analyze the steady state operation of a CSTR used for the production of biomass. Assume the birth and death kinetics of Example 10.13.

Solution A cell balance has the same form as for a molecular reactant:

$$Qc_{in} + V\mathcal{R}_c = Qc_{out} \tag{10.86}$$

For the problem at hand, $c_{in} = 0$ since there is no recycle of live cells. Using the kinetics of Example 10.13,

$$\bar{t}(k_b a_{out} c_{out} - k_d c_{out}) = c_{out} \tag{10.87}$$

A component balance on the substrate A is also needed

$$a_{in} - \beta \bar{t} c_{out} = a_{out} \tag{10.88}$$

where we have used $\mathcal{R}_A = -\beta c_{out}$ in accordance with Equation 10.80. Substituting Equation 10.88 into Equation 10.87 gives

$$\bar{t}(k_b a_{out} c_{out} - k_b \beta \bar{t} c_{out}^2) = c_{out} \tag{10.89}$$

This equation has two roots, one of which is $c_{out} = 0$. The case of $c_{out} = 0$, $a_{in} = a_{out}$ is always one of the steady state solutions to cell growth in a CSTR. The present kinetics admit another solution:

$$c_{out} = \frac{\bar{t}k_b a_{in} - 1 - \bar{t}k_d}{\bar{t}k_b} \tag{10.90}$$

Depending on the parameters, this result appears to allow negative values for c_{out}; but only $c_{out} \geq 0$ is physically realistic. Applying the lower limit $c_{out} = 0$ to Equation 10.90 gives a minimum value for the mean residence time $\bar{t}_{min}$ below which the cells will completely wash out:

$$\bar{t}_{min} = \frac{1}{a_{in}k_b - k_d} \tag{10.91}$$

If $\bar{t} > \bar{t}_{min}$, a steady state with $c_{out} > 0$ is possible but not inevitable.

Example 10.16 Determine the flow rate Q_{opt} that maximizes the production of biomass for the CSTR of Example 10.15.

Solution We begin by showing that an interior maximum exists, $0 < Q_{opt} < \infty$. Biomass production Qc_{out} is zero when $Q = 0$. It is also zero at very high flow rates, $Q > V/\bar{t}_{min}$, since complete washout will occur. There are some intermediate flow rates, $0 < Q < V/\bar{t}_{min}$ where $Qc_{out} > 0$. Thus there is an intermediate optimum for Qc_{out} with Q in the indicated range.

From Equation 10.90,

$$Qc_{out} = Q\left[\frac{\bar{t}k_b a_{in} - 1 - \bar{t}k_d}{\bar{t}k_b}\right] = \left(a_{in} - \frac{k_d}{k_b}\right)Q - \frac{Q^2}{Vk_b} \tag{10.92}$$

Setting $d(Qc_{out})/dQ = 0$ gives

$$\left(a_{in} - \frac{k_d}{k_b}\right) - \frac{2Q_{opt}}{Vk_b} = 0 \tag{10.93}$$

or

$$Q_{opt} = \frac{V}{\bar{t}_{opt}} = \frac{V}{2}(a_{in}k_b - k_d) \tag{10.94}$$

The literature of biochemical engineering refers to a stirred tank reactor as a *chemostat*, this being a device in which the chemical concentrations are static. This statement supposes that a chemostat can indeed achieve a steady state in continuous operation. The dynamic behavior of a chemostat can be determined by the methods of Chapter 8. Microbial kinetics are very complex, particularly when there are two or more microbial species in competition. Multiple steady states are always possible, and oscillatory behavior is quite common.

Continuous fermentations have been designed for the production of single-cell protein, but most nonexcreted products are made in batch fermentations. In particular, proteins such as inteferon and human insulin, made possible by recombinant DNA techniques, are produced in batch fermentations that typically use a genetically altered variety of *Escherichia coli*. As the production volume increases, continuous fermentation becomes more attractive.

Suggestions for Further Reading

A recent and comprehensive treatment of heterogeneous catalysis and reactor design is given by:

H. S. Lee, *Heterogeneous Reactor Design*, Butterworth, Boston, 1984.

The more general books on chemical engineering kinetics and reactor design invariably include some treatment of gas solid–catalyzed reactions. Three texts where this is done particularly well are:

J. J. Carberry, *Chemical and Catalytic Reaction Engineering*, McGraw-Hill, New York, 1976.

J. M. Smith, *Chemical Engineering Kinetics*, 3rd ed., McGraw-Hill, New York, 1983.

J. B. Butt, *Reaction Kinetics and Reactor Design*, Prentice-Hall, Englewood Cliffs, NJ, 1980.

A good introduction to enzyme reactions and fermentations is given in:

J. E. Bailey and D. F. Ollis, *Biochemical Engineering Fundamentals*, McGraw-Hill, New York, 1977.

Interactions between thermodynamics and kinetics are often ignored, and this book has been no exception. The following article will help remedy this omission:

R. Shinnar and C. A. Feng, "Structure of Complex Catalytic Reactions: Thermodynamic Constraints in Kinetic Modeling and Catalyst Evaluation," *I & EC Fundam.*, **24**, 153–170 (1985).

Problems

10.1 The ethylbenzene dehydrogenation catalyst of Example 3.1 has a first order rate constant of 3.752 s^{-1} at 700°C. Could Wenner and Dybdal[9] have used the same catalyst? They reported

$$k = 12,600\, e^{-19,800/T}$$

where $\mathscr{R} = kP_{EB}$ with k in lb-moles per hr per atm per lb of catalyst and T in degrees Rankine. The bulk density of the catalyst is 90 lb/ft^3 and the void fraction is 0.4.

10.2 An observed, gas solid–catalyzed reaction is

$$A + B \rightarrow P$$

Suppose the surface mechanism is

$$A + S \rightleftharpoons AS \qquad K_I = \frac{[AS]}{a[S]} \qquad \text{(I)}$$

$$B + S \rightleftharpoons BS \qquad K_{II} = \frac{[BS]}{b[S]} \qquad \text{(II)}$$

$$AS + BS \xrightarrow{k} PS + S \qquad \text{(III)}$$

$$PS \rightleftharpoons P + S \qquad K_{IV} = \frac{p[S]}{[PS]} \qquad \text{(IV)}$$

Determine the functional form of the rate equation.

[9] R. R. Wenner and F. C. Dybdal, *Chem. Eng. Prog.*, **44**, 275 (1948).

10.3   Alternative rate expressions for the same reaction are

$$\mathscr{R}_A = \frac{1.34a}{1 + 0.63a} \tag{I}$$

$$\mathscr{R}_A = 0.822a^{0.69} \tag{II}$$

Compare their predictions for $0 \leq a \leq 1$. Draw some conclusions.

10.4   The following surface mechanism has been evoked to explain an observed reaction:

$$A_2 + 2S \rightleftharpoons 2AS \tag{I}$$

$$B + S \rightleftharpoons BS \tag{II}$$

$$AS + BS \rightleftharpoons CS + DS \tag{III}$$

$$CS \rightleftharpoons C + S \tag{IV}$$

$$DS \rightleftharpoons D + S \tag{V}$$

(a) What is the observed reaction?

(b) Develop a Hougen and Watson kinetic model assuming Reaction III is rate controlling.

10.5   Repeat Problem 10.4 assuming now that Reaction I is rate controlling.

10.6   The catalytic hydrogenation of butyraldehyde to butanol,

$$H_2 + C_3H_7\overset{\overset{\displaystyle O}{\|}}{C}H \rightleftharpoons C_3H_7CH_2OH \tag{I}$$

has a reported[10] rate equation of the form

$$\mathscr{R}_I = \frac{k_0 e^{-E/R_g T}\left(P_{H_2}P_{BAL} - P_{BOH}/K_{eq}\right)}{\left(1 + K_1 P_{H_2} + K_2 P_{BAL} + K_3 P_{BOH}\right)^2}$$

where P_{H_2}, P_{BAL}, and P_{BOH} are the partial pressures of hydrogen, butyraldehyde, and butanol, respectively.

(a) Develop a surface reaction model to rationalize the observed form of the kinetics.

[10] J. B. Cropley, L. M. Burgess, and R. A. Loke, *Chemtech*, p. 374, June 1984.

(b) Suppose the parallel reaction

$$\underset{\|}{\overset{\text{O}}{}}$$

$$2H_2 + C_3H_7CH \rightarrow C_4H_{10} + H_2O \tag{II}$$

is significant. How might this affect $\mathcal{R}_I$? Give a plausible rate equation for $\mathcal{R}_{II}$.

10.7 The bimolecular reaction

$$A + B \rightleftharpoons P + Q$$

is sometimes catalyzed by using two different metals dispersed on a common support. A mechanism might be

$$A + S_1 \rightleftharpoons AS_1 \tag{I}$$

$$B + S_2 \rightleftharpoons BS_2 \tag{II}$$

$$AS_1 + BS_2 \rightleftharpoons PS_1 + QS_2 \tag{III}$$

$$PS_1 \rightleftharpoons P + S_1 \tag{IV}$$

$$QS_2 \rightleftharpoons Q + S_2 \tag{V}$$

Develop a Hougen and Watson kinetic model assuming Reaction III is rate controlling.

10.8 A platinum catalyst supported on Al_2O_3 is used for the oxidation of sulfur dioxide:

$$SO_2 + \tfrac{1}{2}O_2 \rightarrow SO_3 \qquad \Delta H_{\mathcal{R}} = -95 \text{ kJ mol}^{-1}$$

The catalyst consists of $\frac{1}{8}''$ pellets that pack to a bulk density of 70 lb/ft^3 and $\varepsilon = 0.5$. Mercury porosimetry has found $R_p = 50$ Å. The feed mixture to a differential reactor consisted of 5 mole-% SO_2 and 95 mole-% air. At approximately atmospheric pressure, the following *initial* rate data were obtained:

T, K	$\mathcal{R}$, mol hr^{-1} (gm catalyst)$^{-1}$
653	0.031
673	0.053
693	0.078
713	0.107

Do an order-of-magnitude calculation for the nonisothermal effectiveness factor. *Hint*: Use the pore model to estimate an isothermal effectiveness factor and obtain D_{eff} from that. Assume $k_{\text{eff}} = 0.15$ J m^{-1} s^{-1} K^{-1}.

10.9 Consider a nonporous catalyst particle where the active surface is all external. There is obviously no pore resistance, but a film resistance to mass transfer can still exist. Determine the isothermal effectiveness factor for first order kinetics, $\mathcal{R}_A = -ka_s$.

10.10 It has been proposed that some enzymes exist in active and inactive forms that are in equilibrium. The active form binds substrate molecules for subsequent reaction while the inactive form does not. The overall reaction mechanism might be

$$I \rightleftharpoons E \qquad \frac{[E]}{[I]} = K_i \qquad\qquad\qquad (I)$$

$$A + E \rightleftharpoons AE \qquad \frac{[AE]}{a[E]} = K \qquad\qquad\qquad (II)$$

$$AE \rightarrow P + E \qquad \mathscr{R}_p = k[AE] \qquad\qquad\qquad (III)$$

Derive a Michaelis–Menten kinetic model for this situation.

10.11 The cell growth model with lag phase, Equations 10.79 through 10.83, allows spontaneous deaths to occur throughout the batch cycle. This means that dc/dt is initially negative. It is possible to lose the inoculum completely if the induction period is too long? Long induction periods correspond to small values for $\alpha/(k_\infty a_0)$. Find the critical value for $\alpha/(k_\infty a_0)$ for which C/c_0 never exceeds 1.0. What fraction of the substrate is consumed in this batch cycle? Use $k_d/(K_\infty a_0) = 0.1$ and $\beta c_0 k_\infty a_0 = 0.05$ as in Figure 10.6.

10.12 Cell populations can be limited by depletion of substrate or by accumulation of toxic products. Extend the batch growth models in Section 10.4.2 to account for *product inhibition*. One means of doing this is to assume that product accumulation causes cell deaths at a rate proportional to the existing cell concentration and to the square of the product concentration. Exercise your model numerically, taking the results in Figure 10.6 as a base case. Is it possible for product inhibition to prevent complete consumption of the substrate or is the rate of consumption merely slowed? This question has considerable importance in vinification, and your knowledge of wines may guide you to the solution if your mathematics will not.

10.13 The Eley–Rideal mechanism for gas–solid heterogeneous catalysis envisions reaction between a molecule adsorbed on the solid surface and one that is still in the gas phase. Consider a reaction of the form

$$A + B \rightarrow P$$

There are two logical possibilities for the reaction mechanism:
(a) A(gas) + BS(solid) → PS(solid)
(b) A(gas) + BS(solid) → P(gas) + S(solid)
Determine the form of the pseudohomogeneous, intrinsic kinetics for each of these cases. Assume that the reaction step shown above is rate limiting.

11

Multiphase Reactors

The microbial reactors discussed in Chapter 10 and the fixed-bed reactors discussed in Chapters 7 and 10 are examples of multiphase, or heterogeneous, systems. The current chapter describes other reactor designs where interfacial transport is an important step in the overall reaction sequence. The reaction may be catalytic or noncatalytic; the unifying concept of this chapter is not reaction mechanisms but mass transfer resistances.

When two or more phases are present, it is rarely possible to design a reactor on a strictly first principles basis. Rather than starting with the mass, energy, and momentum transport equations as was done for the laminar flow systems in Chapter 6, we tend to use simplified flow models with empirical correlations for mass transfer coefficients and interfacial areas. The approach is conceptually similar to that used for friction factors and heat transfer coefficients in turbulent flow systems. Typically, it provides a good engineering basis for design and scaleup.

11.1   Gas – Liquid and Liquid – Liquid Reactors

11.1.1   Two-Phase, Stirred Tank Reactors

A simple but effective way of modeling many two-phase reactors is to assume perfect mixing within each phase. Figure 11.1 gives a conceptual view of a *two-phase continuous stirred tank reactor*. For convenience we refer to one phase as being liquid and to the other as being gas, but the mixing and contacting scheme shown in Figure 11.1 obviously can apply to liquid–liquid systems as well. As suggested in the figure, the liquid phase is internally homogeneous, as is the gas. The two phases contact each other through an interface that has area A_i

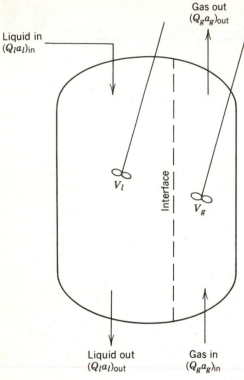

Gas out
$(Q_g a_g)_{out}$

Liquid in
$(Q_l a_l)_{in}$

V_l

Interface

V_g

Liquid out
$(Q_l a_l)_{out}$

Gas in
$(Q_g a_g)_{in}$

Figure 11.1 The two-phase continuous stirred tank reactor.

per unit volume of reactor. Thus the total interfacial area is $A_i V$, where $V = V_l + V_g$ is the total reactor volume. Mass transfer occurs across the interface with rate

$$k_g A_i V(a_g - a_g^*) = k_l A_i V(a_l^* - a_l) \quad \text{moles per unit time} \tag{11.1}$$

where k_g and k_l are the gas and liquid mass transfer coefficients. They have units of moles per time per area per concentration driving force = length per time. The concentrations a_g^* and a_l^* occur at the interface. They differ from the bulk concentrations, and the differences $a_g - a_g^*$ and $a_l^* - a_l$ are the driving forces for mass transfer. If the differences are positive as written, species A will be transferred from the gas to the liquid. Equilibrium is assumed to exist at the interface:

$$a_l^* = k_H a_g^* \tag{11.2}$$

where k_H is a constant that we (loosely) refer to as the Henry's law constant.[1] The rate of mass transfer between the phases can also be expressed in terms of the *overall driving force* $k_H a_g - a_l$:

$$U_m A_i V(k_H a_g - a_l) = k_g A_i V(a_g - a_g^*) = k_l A_i V(a_l^* - a_l) \qquad (11.3)$$

where U_m is the overall mass transfer coefficient:

$$U_m = \frac{k_g k_l}{k_g + k_l k_H} = \frac{1}{(1/k_l) + (k_H/k_g)} \qquad (11.4)$$

(Note the analogy with an overall heat transfer coefficient.)

Material balances can be written for each phase. For the general case of variable physical properties and possibly unsteady operation, the liquid phase balance is

$$(Q_l a_l)_{in} + k_l A_i V(a_l^* - a_l) + V_l(\mathscr{R}_A)_l = Q_l a_l + \frac{d(V_l a_l)}{d\theta} \qquad (11.5)$$

and for the gas phase,

$$(Q_g a_g)_{in} - k_g A_i V(a_g - a_g^*) + V_g(\mathscr{R}_A)_g = Q_g a_g + \frac{d(V_g a_g)}{d\theta} \qquad (11.6)$$

The time derivatives are dropped for steady state, continuous flow, and the flow terms are dropped for batch reactions. Semibatch operation with the liquid charged initially and the gas sparged in continuously is quite common for gas–liquid reactions. For this situation we set $(Q_l)_{in} = (Q_l)_{out} = 0$, and $d(V_g a_g)/d\theta = 0$. Typically $(\mathscr{R}_A)_g = 0$ as well. Figures 11.2 and 11.3 show two common reactor types that are usually modeled as two-phase stirred tanks and that are often operated in the semibatch mode.

The aerated fermentor in Figure 11.2 has water as the continuous phase and air, in the form of small bubbles, as the dispersed phase. The agitator controls the bubble size and circulates the liquid and gas phases. In the alkylation reactor, a gaseous alkene is contacted with a liquid hydrocarbon. A specially designed injection nozzle controls the gas dispersion while circulation is achieved by the recycle pump. Both types of reactors can be converted to continuous operation

---

[1]Strictly speaking, Henry's law is expressed in terms of the gas phase partial pressure, not the gas phase concentration. The literature frequently bases mass transfer coefficients on partial pressures rather than concentrations and occasionally confuses k_g with U_m or k_l when doing so. See Example 11.1.

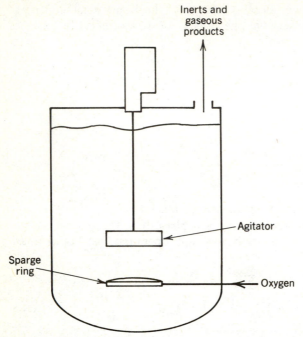

Figure 11.2   Aeration in an agitated fermentation vessel.

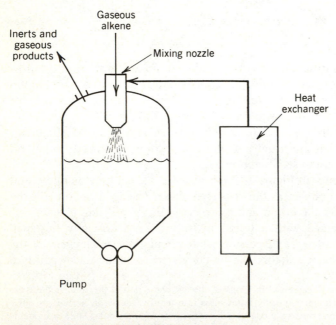

Figure 11.3   Semibatch alkylation reactor.

by adding input and output streams for the liquid phase. Under typical operating conditions, the liquid phase will be nearly homogeneous and, when operated continuously, will have an exponential distribution of residence times. Thus the liquid phase in the example reactors can be modeled as perfectly mixed, and a similar assumption is usually used for the gas phase as well.

Since the gas phase is dispersed, one might prefer to assume an exponential distribution of residence times but with segregated flow rather than perfect mixing in the gas phase. This will make a difference when the gas contains inerts, as when air is used as an oxygen source, since depletion of the reactive component can be appreciable for bubbles that have remained in the system a long time. It makes no difference when the reactive gas is pure. Use of a pure gas also ensures a negligible resistance to mass transfer on the gas side since $a_g^* = a_g$. In general, the high diffusivity of gases causes the gas side resistance to be much smaller than that on the liquid side. Often, k_g is an order of magnitude larger than k_l and is ignored.

Example 11.1 An article in the literature reports the absorption rate of pure oxygen into a sodium sulfite solution at 20°C using an agitated stirred tank having a liquid depth of 3 ft. A large excess of oxygen was continuously injected into the tank through a sparge ring located just under the agitator. The liquid reaction (sulfite oxidation) was semibatch, but there was sufficient sodium sulfite present so that dissolved oxygen concentrations were nearly zero throughout the experiment. The oxygen consumption was measured using gas flow rates. For a particular set of operating conditions, the result was reported as $k_g A_i = 0.04$ lb-moles hr^{-1} ft^{-3} atm^{-1}. What was actually measured and what is its value in SI units?

Solution Since pure oxygen was sparged $a_g = a_g^*$; and since $a_1 \approx 0$ due to the fast sulfite reaction, it is apparent this experiment actually measured $Q_g a_g = k_l A_i V a_l^*$ from which a value for $k_l A_i$ was determined. It was reported as $k_g A_i$ using an earlier notation which uses the symbol k_g to denote the overall mass transfer coefficient when the driving force is expressed in partial pressure units rather than concentration units. It is perfectly acceptable, although now unconventional, to express the liquid side driving force in terms of partial pressures. Thus in the current notation, $k_l A_i = 0.04$ lb-moles hr^{-1} ft^{-3} atm^{-1} = 0.178 mol^{-1} s^{-1} m^{-3} atm^{-1} = 1.76×10^{-6} mol s^{-1} m^{-3} Pa^{-1}.

The mass transfer rate actually measured was $0.04\Delta P$, lb-moles hr^{-1} ft^{-3}, where $\Delta P = P_g - P_l$ is the oxygen driving force measured in atmospheres. In this study $P_l \approx 0$ since $a_l \approx 0$. The oxygen pressure on the gas side does not vary due to reaction but does vary due to the liquid head. Assume that the pressure at the top of the tank was 1 atm. Then $P_g = 0.975$ atm since the vapor pressure of water at 20°C should be subtracted. At the bottom of the tank, $P_g = 1.0635$ atm. The logarithmic mean is appropriate: $\Delta P = 1.0183$ atm. Thus the transfer rate was $Q_g a_g/V = 0.0407$ lb-moles hr^{-1} ft^{-3}, or 0.181 mol s^{-1} m^{-3}, and

$$0.181 = k_l A_i (a_l^* - a_l) = k_l A_i a_l^*$$

where a_l^* is the solubility of oxygen in water at 20°C and 1.018 atm. A literature source expresses Henry's law for oxygen in water at 20°C as

$$2.95 \times 10^{+7} = \frac{P}{x} \text{ mm Hg/mole fraction}$$

where x is the mole fraction of oxygen. This gives $a^* = 1.46$ mol/m^3. Thus

$$k_l A_i = \frac{0.181}{1.46} = 0.124 \text{ s}^{-1}$$

For semibatch operation with reaction only on the liquid side, Equation 11.5 reduces to

$$k_l A_i V (a_l^* - a_l) + V_l \mathscr{R}_A = \frac{d(V_l a_l)}{d\theta} \tag{11.7}$$

In many cases the gas is sparingly soluble; the liquid would quickly saturate in the absence of reaction; and the accumulation term can be ignored. If the sparingly soluble gas is consumed by an nth order reaction,

$$k_l A_i V (a_l^* - a_l) - V_l k a_l^n = 0 \tag{11.8}$$

which can be solved for the liquid phase concentration. For $n = 1$,

$$a_l = \frac{k_l A_i V a_l^*}{k_l A_i V + V_l k} \tag{11.9}$$

Dropping the accumulation term from Equation 11.7 is a kind of quasi-steady state hypothesis similar to those in Section 2.3.2. The liquid phase composition adjusts itself so that the mass transfer rate is matched to the reaction rate. These rates then remain constant until the vessel overflows with accumulated product or until depletion of some reactive component in the liquid phase becomes significant. If $k_l A_i V \gg V_l k$, $a_l \approx a_l^*$ and the system is said to be **reaction limited**. If $k_l A_i V \ll V_l k$, $a_l \approx 0$ and the system is said to be **mass transfer limited**. Table 11.1 gives typical values of V_l/V and $k_l A_i$ for air–water contacting.

All the devices in this table operate with continuous gas flow. The columns usually operate with continuous liquid flow as well, but a batch reaction on the liquid side can be achieved by external recycle of the liquid.

Table 11.1 **Typical Operating Ranges for Gas – Liquid Contacting Devices**[a]

Device	Fractional Liquid Holdup, V_l / V	$k_l A_i$, s^{-1}
Mechanically agitated tanks	0.9	0.02–0.2
Tray columns	0.15	0.01–0.05
Packed columns	0.05	.005–.02
Bubble columns	0.95	.005–.01

[a]Data from N. Harnby, M. F. Edwards, and A. W. Nienow, *Mixing in the Process Industries*, Butterworth, London, 1985.

Example 11.2 Define ranges for pseudo-first order rate constants such that the various devices in Table 11.1 will be limited by mass transfer.

Solution The requirement for mass transfer being strictly limiting is that $k_l A_i V \ll V_l k$. Suppose we insist on an order-of-magnitude difference between these terms. Then

$$10 k_l A_i \frac{V}{V_l} < k$$

Use $V_l/V = 0.9$ and the upper limit of $k_l A_i = 0.2$ s^{-1} for a mechanically agitated tank. This gives $k > 2.2$ s^{-1}. Thus a reaction with pseudo-first order rate constant above about 2 s^{-1} is almost certain to be mass transfer limited. Repeating this logic for the other devices in Table 11.1 gives

Device	Minimum k for Strict Mass Transfer Limitation
Agitated tank	2 s^{-1}
Tray columns	3 s^{-1}
Packed columns	4 s^{-1}
Bubble columns	0.1 s^{-1}

We consider now two examples of semibatch operation where the change in liquid phase composition has a significant effect on reaction rate and selectivity.

Example 11.3 Suppose the liquid phase reaction is

$$A + B \overset{k}{\rightarrow} \text{Products}$$

where A is a sparingly soluble gas that is sparged continuously and where all the liquid B was charged to the vessel initially. Determine $a_l(t)$ and $b(t)$.

Solution   The accumulation term in Equation 11.7 is dropped to give

$$k_l A_i V(a_l^* - a_l) = V_l ka_l b \tag{11.10}$$

and

$$a_l = \frac{k_l A_i V a_l^*}{k_l A_i V + V_l kb} \tag{11.11}$$

Since B is not transferred between phases, its material balance has the usual form for a batch reactor:

$$\frac{db}{d\theta} = -ka_l b = \frac{-kk_l A_i V a_l^* b}{k_l A_i V + V_l kb} = \frac{-k'b}{1 + k_B b} \tag{11.12}$$

The appearance of this "heterogeneous" form for the rate expression reflects the presence of a mass transfer step in series with the reaction step. Equation 11.12 can be integrated subject to the initial condition that $b = b_0$ at $\theta = 0$. This integration gives $b(\theta)$, which can be substituted into Equation 11.11 to give $a_1(\theta)$. Finally, one might evaluate $d(a_l V_l)/d\theta$ to see if the accumulation term for A really was negligible in Equation 11.7.

Example 11.4   Consider the alkylation of phenol in a semibatch, two-phase stirred tank. Suppose the liquid phase contains a homogeneous, ortho-directing catalyst so that alkylation occurs only at the ortho positions. The reactions are

$$A + P \xrightarrow{k_P} M$$

$$\tag{11.13}$$

$$A + M \xrightarrow{k_M} N$$

where M is the monosubstituted product (2-alkylphenol) and N is the disubstituted product (2,6-dialkylphenol). Determine the distribution of phenol, monoalkylate, and dialkylate as a function of conversion. The alkene is pure and sparingly soluble in the liquid phase.

Solution   The liquid phase components react according to

$$\frac{dp}{dt} = -k_P ap \tag{11.14}$$

$$\frac{dm}{dt} = +k_P ap - k_M am \tag{11.15}$$

$$\frac{dn}{dt} = k_M am \tag{11.16}$$

from which it follows that

$$p + m + n = p_0 \tag{11.17}$$

where we have assumed $m_0 = n_0 = 0$. The alkene concentration in the liquid phase is found from Equation 11.7 with $d(V_l a_l)/dt \approx 0$ and $\mathscr{R}_A = -k_P a_l p - k_M a_l m$. This gives

$$a_l = \frac{k_l A_i V a_l^*}{k_l A_i V + V_l k_P p + V_l k_M m} \tag{11.18}$$

Using $a_l^* = k_H a_g$, these equations can be solved numerically for a_l, p, m, and n as functions of time. However, a more elegant approach is possible in this and similar selectivity problems. The basic idea is to eliminate time as a parameter. Divide Equations 11.15 by 11.14 to obtain

$$\frac{dm}{dp} = -1 + \frac{k_M}{k_P} \frac{m}{p} \tag{11.19}$$

This eliminates t and, as it happens, a_l as well. Solving Equation 11.19 gives

$$\frac{m}{p_0} = \frac{(p/p_0)^{k_M/k_P} - p/p_0}{1 - k_M/k_P} \tag{11.20}$$

which incorporates the initial condition that $m = 0$ when $p = p_0$. Dividing Equation 11.16 by Equation 11.14 gives a differential equation for n. The solution is

$$\frac{n}{p_0} = 1 - \left[\frac{(p/p_0)^{k_M/k_P} - (k_M/k_P)(p/p_0)}{1 - k_M/k_P} \right] \tag{11.21}$$

Note that p/p_0 is the fraction unreacted of phenol. Knowing the fraction unreacted and the ratio of rate constants k_M/k_P, we can calculate m/p_0 and n/p_0. The product distribution can thus be found as a function of phenol conversion without explicitly knowing how the conversion varies with time.

The usual case in these ortho-alkylations is $k_M/k_P = 0.5$. This corresponds to equal reactivity of the ortho sites. Specifically, the probability of alkylation at a given ortho site is independent of whether the other site on that molecule has been alkylated. Some values for p/p_0, m/p_0, n/p_0 with $k_M/k_P = 0.5$ are:

Time	p/p_0	m/p_0	n/p_0
$t = 0$	1.000	0	0
	0.750	0.232	0.018
	0.500	0.414	0.086
	0.250	0.500	0.250
	0.125	0.457	0.418
$t = \infty$	0	0	1.000

This treatment of two-phase reactors has introduced a number of phenomenological parameters that must be determined: k_H, $k_g A_i$, $k_l A_i$, V_l, and V_g. The phase volumes V_l and V_g can be measured from experiments with and without aeration. The Henry's law constant can be measured under equilibrium conditions if the gas and liquid phases do not react. For example, phenol and alkenes react slowly in the absence of a catalyst so that reasonable measurements of k_H can be made under what would otherwise be reacting conditions just by leaving out the catalyst. In essence, these are solubility measurements where the alkene partial pressure is set and its pseudoequilibrium[2] concentration in the liquid phase is measured. Note that the alkene solubility will depend on the liquid phase composition, increasing with the extent of alkylation. Thus measurements must be made at a variety of compositions.

Measurements of $k_g A_i$ and $k_l A_i$ can also be done under nonreacting conditions. Example 11.5 describes a standard technique for measuring $k_l A_i$ which is similar in spirit to the residence time measurements of Chapter 9.

When significant reaction between the gas and liquid phases cannot be avoided, extrapolations from nonreacting conditions or extrapolations using analogous but nonreacting species are sometimes possible, but see the discussion on enhancement factors in Section 11.1.3. It is also possible to use parameter estimation techniques to fit the experimental data. These are similar in concept to the data fitting techniques discussed in Section 4.3.

Example 11.5 Air is sparged into a batch vessel until the liquid phase is saturated with respect to oxygen. The air supply is suddenly turned off and replaced with nitrogen at the same volumetric flow rate. A dissolved oxygen meter is used to monitor the oxygen content in the liquid phase. Show how this can be used to find $k_l A_i$ assuming that the gas phase mass transfer resistance is negligible and that both phases are perfectly mixed.

Solution Let a and b denote the gas phase and liquid phase oxygen concentrations, respectively. Then for $t > 0$, the liquid phase material balance gives

$$V_l \frac{db}{dt} = k_l A_i V(b^* - b) \tag{11.22}$$

while for the gas phase we have

$$V_g \frac{da}{dt} = -k_l A_i V(b^* - b) - Q_g a \tag{11.23}$$

---

[2] The true thermodynamic equilibrium is nearly complete alkylation. In the absence of a catalyst, the reaction equilibrium is established relatively slowly compared to the mass transfer steps. Thus solubility measurements can be made before any significantly reaction occurs in the liquid phase.

Using Henry's law $b^* = k_H a$ gives a pair of equations containing a and b as unknowns. They can be combined to yield a second order differential equation in b:

$$\frac{d^2 b}{dt^2} + \left(\frac{k_l V A_i}{V_l} + \frac{k_l A_i V k_H}{V_g} + \frac{1}{\bar{t}_g} \right) \frac{db}{dt} + \frac{k_l A_i V}{V_l \bar{t}_g} b = 0 \qquad (11.24)$$

where $\bar{t}_g = V_g / Q_g$. This equation can be used to estimate $k_l A_i$ from an experimental $b(t)$ curve in at least three ways. They are:

1. *Initial second derivative method*   At $t = 0$, $db/dt = 0$. Therefore,

$$\frac{1}{b} \frac{d^2 b}{dt^2} \Big|_0 = -\frac{k_l A_i V}{V_l \bar{t}_g} \qquad (11.25)$$

so that $k_l A_i$ can be calculated assuming the other parameters are known. This method suffers the obvious difficulty of measuring a second derivative.

2. *The inflection point method*   At the inflection point, $d^2 b/dt^2 = 0$ and

$$\frac{1}{b} \frac{db}{dt} \Big|_{\text{inflect}} = \frac{d \ln b}{dt} \Big|_{\text{inflect}} = \frac{-k_l A_i V}{k_l A_i V \bar{t}_g + \dfrac{k_l A_i V \bar{t}_g k_H V_l}{V_g} + V_l} \qquad (11.26)$$

3. *The asymptotic method*   For most systems, $V_l / (k_l A_i) \gg \bar{t}_g$; and for large values of t,

$$b(t) \approx \exp \left(\frac{-k_l A_i V t}{V_l} \right) \qquad (11.27)$$

When applicable, this method is the least demanding in terms of experimental accuracy. It is merely necessary to estimate the slope of what should be a straight line when $\ln b$ is plotted versus t. By comparison, the inflection point method requires estimating the slope at an earlier time before it is constant.

The above examples treated semibatch, two-phase stirred tank reactors. We turn now to the fully continuous case. Begin by supposing the phenol alkylation reactor in Example 11.4 has been converted to continuous operation.

Example 11.6   The alkylation reactor of Figure 11.3 is to be operated continuously with a liquid feed of essentially pure phenol and a gas feed of high-purity propylene. An ortho-directing, homogeneous catalyst is employed so that the chemistry of Equation 11.13 applies. Find the conversion of phenol and the distribution of products.

Solution With a gas feed of high-purity propylene, it is possible to operate with $(Q_g)_{out} = 0$. (Note that trace amounts of gas phase inerts may or may not accumulate depending on their solubility in the liquid phase.) There is no reaction in the gas phase. Thus Equations 11.3 and 11.6 can be combined to give

$$(Q_g a_g)_{in} = k_l A_i V(a_l^* - a_l) \tag{11.28}$$

which merely tells us that the propylene feed rate must match its transfer rate into the liquid phase. This same result is found for the quasi-steady operation of a semibatch alkylation. The liquid phase balance, Equation 11.5, gives more useful information. For propylene,

$$k_l A_i V(a_l^* - a_l) - V_l(k_P a_l p + k_M a_l m) = Q_l a_l \tag{11.29}$$

The phenol balance gives

$$Q_l p_{in} - V_l k_P a_l p = Q_l p \tag{11.30}$$

where we have assumed constant liquid density with $(Q_l)_{in} = (Q_l)_{out} = Q_l$. For the monoalkylate,

$$V_l(k_P a_l p - k_M a_l m) = Q_l m \tag{11.31}$$

and for the dialkylate,

$$V_l k_M a_l m = Q_l n \tag{11.32}$$

Equations 11.29 to 11.32 contain the four liquid phase compositions (a_l, p, m, and n) as unknowns. These are readily solvable when a_l^* is known. In practice, however, a_l^* will depend on p, m, and n, since the solubility of propylene will depend on the extent of alkylation as well as the temperature and pressure in the reactor. A trial-and-error solution is then required to find a_l and the conversion of phenol. The product selectivity at a given level of phenol conversion can be found by simpler means. The fraction of phenol unreacted is found from Equation 11.30 as

$$\frac{p}{p_{in}} = \frac{1}{1 + k_P \bar{t}_l a_l} \tag{11.33}$$

where $\bar{t}_l = V_l/Q_l$. Suppose p/p_{in} and k_P/k_M are given. Then $k_M \bar{t}_l a_l$ can be found once $k_P \bar{t}_l a_l$ is calculated from Equation 11.33. With $k_P \bar{t}_l a_l$ and $k_M \bar{t}_l a_l$

known, the concentrations of mono- and dialkylate are

$$\frac{m}{p_{in}} = \frac{k_p \bar{t}_l a_l}{(1 + k_p \bar{t}_l a_l)(1 + k_M \bar{t}_l a_l)}$$

(11.34)

$$\frac{n}{p_{in}} = \frac{k_p k_M (\bar{t}_l)^2 a_l^2}{(1 + k_p \bar{t}_l a_l)(1 + k_M \bar{t}_l a_l)}$$

The usual case is $k_M/k_P = 0.5$. Some product distributions for the continuous flow reactor are:

$\bar{t}$	p/p_{in}	$\dfrac{m}{p_{in}}$	$\dfrac{n}{p_{in}}$
0	1.000	0	0
	0.750	0.214	0.036
	0.500	0.333	0.167
	0.250	0.300	0.450
	0.125	0.194	0.681
$t = \infty$	0	0	1.000

Not surprisingly, the continuous-flow reactor gives poorer selectivity for the monoalkylate than did the semibatch reactor considered previously. This, of course, is due to the entire reaction in a perfect mixer occurring at the exit concentrations where an appreciable concentration of monoalkylate is available for further reaction.

Example 11.7 Carbon dioxide is sometimes removed from natural gas by reactive absorption in a tray column. The absorbant, typically an amine, is fed to the top of the column and gas is fed at the bottom. Liquid and gas flow patterns are similar to those in a distillation column with gas rising, liquid falling, and gas–liquid contacting occurring on the trays. Develop a model for a multitray CO_2 absorber assuming that individual trays behave as two-phase, stirred tanks.

Solution The reaction is

$$C + A \rightarrow P$$

which we assume to be elementary with rate constant k. There are J trays in the column and they are numbered $j = 1, 2, \ldots, J$ starting from the bottom. Figure 11.4 shows a typical tray and indicates the notation. Since the column has many trays (typically 20 or more), composition changes on each tray are small, and it is reasonable to assume perfect mixing within both phases on an individual tray. Thus the gas phase balance for CO_2 is

$$F_{j-1} c_{j-1} = F_j c_j + k_g \Delta A_i (c_j - c_j^*)$$

(11.35)

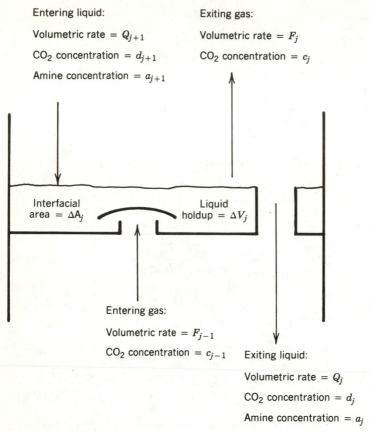

Entering liquid:

Volumetric rate = Q_{j+1}

CO_2 concentration = d_{j+1}

Amine concentration = a_{j+1}

Exiting gas:

Volumetric rate = F_j

CO_2 concentration = c_j

Interfacial area = ΔA_j

Liquid holdup = ΔV_j

Entering gas:

Volumetric rate = F_{j-1}

CO_2 concentration = c_{j-1}

Exiting liquid:

Volumetric rate = Q_j

CO_2 concentration = d_j

Amine concentration = a_j

Figure 11.4   Typical tray in a tray column reactor for acid gas scrubbing.

where ΔA_i is the interfacial area per tray. For the dissolved but unreacted CO_2 in the liquid phase,

$$Q_{j+1}d_{j+1} + k_l \Delta A_i \left(d_j^* - d_j \right) = \Delta V_j \, ka_j d_j + Q_j d_j \tag{11.36}$$

For the amine

$$Q_{j+1}a_{j+1} = \Delta V_j \, ka_j d_j + Q_j d_j \tag{11.37}$$

Since a single tray gives a small increment of conversion, it would be reasonable to assume F and Q constant for that tray. However, there is likely to be a significant change in these flow rates for the column as a whole.

The inputs to the jth tray are $c_{j-1}, d_{j+1}, a_{j+1}, F_{j-1}, Q_{j+1}$. Suppose these are known. Then Equations 11.35 to 11.37 contain five compositional unknowns: c_j, c_j^*, d_j^*, d_j, and a_j. There are two unknown flow rates: F_j and Q_j. The flow rates can be calculated from suitable equations of state, but we still need two additional equations to determine the concentrations. These are of course Equations 11.2 and 11.3. In the current notation,

$$d_j^* = k_H c_j^* \tag{11.38}$$

and

$$k_g \Delta A_i \left(c_j - c_j^* \right) = k_l \Delta A_i \left(d_j^* - d_j \right) \tag{11.39}$$

Equations 11.35 to 11.39 are the minimum set of equations that must be solved per tray. Two additional equations are required for nonisothermal absorbers, and most acid gas scrubbing reactors will have an additional reactive component such as H_2S or SO_2. Thus five to eight nonlinear algebraic equations must be solved for each tray so that a typical column will require solution of several hundred simultaneous equations. This is computationally feasible. The multidimensional Newton's method described in Appendix 3.1 is usually used. An alternative approach is to guess the composition of, say, the exiting liquid stream at the bottom of the column. With this initial guess, a sequential, tray-by-tray calculation is possible which involves solving only the five to eight equations simultaneously. This approach is conceptually similar to the method of shooting described in Chapter 7. It presents similar difficulties, and the fully implicit scheme is usually preferred.

Example 11.8   With highly reactive absorbants, the mass transfer resistance in the gas phase can be controlling. Determine the number of trays needed to reduce the CO_2 concentration in a methane stream from 5 percent to 100 ppm (by volume) assuming the liquid mass transfer and reaction steps are fast. A 0.9-m diameter column is to be operated at 8 atm and 50°C with a gas feed rate of 0.2 m^3/s. The trays are bubble caps operated with a 0.1-m liquid holdup. Literature correlations suggest $k_g = 0.002$ m/s and $\Delta A_i' = 20$ m^2 per m^2 of tray area.

Solution   Ideal gas behavior is a reasonable approximation for the feed stream. The inlet concentrations are 287 mol/m^3 of methane and 15 mol/m^3 of carbon dioxide. The column pressure drop is mainly due to the liquid head on the trays and will be negligible compared to 8 atm unless there is an enormous number of trays. Thus the gas flow rate F will be approximately constant for the column as a whole. With the gas side resistance controlling, we only need Equation 11.35 with $c_j^* \approx 0$:

$$F c_{j-1} = F c_j + k_g \Delta A_i c_j \tag{11.40}$$

Substituting known values,

$$\frac{c_{j-1}}{c_j} = 1 + \frac{k_g \Delta A_i}{F} = 1.13 \tag{11.41}$$

Solution gives

$$\frac{c_0}{c_J} = \frac{0.05}{10^{-4}} = (1.13)^J \tag{11.42}$$

and $J = 51$ trays. The indicated separation appears feasible in a bubble cap column although the design engineer should not be content with the glib assumption of negligible liquid side resistance.

Equations 11.5 and 11.6 can be written for each of N reactive components, giving a set of $2N$ algebraic equations for steady state (flow) operation or $2N$ ordinary differential equations for unsteady state operation. When the reaction temperatures are unknown, two heat balances are also needed:

$$\begin{aligned}
(Q_l \rho_l H_l)_{\text{in}} - (\Delta H_{\mathscr{R}} \mathscr{R})_l V_l &= (Q_l \rho_l H_l)_{\text{out}} + UA_{\text{ext}}(T - T_{\text{ext}}) \\
&+ h_i A_i V(T_l - T_g) + \frac{d(V_l \rho_l H_l)}{dt}
\end{aligned} \tag{11.43}$$

and

$$\begin{aligned}
(Q_g \rho_g H_g)_{\text{in}} - (\Delta H_{\mathscr{R}} \mathscr{R})_g V_g &= (Q_g \rho_g H_g)_{\text{out}} - h_i A_i V(T_l - T_g) \\
&+ \frac{d(V_g \rho_g H_g)}{dt}
\end{aligned} \tag{11.44}$$

These equations allow for interphase heat transfer with h_i as the heat transfer coefficient. As written, they envision heat transfer to the environment to occur only through the liquid phase since we are considering it to be the continuous phase.

Equations 11.2, 11.3, 11.5, 11.6, 11.43, and 11.44 are the complete design equations for a two-phase reactor with perfect mixing in each phase. Just by deleting one phase, they govern the performance of a perfect mixer (Chapter 3) which may be nonisothermal (Chapter 4) and unsteady (Chapter 8). The reader will appreciate that these equations are rarely used in their full form even for single-phase systems.

This section has treated the situation where both the liquid and gas phases are perfectly mixed. Thus parameters such as $k_l A_i$ were taken as constant throughout the vessel, as were concentrations and temperatures within a phase. The resulting reactor models are of the lumped variety. We turn now to situations where both phases are in piston flow. This gives rise to distributed models.

11.1.2   Fluid – Fluid Contacting in Piston Flow

Some contacting devices operate with approximately piston flow in both phases. The rotating discs column in Figure 11.5 illustrates equipment commonly used for liquid–liquid extraction and for reactive contacting of a liquid phase with another fluid phase. The discs rotate to create and maintain the dispersion of the upward-flowing, discontinuous phase in the downward-flowing, continuous liquid phase. The stators on the walls of the column prevent channeling and also compartmentalize the column. An accurate model for such a device might treat this situation as J tanks in series, where J is the number of rotating discs. We

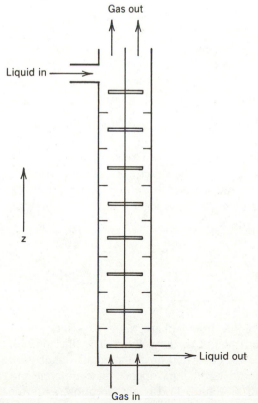

Figure 11.5 Rotating disc contactor with countercurrent, downward flow of a liquid and upward flow of a gas or lower-density liquid.

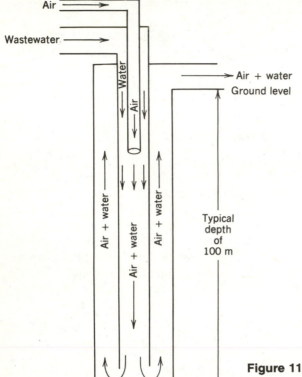

Figure 11.6 A deep shaft fermentor for wastewater treatment.

now suppose J is large enough so that piston flow is a reasonable approximation for both phases.

For simplicity of discussion and notation, we will refer to one phase as being liquid and the other phase as being gas. The gas phase is upward flowing. It flows in the $+z$-direction. The liquid phase may be either upward flowing (cocurrent) or downward flowing (countercurrent). A steady state but otherwise general material balance gives

$$\pm \frac{d(A_l \bar{u}_l a_l)}{dz} = A_l(\mathscr{R}_A)_l + k_l A_i'(a_l^* - a_l) \tag{11.45}$$

where A_l is the cross-sectional area of the liquid phase, $\bar{u}_l$ is its velocity, and where A_i' is the interfacial area per unit height of the column. The plus sign on the derivative in Equation 11.39 is used for cocurrent flow; the minus sign is used for countercurrent flow. Most gas–liquid contactors operate in countercurrent flow. The deep shaft fermentor shown in Figure 11.6 is an exception, as is the

Table 11.2 **Simplifying Assumptions for Gas – Liquid Reactors**

Assumption	Possible Rationale
$\left\|\dfrac{dP}{dz}\right\| = \rho_l g$	Gas lift is negligible so that the liquid exerts its normal static head
$\dfrac{dP}{dz} = 0$	Short column or column operated at high pressure or with the liquid phase discontinuous
a_g and a_g^* are constant	Reactant gas is pure or is fed at a high rate in a constant-pressure column
$a_l \bar{u}_l$ is constant	Above plus constant liquid density
$a_g u_g \rho_g$ is constant	Mass transfer has a negligible effect on the mass flow of gas
$a_g u_g$ is constant	As above in a constant-pressure column
$k_l A_i'$ is constant	Redispersion of gas compensates for coalescence and pressure effects; negligible depletion of the reactive component or else a high level of inerts are present in the gas phase
A_g and A_l are constant	Consistent with A_i' being constant; note that constancy of A_g implies constancy of A_l since $A_l + A_g = A_c$
$\dfrac{d(A_l a_l)}{dz} = 0$	Negligible solubility of the reactive gas

trickle bed reactor discussed in Section 11.4. The gas phase material balance is

$$\frac{d(A_g \bar{u}_g a_g)}{dz} = A_g (\mathscr{R}_A)_g - k_g A_i'(a_g - a_g^*) \tag{11.46}$$

Equations 11.2 and 11.3 apply as written although it must now be recognized that a_g, a_g^*, a_l^*, and a_l will be functions of z. In general, we also expect k_g, k_l and A_i' to vary with z. Bubbles will grow (and a_g will decrease) due to pressure changes as they rise in the column. Coalescence and dispersion may be important. Depletion or enrichment of the gas phase due to reaction and mass transfer may be important. These effects could be quite major in a device such as the deep shaft fermentor. Bubbles breaking the surface could be quite large due to coalescence and pressure changes and might contain little oxygen. Thus little or no mass transfer may be occurring in the upper part of the column. On the other hand, oxygen partial pressures of 2 atm (total pressure of 10 atm) will give a very high concentration driving force at the bottom of the column.

A rigorous analysis of two-phase, tubular reactors is a difficult problem in two-phase fluid mechanics. As a practical matter, simplifying assumptions such as constant A_g and A_i' are usually necessary. Table 11.2 lists a number of typical simplifications.

Example 11.9 A rotating discs column is used to obtain kinetic data on the reaction

$$A(\text{gas}) + B(\text{liquid}) \rightleftharpoons C(\text{liquid}) \tag{11.47}$$

The gas is sparingly soluble and only the liquid phase concentrations of B and C can be measured. Determine a plausible form for correlating the data.

Solution Assume the elementary liquid phase reaction is

$$A + B \underset{k_r}{\overset{k_f}{\rightleftharpoons}} C \tag{11.48}$$

The accumulation of A in the liquid phase is assumed to be small compared to its reaction rate. Thus Equation 11.45 becomes

$$k_l A_i'(a_l^* - a_l) = -A_l(-k_f a_l b + k_r c) \tag{11.49}$$

Solving for a_l,

$$a_l = \frac{k_l a_i' a_l' + A_l k_r c}{k_l A_i' + A_l k_f b} \tag{11.50}$$

Since B is not transferred between phases,

$$\frac{1}{A_l} \frac{d(A_l \bar{u}_l b)}{dz} = -k_f a_l b + k_r c = \frac{k_l A_i'(-k_r a_l^* b + k_r c)}{k_l A_i' + A_l k_f b} \tag{11.51}$$

This result remains quite general since $k_l A_i'$, $a_l^* A_l$, and $\bar{u}_l$ can all vary with z. Suppose, however, that the gas is sparged at a high rate and that the reactor is operated under pressure. This allows gas phase concentrations and bubble sizes to be approximately constant, and thus the various parameters can be independent of z. With redefined constants, Equation 11.51 becomes

$$\bar{u}_l \frac{db}{dz} = \frac{1}{\bar{t}_l} \frac{db_l}{dz} = \frac{-k_{\rm I} b + k_{\rm II} c}{1 + k_{\rm B} b} \tag{11.52}$$

where $z = z/L$. Equation 11.52 corresponds to pseudohomogeneous piston flow with reaction according to a Hougen and Watson kinetic scheme. It can be integrated using $c = c_{\rm in} + b_{\rm in} - b$. The three constants $k_{\rm I}$, $k_{\rm II}$, and $k_{\rm B}$ can then be estimated from measurements of $b_{\rm out}$ and $c_{\rm out}$.

The above example eliminated all gas phase concentration gradients and, in effect, treated the gas phase as though it were perfectly mixed. The solubility of the gas was also ignored. These assumptions resulted in the simplest possible model (for the given elementary reaction) of a gas–liquid reactor in which the liquid phase is in piston flow. The next example treats a more realistic situation but remains far from a general case.

Example 11.10 A column fermentor is to be used for the continuous, aerobic fermentation of wastewater. In this application, the reactor is more commonly known as a **digester**. Air pumped into the bottom of the column serves as the oxygen source. The flow is cocurrent. Develop a simplified reactor model.

Solution Digesters normally operate with the discharge at atmospheric pressure; and unless the column is very short indeed, the liquid head will have a significant effect on the oxygen partial pressure. Thus we use

$$\frac{dP}{dz} = -\rho_l g \tag{11.53}$$

with $z = 0$ at the bottom of the reactor. Assume $A_l \bar{u}_l$ to be constant since the fermentation will have no significant effect on the mass flow or density of the liquid phase. Assuming $A_g \bar{u}_g \rho_g$ to be constant is also reasonable since much of the gas phase is inert and since oxygen consumption will be partially compensated by transfer of product CO_2 into the gas phase. Since pressure varies with z, we cannot assume $A_g \bar{u}_g$ is constant. Instead, the ideal gas law gives

$$A_g \bar{u}_g = \frac{(A_g \bar{u}_g)_{in} P_{in}}{P(z)} \tag{11.54}$$

Substituting Equations 11.3 (with A_i' replacing $A_i V$), 11.53, and 11.54 into Equation 11.46 gives

$$\frac{da_g}{dz} = \frac{[A_g(\mathscr{R}_A)_g - U_m A_i'(k_H a_g - a_l)] P}{(A_g \bar{u} P)_{in}} - \frac{\rho_l g}{P} a_g \tag{11.55}$$

This is the gas side balance for component A (oxygen). The liquid side balance for A gives

$$\frac{da_l}{dz} = \frac{\mathscr{R}_A}{\bar{u}_l} + \frac{U_m A_i'(k_H a_g - a_l)}{A_l \bar{u}_l} \tag{11.56}$$

A liquid side balance for the substrate (say, a waste hydrocarbon) is

$$\frac{db}{dz} = \frac{\mathscr{R}_B}{\bar{u}_l} \tag{11.57}$$

Equations 11.55 to 11.57 constitute the simplified model. They can be solved numerically to give a_g, a_l, and b_l as functions of z. Necessary parameters will be known from initial conditions except for $U_m A_i'$ and $\bar{u}_l$. Both of these could be treated as functions of z as far as the numerical integration is concerned. In practice, they will probably be assumed constant. The values of these "lumped" parameters can be estimated from actual reaction data or from literature correlations. See suggestions for further reading the end of this chapter.

Equations 11.2, 11.3 (with A_i' replacing A_iV), 11.45, and 11.46 constitute the basic set of design equations for two-phase, piston flow reactors. If the reactors are nonisothermal, two energy balances are also needed:

$$\pm \frac{d(A_l \bar{u}_l \rho_l H_l)}{dz} = -A_l(\Delta H_{\mathscr{R}}\mathscr{R})_l + h_i A_i'(T_g - T_l) + UA_{\text{ext}}'(T_{\text{ext}} - T_l) \qquad (11.58)$$

and

$$\frac{d(A_g \bar{u}_g \rho_g H_g)}{dz} = -A_g(\Delta H_{\mathscr{R}}\mathscr{R})_g + h_i A_i'(T_l - T_g) \qquad (11.59)$$

In writing these equations we have again assumed that the liquid phase is continuous and that any external heat transfer occurs through that phase.

This section has treated two-phase, piston flow reactors while the previous section treated two-phase perfect mixers. Combinations such as piston flow in the liquid phase and perfect mixing in the gas phase (e.g., a spray reactor) can be treated within the same theoretical framework. Table 11.3 lists a variety of

Table 11.3 Typical Flow and Mixing Regimes for Gas – Liquid reactors

Type of Reactor	Liquid Phase	Gas Phase
Stirred tank with sparger	Continuous, well mixed	Discontinuous, but usually assumed well mixed
RDC and pulsed columns	Continuous, piston flow	Dispersed, piston flow
Bubble columns	Continuous, piston flow	Dispersed, piston flow
Packed columns	Continuous or trickle, piston flow	Continuous, trickle, or dispersed; piston flow
Tray columns	Continuous, well mixed on trays	Discontinuous, piston flow but often assumed well mixed on an individual tray
Spray towers	Discontinuous, piston flow	Continuous, typically well mixed

reactor types together with the flow and mixing assumptions usually used to model them. Nonideal flow models are becoming possible but are not yet a standard design tool.

11.1.3 Surface Renewal and Enhancement Factors

The *film model* for interphase mass transfer envisions a stagnant film of liquid adjacent to the interface. A similar film may also exist on the gas side. These films, which act much like membranes, cause diffusional resistances to mass transfer. The concentration on the gas side of the liquid film is a_i^*; that on the bulk liquid side is a_i; and concentrations within the film are governed by the one-dimensional, steady state diffusion equation

$$\mathcal{D}_A \frac{d^2 a}{dx^2} = 0 \tag{11.60}$$

This is subject to the boundary conditions that $a = a_i^*$ at $x = 0$ and $a = a_l$ at $x = \delta$. The solution is

$$a(x) = a_i^* + (a_l - a_i^*)\frac{x}{\delta} \tag{11.61}$$

The flux through the film is given by

$$-\mathcal{D}_A \frac{da}{dx} = \frac{\mathcal{D}_A}{\delta}(a_i^* - a_l) \tag{11.62}$$

Multiplying this result by A_i and comparing it to Equation 11.1 gives

$$k_l = \frac{\mathcal{D}_A}{\delta} \tag{11.63}$$

This is the central result of film theory; and, as is discussed in any good text on mass transfer, it happens to be wrong. Experimental measurements show k_l proportional to $\sqrt{\mathcal{D}_A}$ rather than to $\mathcal{D}_A$, at least in turbulent flow systems.

Two rather similar models have been devised to remedy the problems of simple film theory. Both the *penetration theory* of Higbie and the *surface renewal theory* of Danckwerts replace the idea of steady state diffusion across a film with transient diffusion into a semi-infinite medium. We give here a brief account of surface renewal theory. It will be applied to the important case of absorbing a highly reactive gas.

Surface renewal theory envisions a continuous exchange of material between the bulk fluid and the interface. A mechanism such as eddy diffusion brings material, of uniform composition a_l, to the interface and exposes it to the gas phase for a period of time t. The exposed fluid is then replaced with fresh fluid, and so on. Diffusion into the freshly exposed surface is governed by

$$\frac{\partial a}{\partial t} = \mathcal{D}_A \frac{\partial^2 a}{\partial x^2} \tag{11.64}$$

subject to the initial condition that $a = a_l$ at $t = 0$ and boundary conditions that $a = a_l^*$ at $x = 0$ and $a = a_l$ at $x = \infty$. The solution to this equation is differentiated to give the flux as in Equation 11.62. Unlike film theory, the flux into the surface varies with the exposure time t, being high at first but gradually declining as the concentration gradient at $x = 0$ decreases. For short exposure times,

$$-\mathcal{D}_A \frac{\partial a}{\partial x}\bigg|_{x=0} = (a_l^* - a_l)\sqrt{\frac{\mathcal{D}_A}{\pi t}} \tag{11.65}$$

This result gives the flux for a small portion of the surface that has been exposed for exactly t seconds. Other portions of the surface will have been exposed for different times and thus will have different instantaneous fluxes. To find the average flux, we need the differential distribution of exposure times, $f(t)$. Danckwerts assumed this to be an exponential distribution having mean τ:

$$f(t) = \frac{1}{\tau} \exp\left(-\frac{t}{\tau}\right) \tag{11.66}$$

where $f(t)\, dt$ is the fraction of the interfacial area that has been exposed from t to $t + dt$ seconds. The average flux is

$$\int_0^\infty \left[\frac{1}{\tau} \exp\left(-\frac{t}{\tau}\right)\right](a_l^* - a_l)\sqrt{\frac{\mathcal{D}_A}{\tau}}\, dt = (a_l^* - a_l)\sqrt{\frac{\mathcal{D}_A}{\tau}} \tag{11.67}$$

$$k_l = \sqrt{\frac{\mathcal{D}_A}{\tau}} \tag{11.68}$$

which has the required proportionality to $\sqrt{\mathcal{D}_A}$.

We turn now to the case of gas absorption with chemical reaction. Even a slow reaction in the bulk liquid can give $a_l \approx 0$ since the volume of the bulk liquid can be quite large. Thus the existence of the reaction can increase mass transfer rates by increasing the overall driving force, but the slow reaction will not affect k_l. With a faster reaction, however, component A will be consumed on a time scale commensurate with the surface renewal time τ. This consumption will increase the concentration gradient and thus the diffusive flux. The governing equation is now

$$\frac{\partial a}{\partial t} = \mathscr{D}_A \frac{\partial^2 a}{\partial z^2} + \mathscr{R}_A \tag{11.69}$$

with initial and boundary conditions identical to those for Equation 11.64. For a first order reaction, $\mathscr{R}_A = -ka$, and the average flux is

$$\int_0^\infty \left[\frac{1}{\tau} \exp\left(-\frac{t}{\tau}\right) \right]\left[-\mathscr{D}_A \frac{\partial a}{\partial x}\Big|_0 \right] dt = (a_l^* - a_l)\sqrt{1 + (k_\tau)^2}\sqrt{\frac{\mathscr{D}_A}{\tau}} \tag{11.70}$$

Thus

$$k_l = \sqrt{1 + (k\tau)^2}\sqrt{\frac{\mathscr{D}_A}{\tau}} = (k_l)_0 \mathscr{E} \tag{11.71}$$

where $\mathscr{E} = k_l/(k_l)_0$ is known as the **enhancement factor** and $(k_l)_0$ denotes what the mass transfer coefficient would be were there no reaction. Measurements of k_l and $(k_l)_0$ have been used to estimate τ. It is typically on the order of 10^{-2} or 10^{-3} seconds so that quite fast reactions are needed to give significant enhancement of mass transfer coefficients.

Consider a very fast, second order reaction between the gas phase component A and a liquid component B. The concentration of B will quickly fall to zero in the vicinity of the freshly exposed surface; and a reaction plane, within which $b = 0$, will gradually move away from the surface. The enhancement factor for this situation is

$$\mathscr{E} = 1 + \frac{b_{\text{bulk}}}{a_l^*} \tag{11.72}$$

assuming A and B have similar liquid phase diffusivities. Since a_l^* is small for sparingly soluble gases, the enhancement factor can be quite large.

11.2   Moving Solids Reactors

Fixed-bed reactors are ideal for many solid-catalyzed gas reactions. The contacting of the solid by the gas tends to be quite uniform, and long contact times (see Equation 10.13) are possible. However, packed beds have severe heat transfer limitations, and scaleup must often be done using many small-diameter tubes in parallel rather than a single, large-diameter bed. Also, the large particle sizes needed to minimize pressure drop lead to diffusional resistances within the catalyst particles. If catalyst deactivation is rapid, the fixed-bed geometry may cause problems in regeneration. For gas–solid noncatalytic reactions, the solid particles may shrink or grow as the reaction proceeds. This too is not easily accommodated in a fixed bed.

Many types of gas–solid reactors have been designed to allow motion of the solid relative to the fixed walls of the reactors. This motion is desired for one of the following reasons:

1. To enhance heat transfer between the particle and the environment.
2. To enable use of small particles.
3. To enable continuous regeneration of catalyst particles.
4. To facilitate continuous removal of ash and slag.
5. To accommodate size changes of the particles concurrent with reaction.

The particle motion can be accomplished by purely mechanical means—perhaps aided by gravity—as in rotary cement kilns and fireplace grates. Chemical engineers usually prefer designs where the particle motion is brought about through hydrodynamic forces that are generated by a fluid phase that also participates in the reaction. Such designs tend to be more controllable and

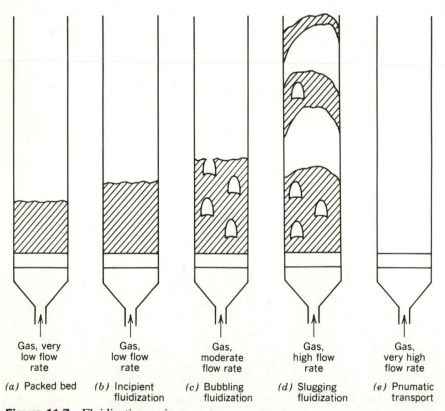

Gas, very low flow rate	Gas, low flow rate	Gas, moderate flow rate	Gas, high flow rate	Gas, very high flow rate
(a) Packed bed	(b) Incipient fluidization	(c) Bubbling fluidization	(d) Slugging fluidization	(e) Pnumatic transport

Figure 11.7 Fluidization regimes.

scalable, although scalability can be a problem. Sophisticated pilot plant and modeling efforts are usually necessary for any form of *fluidized-bed reactor*.

Figure 11.7 illustrates various regimes of fluidization for small-diameter catalyst particles in a gas-fluidized bed. Typically, the particles will be roughly spherical with a mean size of about 60 μm. The size distribution will be fairly broad with 95 weight-% in the range of 30 to 100 μm. The particle density will be just over 1 gm/cm^3. The particles rest on a microporous distributor plate and, at low flow rates, act as a packed bed. At high flow rates, the drag on the particles will act to lift them off the surface. The pressure drop across the bed will increase with increasing flow rate and at some point will exactly equal the weight of the bed. The linear velocity at this point is called the *minimum fluidization velocity*, u_{min}. The particles will be mobile in the sense that they are suspended and could be mechanically stirred, but relatively little solids motion will be caused just by the gas flow. Incipient fluidization is illustrated in Figure 11.7b. Note that no gas bubbles exist at this point.

As the gas flow is increased beyond the minimum fluidization velocity, voids that are essentially free of solid particles begin to form and move upward. These voids act much like bubbles in a gas–liquid system. The voids grow as they rise in the bed due to pressure reduction and coalescence and provide a stirring mechanism for the suspended particles. Figure 11.7c illustrates the bubbling regime of fluidization which lasts until the superficial gas velocity is many times higher than u_{min}. At this point, the bubbles are so large that they span the reactor and cause a qualitatively different form of fluidization known as slugging. Solids at the center of the bed are conveyed upward but rain back down near the walls. Figure 11.7d depicts the slugging regime. There is no clear upper boundary to this regime. It gradually merges into the regime of pneumatic transport as the gas velocity exceeds the terminal velocity for the smaller particles. A batch fluidized bed operating in the slugging regime will gradually coarsen as smaller particles are elutriated.

11.2.1 Bubbling Fluidization

Fluidized-bed reactors have received substantial attention from academic and industrial researchers. The dynamics of bubble formation and growth and of solids movement within the bed have been analyzed in great detail, and elaborate computer simulations have been developed for all regimes of fluidization. The reader must be referred to the specialized (and sometimes proprietary) literature for details on such models. Here we describe a fairly simple model that is applicable to the bubbling regime and that treats a catalytic fluidized bed much like a gas–liquid reactor. The bubbles play the role of the gas phase. Acting much like a liquid is an *emulsion phase*, which consists of solid particles and suspending gas at conditions similar to those of incipient fluidization. The quasi-phases are in cocurrent flow with reaction occurring only in the emulsion phase. The downward flow of solids that occurs near the walls is not explicitly considered in this simplified model.

For the emulsion phase,

$$A_e u_e \frac{da_e}{dz} = A_e \mathcal{R}_A + U_m A_i'(a_b - a_e) + A_e D_e \frac{d^2 a_e}{dz^2} \tag{11.73}$$

and for the bubble phase,

$$A_b u_b \frac{da_b}{dz} = -U_m A_i (a_b - a_e) \tag{11.74}$$

These are versions of Equations 11.45 and 11.46, except that we have used overall mass transfer coefficients and overall driving forces and have added an axial dispersion term for the emulsion phase. The various parameters $(A_e, u_e, U_m, A_i, D_e, A_b, u_b)$ are assumed independent of position in the bed although they will generally depend on operating conditions.

Converting Equations 11.73 and 11.74 to dimensionless form gives

$$\frac{da_e}{dz} = \bar{t}_e \mathcal{R}_A + \left(\frac{U_m A_i}{u_e A_e}\right)(a_b - a_e) + \frac{1}{P_e} \frac{d^2 a_e}{dz^2} \tag{11.75}$$

and

$$\frac{da_b}{dz} = -\left(\frac{U_m A_i}{u_e A_e}\right)\left(\frac{u_e A_e}{u_b A_b}\right)(a_b - a_e) \tag{11.76}$$

The boundary conditions associated with Equation 11.75 are the closed, or Danckwerts, variety:

$$a_{in} = a_e(0^+) - \frac{1}{P_e} \frac{da_e}{dz}\bigg|_{0^+}$$

$$\frac{da_e}{dz}\bigg|_1 = 0 \tag{11.77}$$

Equation 11.76 governs the bubble phase and has the usual boundary condition for piston flow:

$$a_b(0) = a_{in} \tag{11.78}$$

The exit concentration is an average for both phases:

$$a_{out} = \frac{Q_e}{Q}(a_e)_{out} + \frac{Q_b}{Q}(a_b)_{out} \tag{11.79}$$

The intrinsic kinetics will usually be expressed in terms of catalyst mass, that is, moles reacted per time per mass of catalyst. Then the pseudohomogeneous rate to be used in Equation 11.75 is

$$[\mathscr{R}_A]_{\text{pseudohomogeneous}} = \frac{\rho_e}{\varepsilon_e}[\mathscr{R}_A]_{\text{catalyst mass}} \qquad (11.80)$$

which follows from Equation 10.14 for the case of heterogeneous rates based on catalyst mass rather than surface area.

Equations 11.75 to 11.80 constitute a fairly simple model for a gas-fluidized, catalytic reactor. These equations contain three dimensionless groups: $(U_m A_i)/(u_e A_e)$, $P_e = u_e L/D_e$, and $(u_e A_e)/(u_b A_b) = Q_e/Q_b$. Additionally, one must know the emulsion phase contact time, $\rho_e \bar{t}_e/\varepsilon_e$, which becomes a fourth dimensionless group when combined with the rate constant, for example, $\rho_e \bar{t}_e k/\varepsilon_e$ for a first order reaction. Values for the various parameters can be estimated from published correlations—see Suggestions for Further Reading. These should be consulted for any serious design calculations. However, qualitative insight into the use of bubbling fluidized beds as catalytic reactors can be obtained from ad hoc models based on additional simplifying assumptions.

Example 11.11   To a first approximation, the properties of the emulsion phase remain constant as the flow rate through the bed is increased. Thus u_e, ρ_e, ε_e, and V_e are independent of Q in the bubbling regime of fluidization. The area of the emulsion phase decreases as Q increases since bubbles occupy part of the reactor cross section. This causes the bed to expand since $LA_e = V_e$ is constant, and the volumetric flow rate through the emulsion phase, $Q_e = A_e u_{\min}$, actually decreases. The area of the bubble phase starts at zero when $Q = Q_{\min}$ and increases with increasing flow rate until $A_b = A_c$ at $Q = Q_{\text{slug}}$. Suppose the variation of A_b with Q is approximately linear over $Q_{\min} < Q < Q_{\text{slug}}$. Then

$$A_b = \frac{A_c(Q - Q_{\min})}{Q_{\text{slug}} - Q_{\min}}$$

$$A_e = \frac{A_c(Q_{\text{slug}} - Q)}{Q_{\text{slug}} - Q_{\min}}$$

$$Q_e = A_e u_{\min} = \frac{Q_{\min}(Q_{\text{slug}} - Q)}{Q_{\text{slug}} - Q_{\min}} \qquad (11.81)$$

$$Q_b = A_b u_b = \frac{Q_{\text{slug}}(Q - Q_{\min})}{Q_{\text{slug}} - Q_{\min}}$$

$$\frac{\rho_e \bar{t}_e}{\varepsilon_e} = \frac{\rho_{\min} L}{\varepsilon_{\min} u_{\min}} = \bar{t}_{\min} \frac{L}{L_{\min}} = \frac{\bar{t}_{\min}(Q_{\text{slug}} - Q_{\min})}{Q_{\text{slug}} - Q}$$

This set of equations defines the phase volumes and velocities given bed properties at incipient fluidization and Q_{slug}.

Now suppose that a first order, solid-catalyzed reaction is occurring and that the following parameters are known from literature sources or have been measured:

$$Q_{slug} = 15 Q_{min} \tag{11.82}$$

$$P_e = \frac{5 Q_{min}}{Q - Q_{min}} \tag{11.83}$$

$$\frac{U_m A_i}{u_e A_e} = \frac{5.7 Q}{Q_{min}} \tag{11.84}$$

$$\frac{\rho_e \bar{t}_e k}{\varepsilon_e} = \frac{2(Q_{slug} - Q_{min})}{Q_{slug} - Q} \tag{11.85}$$

The first three of these assumptions represent typical behavior for small-diameter, fluidized beds. Note that Equation 11.83 predicts piston flow ($P_e = \infty$) at $Q = Q_{min}$ and a close approach to perfect mixing at $Q > 10 Q_{min}$. The final assumption supposes that $\rho_e \bar{t}_e k / \varepsilon_e = 2$ at incipient fluidization so that $a_{out}/a_{in} = \exp(-2) = 0.135$ at $Q = Q_{min}$. The model now consists of Equations 11.75 through 11.85 and allows a_{out}/a_{in} to be predicted for flow rates in the range $Q_{min} < Q < Q_{slug}$.

The numerical evaluation of the model uses the shooting method and is similar to the axial dispersion example in Section 7.1.1. The marching-ahead equations are

$$a_{j+1} = \left[2 + P_e \Delta z + P_e \left(\frac{\rho_e \bar{t} k}{\varepsilon_e} \right) \Delta z^2 + \left(\frac{U_m A_i}{u_e A_e} \right) P_e \Delta z^2 \right] a_j$$
$$\tag{11.86}$$
$$- (1 + P_e \Delta z) a_{j-1} - \left(\frac{U_m A_i}{u_e A_e} \right) P_e \Delta z^2 b_j$$

and

$$b_{j+1} = \left[1 - \left(\frac{U_m A_i}{u_e A_e} \right) \left(\frac{Q_e}{Q_b} \right) \Delta z \right] b_j + \left(\frac{U_m A_i}{U_e A_e} \right) \left(\frac{Q_e}{Q_b} \right) \Delta z \, a_j \tag{11.87}$$

where $a = a_e$ and $b = a_b$. The initial condition is

$$a_1 = (1 + P_e \Delta z) a_0 - P_e \Delta z \, a_{in} \tag{11.88}$$

For $Q/Q_{min} = 10$, the various parameters are $P_e = 0.556$, $(U_m A_i)/(u_e A_e) = 57$, $\rho_e \bar{t} k / \varepsilon_e = 5.6$, and $Q_e/Q_b = 0.037$. For $\Delta z = 0.0625$ the marching-ahead equa-

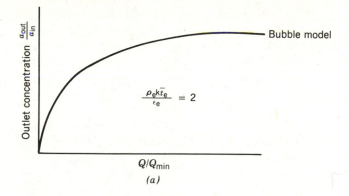

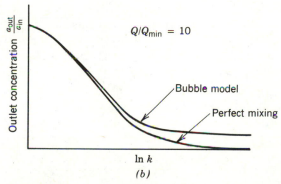

Figure 11.8   Predictions of simplified model of a bubbling fluidized bed.

tions are

$$a_1 = 1.034729a_0 - 0.03472$$

$$a_{j+1} = 2.05056a_j - 1.03472a_{j-1} - 0.00369b_j \qquad \text{(11.89)}$$

$$b_{j+1} = 0.99606b_j + 0.00369a_j$$

A few results are:

Δz	$a_E(0^+)$	$(a_E)_{out}$	$(a_B)_{out}$	a_{out}
$\frac{1}{16}$	0.8983	0.7861	0.8344	0.8327
$\frac{1}{32}$	0.8953	0.7842	0.8349	0.8331
$\frac{1}{64}$	0.8938	0.7834	0.8352	0.8333
$\frac{1}{128}$	0.8930	0.7829	0.8353	0.8334

Convergence is $0(\Delta z)$. This example is not as ill-conditioned as the one in Section 7.1.1 but becomes worse as $Q/Q_{min} \to 1$ and $Pe \to \infty$. Figure 11.8 shows a_{out}/a_{in} as a function of Q/Q_{min}.

The model predictions in Figure 11.8 illustrate the qualitative behavior of gas-fluidized beds operating in the bubbling regime. Figure 11.8a shows the response of the outlet concentration to changes in gas flow rate. The fraction unreacted rises sharply as flow rate is increased. This rise is much faster than would occur in a perfect mixer and is slightly faster than would occur in a piston flow reactor (which also responds poorly to upward demand changes). Figure 11.8b shows the model response to a change in the first order rate constant at a fixed flow rate. Amounts of gas remain unreacted despite very high values for k. The bed behaves quite badly if high conversions are desired. Part of the entering gas remains in the bubble phase and effectively bypasses the emulsion phase where reaction occurs. In this respect the bed behaves worse than a perfect mixer (which is also poor at achieving high conversions).

Example 11.12   Determine the fraction of gas bypassing the emulsion phase for the model of the previous example. Define bypassing as the fraction unreacted for a first order reaction in the limit of high rate constant.

Solution   As $k \to \infty$, $a_e \to 0$ and Equation 11.76 becomes

$$\frac{da_b}{dz} = -\left(\frac{U_m A_i}{u_e A_e}\right)\left(\frac{u_e A_e}{u_b A_b}\right) a_b \tag{11.90}$$

which has the solution

$$(a_b)_{\text{out}} = a_{\text{in}} \exp\left[-\left(\frac{U_m A_i}{u_b A_b}\right)\frac{Q_e}{Q_b}\right] \tag{11.91}$$

Combining this with Equations 11.79 and 11.81 gives

$$\frac{a_{\text{out}}}{a_{\text{in}}} = \frac{Q_{\text{slug}}(Q - Q_{\text{min}})}{Q(Q_{\text{slug}} - Q_{\text{min}})} \exp\left[-\left(\frac{U_m A_i}{u_b A_b}\right)\frac{Q_{\text{min}}(Q_{\text{slug}} - Q)}{Q_{\text{slug}}(Q - Q_{\text{min}})}\right] \tag{11.92}$$

Using the model parameters in Equations 11.82 to 11.85 gives the following results:

Q/Q_{min}	Fraction Bypassed
1	0
2	2.7×10^{-5}
4	3.1×10^{-3}
8	4.5×10^{-2}
12	0.28
14	0.66

11.2.2 Other Fluidization Regimes

Many industrial fluidized beds operate with flow rates substantially above the bubbling regime. They have gas velocities higher than the entrainment velocity of at least the smaller particles and sometimes higher than the entrainment velocity of all particles. This is called *fast fluidization*, or *transport-line* fluidization. Elutriated particles are collected in a cyclone and continuously recycled back to the inlet of the reactor. (Particle recycle is often combined with catalyst regeneration.) A well-defined bed of particles may not exist in such systems. Instead, the particles are distributed more or less uniformly throughout the reactor. The two-phase model does not apply. Typically, the reactor is described with a pseudohomogeneous, axial dispersion model. Gas–solids contacting tends to be fairly uniform, and heat transfer between the particles and the gas stream is quite good. The small particles give effectiveness factors near 1. This reaction environment appears ideal: It is nearly isothermal and near to piston flow, and the intrinsic kinetics are applicable. However, the maximum contact time in such a reactor is quite limited because of the low catalyst densities and high gas velocities that prevail in a fast-fluidized or transport-line reactor. Thus the reaction must be fast, or low conversions must be acceptable. Also, the catalyst must be quite robust to minimize particle attrition.

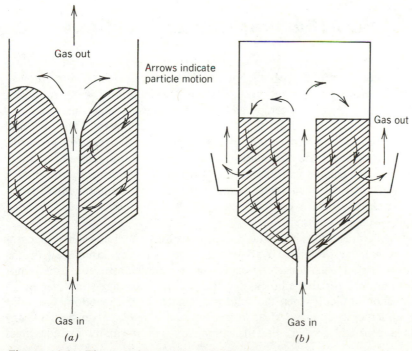

Figure 11.9 The spouting regime of fluidization.

A very different regime of fluidization is called *spouting*. Spouting can occur —and is usually undesirable—in a normal fluidized bed if the gas is introduced at localized points rather than being distributed evenly over the rector cross section. See Figure 11.9a. Spouting provides uniform gas–solid contacting, good fluid-particle heat transfer, and relatively long contact times when used in a draft-tube, side exit reactor as shown in Figure 11.9b.

The velocity in the spout is high enough to entrain all particles, but they disengage in the low-velocity regions above the bed. This causes circulation of particles with upward movement in the spout but generally downward motion in the bed. Contact times within the spout are quite short, and little reaction occurs there. Thus the freely spouted bed in Figure 11.9a would show an extreme amount of bypassing. In Figure 11.9b, however, the gas is forced to turn around and flow cocurrently with the downward-moving particles. The reaction environment in this region is close to that in a fixed-bed reactor, but the overall reactor is substantially better than a fixed-bed reactor in terms of fluid-particle heat transfer and heat transfer to the reactor walls. To a reasonable approximation, the reactor in Figure 11.9b can be modeled as a (possibly nonisothermal) piston flow reactor with recycle. The fluid mechanics of spouting have been examined in detail so that model variables such as pressure drop, gas recycle rate, and solids circulation rate can be estimated. Spouted bed reactors use rather larger particles than the usual fluid bed. One-millimeter particles are typical, compared to 50 to 100 μm for most fluidizable catalysts.

11.3 Fluid–Solid Noncatalytic Reactions

Consider now the case where a solid directly participates in an overall reaction. Examples include the burning of solid fuels, the decoking of cracking catalyst, the reduction of iron ore with hydrogen, and the purification of water in an ion exchange bed. A unifying aspect of all these examples is that the solid participates directly and will ultimately be consumed or exhausted. The size and shape of the fluid–solid interface will usually change as the reaction proceeds. Mass transfer resistances are frequently important, and the magnitude of these resistances may also change with the extent of reaction. The diversity of possible chemistries and physical phenomena is too great to allow comprehensive treatment. We necessarily take a limited view and refer the reader to the research literature on specific processes.

A glib generalization is that the design equations for fluid–solid noncatalytic reactors can be obtained by combining the intrinsic kinetics with the appropriate transport equations. The experienced reader knows that this is not always possible even for the solid-catalyzed reactions considered in Chapter 10 and is much more difficult when the solid participates in the reaction. The solid surface is undergoing change. Measurements usually require transient experiments.[3] As a practical matter, the measurements will normally include mass

[3] Steady state experiments are possible in fluidized beds and other moving solids reactors. One must then deal with a population of particles, no two of which may be quite alike.

transfer effects and are often made in pilot-scale equipment intended to simulate a full-scale reactor.

Consider a gas–solid reaction of the general form

$$\nu_A A + \nu_S S \rightarrow \text{Products} \tag{11.93}$$

Any of the following mass transfer resistances can be important

Film diffusion   With a fast surface reaction on a nonporous particle, fluid phase mass transfer limitations can arise.

Pore diffusion   With porous particles, pore diffusion is likely to limit reaction rates at the internal surface.

Product layer diffusion   Many fluid–solid reactions generate ash or oxide layers that impede further reaction.

Sublimation   Some solids sublime before they react in the gas phase. Heat transfer can then become the rate-limiting step.

Finally, of course, the reaction itself can be rate limiting.

A useful, semiempirical approach to noncatalytic surface reactions is to postulate the following rate equation:

$$\mathscr{R}_A = \nu_A k a A_i \tag{11.94}$$

This does not address the mechanism of the surface reaction but supposes that the rate will be proportional to the exposed area. For $\mathscr{R}_A$ to have its usual interpretation as moles formed per unit volume of reactor per unit time, A_i should be the surface area of the fluid–solid interface per unit volume of reactor. For single-particle experiments, A_i will be the surface area and $\mathscr{R}_A$ will be in moles reacted per unit time.

Example 11.13   Explore the suitability of Equation 11.94 for reflecting various forms of mass transfer and kinetic limitations.

Solution   We consider each possibility in turn as being rate limiting. For film diffusion control,

$$\mathscr{R}_A = k_g A_i (a_s - a) \tag{11.95}$$

But if mass transfer is strictly limiting, $a_s \approx 0$. Thus with a redefinition of constants, Equation 11.95 agrees with Equation 11.94.

If pore diffusion is controlling, we repeat the effectiveness factor calculations of Section 10.3.1 to obtain

$$\mathscr{R}_A = -\eta k a A_i \tag{11.96}$$

which also has the functional form of Equation 11.94.

The effective diffusivity model of Section 10.1.3 can be used for diffusion through a product layer. For spherical particles,

$$\frac{\partial a_r}{\partial \theta} = \mathscr{D}_{\text{eff}} \left(\frac{\partial^2 a_r}{\partial r_p} + \frac{2}{r_p} \frac{\partial a_r}{\partial r_p} \right) = 0 \tag{11.97}$$

The boundary and initial conditions associated with this Equation are

$$a_r = a \qquad \text{at} \qquad r_p = d_p/2$$

$$a_r = 0 \qquad \text{at} \qquad r_p = d_s/2 \tag{11.98}$$

$$d_p = d_s = d_0 \qquad \text{at} \qquad \theta = 0$$

where d_p is the diameter of the particle including the product layer and d_s is the diameter of the unreacted core. Note that there is no reaction within the product layer and thus no reaction term in Equation 11.97. There is a sharp reaction front located at $r_p = d_s/2$, and the concentration of A is zero at that point. The flux, however, is not zero and determines the reaction rate:

$$\mathscr{R}_A = A_i \mathscr{D}_{\text{eff}} \frac{\partial a_r}{\partial r_p} \bigg|_{d_s/2} \tag{11.99}$$

where we are adopting the single-particle viewpoint, with $\mathscr{R}_A$ in moles per unit time. Equation 11.97 explicitly recognizes the transient nature of single-particle experiments. Both d_p and d_s will be functions of time, and thus the diffusion path length will vary with time. However, the rate of change of $(d_p - d_s)/2$ will usually be quite small compared to the net diffusion velocity:

$$\frac{d(d_p - d_s)}{d\theta} \ll \frac{D_{\text{eff}}}{\rho_A} \frac{\partial a_r}{\partial r_p} \bigg|_{d_s/2} \tag{11.100}$$

This allows a type of pseudo-steady state assumption, and Equation 11.97 reduces to

$$\frac{d^2 a_r}{dr_p^2} + \frac{2}{r_p} \frac{da_r}{dr_p} = 0 \tag{11.101}$$

We leave it to the mathematically inclined reader to solve this equation subject to the boundary conditions of Equation 11.98 and then to show that the flux is directly proportional to the fluid phase concentration a. Thus Equation 11.99 becomes equivalent in form to Equation 11.94.

Table 11.4   Examples of Fluid – Solid Reactions

Particle Geometry Largely Unaffected by Reaction	Particle Geometry Strongly Affected by Reaction
Decoking of catalyst pellets	Combustion of coal
Ion exchange reactions	Reduction of ore
Hydrogen storage in a metal lattice	Production of acetylene from CaC_2
Semiconductor doping	Semiconductor etching

We expect the sublimation rate to depend on the surface area of the solid and on the temperature. Thus if sublimation is rate controlling,

$$\mathscr{R}_A = \nu_A k' A_i \tag{11.102}$$

where k' is a temperature-dependent constant. The fluid phase concentration of A will be constant at its bulk value a. Equation 11.102 will agree with Equation 11.94 provided $k' = ka$. Thus sublimation represents a special case where the reaction becomes zero order with respect to A.

For the final situation, that of a strict rate limitation, we should use a rate equation that agrees with the intrinsic kinetics. No sweeping claims can be made for Equation 11.94. However, experience shows that first order rate expressions often provide an excellent fit to experimental data regardless of the underlying reaction mechanism.

The analysis of fluid–solid reactions is easier when the particle geometry is independent of the reaction. Table 11.4 lists situations where this assumption is or is not reasonable. However, even when the reaction geometry is fixed, moving boundary problems and sharp reaction wave fronts are the general rule for fluid–solid reactions. The next few examples explore this point.

Example 11.14   It is sometimes possible to design systems where the reaction and diffusion steps are both fast compared to bulk transport by convection. This is the design intent for ion exchange columns. The reaction front moves through the bed at a speed dependent only on the supply of fluid phase reactants. Assuming piston flow in a constant-diameter column, the location of the reaction front is given by

$$z_{\mathscr{R}}(\theta) = \frac{1}{C_A} \int_0^\theta a_{in}(\theta) \bar{u}(\theta) \, d\theta \tag{11.103}$$

Here, C_A is the capacity of the ion exchange resin measured in moles of A per unit volume. *Breakthrough* occurs when all the active sites on the ion exchange resin are occupied and $z_{\mathscr{R}} = L$. It will occur somewhat sooner than this, namely, at lower values of the integral in Equation 11.103, if the reaction front becomes diffuse because of mass transfer or reaction rate limitations.

Example 11.15 Coke formation is a major cause of catalyst deactivation. Decoking is accomplished by periodic oxidations in air. Consider a microporous catalyst that has its internal surface covered with a uniform layer of coke. Suppose that the decoking reaction is stopped short of completion. What is the distribution of residual coke under the following circumstances:

(a) The oxidation is reaction rate limited?
(b) The oxidation is pore diffusion limited?

Solution For part (a), oxygen has access to the entire internal surface. We expect a gradual reduction in coke thickness throughout the catalyst pellet. If a completely clean surface is required for catalytic activity, partial decoking will achieve very little.

For part (b), the reaction is fast, and oxygen is consumed as soon as it contacts carbon. Thus there are two zones in the pellet. The outer zone contains oxygen and no carbon. The inner zone contains carbon at its original thickness and no oxygen. The reaction is confined to a narrow front between the zones. The rate at which the front advances is determined by the rate of diffusion of oxygen and the extent of carbon loading in the pores. The diffusion rate in a spherical pellet can be found from Equation 11.101 subject to the boundary conditions of Equation 11.98. Note that we have again needed the pseudo-steady hypothesis since d_s will vary with time.

Partial oxidations in the diffusion-controlled regime give partial restoration of catalyst activity since some of the surface is completely cleaned. The net rate of oxidation is also higher; but because of the high temperatures, there is greater risk of surface loss through sintering. Decoking has been studied extensively because of its importance to the chemical industry. The two cases considered in this example are known as the *uniform* and *shell progressive* models, respectively. See H. H. Lee, *Homogeneous Reactor Design*, Butterworth, Boston, 1985, for further details.

Example 11.16 In the absence of mass transfer limitations, fluid–solid noncatalytic reactions consume the solid at a constant linear rate. To see that this is true, start with Equation 11.94 but note that the reaction will be zero order with respect to A since we are assuming that A is available in great excess. Thus

$$\mathscr{R}_s = -kA_i \quad \text{mol/s} \tag{11.104}$$

Divide by $\rho_s A_i$ to find the linear burning rate

$$\frac{dz}{d\theta} = -\frac{k}{\rho_s} \quad \text{m/s} \tag{11.105}$$

Here, the coordinate z is in the direction normal to the surface being burned, and ρ_s is the density of the solid.

For a sphere of carbon slowly oxidizing in air, $z = r$, and Equation 1.105 predicts that the diameter of the sphere will decrease at a constant rate.

11.4   Electronic Device Fabrication

The fabrication of modern electronic devices, such as large-scale integrated circuits, involves an elaborate sequence of chemical operations. A typical process starts with a wafer of high-purity silicon that has been cut from a single crystal. Electronic functionality is achieved by creating a multilayer structure in and on the surface of the wafer in a precise geometric pattern. The pattern is laid down by a process known as *photolithography* using the following sequence of steps:

1. The surface is coated with a polymer known as a *photoresist*.
2. An image is formed on the surface using UV or visible light. This causes crosslinking of the polymer in those areas exposed to light.
3. The polymer covering those regions *not* exposed to light is removed by a solvent.
4. The surface is treated with a chemical agent, or *dopant*, to modify it in those regions which were not exposed to light.
5. The crosslinked polymer is removed using a more aggressive solvent.

This procedure, with minor variations, is repeated dozens of times in the manufacture of a semiconductor chip. One important variation uses a *negative photoresist* in Step 1. This polymer degrades when exposed to light. Thus the exposed regions are removed by the solvent in Step 3 and are subjected to chemical treatment in Step 4. The chemical treatment can be done using reagents in a liquid phase, but gas phase treatments by a process known as *chemical vapor deposition* is becoming progressively more important as individual features in the integrated circuit become smaller than 1 μm.

Example 11.17   A frequently encountered example of chemical vapor deposition is the formation of polycrystalline silicon by the decomposition of silane:

$$SiH_4 \rightarrow Si + 2H_2 \tag{11.106}$$

The decomposition occurs on the surface and has an observed[4] rate equation of

---

[4] W. A. P. Claassen, et al., *J. Crystal Growth*, **57**, 259 (1982).

the form

$$\mathcal{R} = \frac{k[\mathrm{SiH_4}]}{1 + k_A[\mathrm{SiH_4}] + k_B[\mathrm{H_2}]^{1/2}} \tag{11.107}$$

This form suggests a Hougen and Watson mechanism in which silane and hydrogen atoms occupy sites that must also be used by the silicon being deposited.

The primary reaction of Equation 11.106 can be complemented by dopant reactions involving compounds such as $\mathrm{AsH_3}$, $\mathrm{PH_3}$, and $\mathrm{B_2H_6}$, which deposit trace amounts of the dopant metals in the silicon lattice. Dopant atoms can also be deposited directly from the vapor phase without use of a chemical intermediate such as $\mathrm{AsH_3}$ that decomposes on the surface. Vapor deposition of conductors such as gold and silver is usually called *vacuum metallizing* and allows the internal connections to be made in an integrated circuit.

Example 11.18 Electrical connections are required between the various layers that are deposited on a wafer. As a first step in providing these connections, channels are cut into the surface using photolithography followed by *chemical etching*. Straight-sided channels having a width equal to the opening in the photoresist are desired. See Figure 11.10a. Experience shows that the channel will undercut the photoresist by an appreciable distance. Explain this.

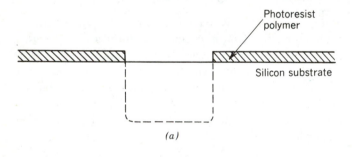

(a)

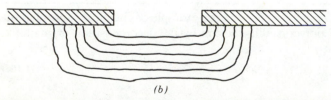

(b)

Figure 11.10 Chemical etching to create interconnections between layers on a silicon wafer.

Solution The chemical etchant is an aggressive fluid that reacts with the solid. The reaction proceeds in a direction everywhere normal to the existing surface. See Example 11.16. The walls of the channel will be attacked at the same rate as the base. Since there are two walls being attacked simultaneously, the minimum possible width-to-depth ratio is 2. See Figure 11.10b.

From a reaction engineering viewpoint, semiconductor device fabrication is a sequence of semibatch reactions interspersed with more strictly mass transfer steps such as polymer dissolution. Mass transfer also plays a role in chemical vapor deposition itself in the sense that adsorption, desorption, and crystal growth must occur as part of the overall reaction of Equation 11.93.

Manufacturing economics require that many devices be fabricated simultaneously in large reactors. Uniformity of treatment from point to point is extremely important, and the possibility of concentration gradients in the gas phase must be considered. For some reactor designs, standard models such as axial dispersion may be suitable for describing mixing in the gas phase.[5] More typically, many vapor deposition reactors have such low L/R ratios that two-dimensional dispersion must be considered:

$$\bar{u}\frac{\partial a}{\partial z} = D_z\frac{\partial^2 a}{\partial z^2} + D_r\left(\frac{1}{r}\frac{\partial a}{\partial r} + \frac{\partial^2 a}{\partial r^2}\right) + \mathscr{R}_{\text{A}} \tag{11.108}$$

This model has been applied to vacuum coaters where the material being vapor deposited is evaporated from one or more point sources. Note that D_z and D_r are empirical parameters that account for both convection and diffusion. Rotary vacuum coaters avoid any dependence in the θ-direction by rotating the substrate as it is coated.

Not all fabrication techniques for electronic materials are restricted to batch or semibatch reactors. The manufacture of magnetic recording tape begins with a dispersion of magnetic particles in a medium that contains a UV-curable prepolymer. The dispersion is coated and cured onto a polyester film substrate in a continuous operation. Continuous coating is also used to manufacture the organic photoconductors used in xerographic copying machines. These examples involve liquid coatings, and the technology is rather similar to that for manufacturing photographic films. However, continuous vapor deposition is used for metallizing plastic films, and chemical vapor deposition is used to manufacture photovoltaic cells.[6]

[5] See K. F. Jensen and K. F. Roenigh, "Modelling of Reactors for Chemical Vapor Deposition of Microelectronic Materials," *Insty. Chem. Eng. Symp. Ser. No. 87*, 255–262, 1984.

[6] T. W. F. Russell, "Chemical Reaction Engineering in Photovoltaic Cell Processing," *Instn. Chem. E. Symp. Ser. No. 87*, 271–277, 1984.

11.5 Three-Phase Reactors

Some reactor types and operating regimes involve three or even more phases. Attention will be restricted to the fairly specific situation of a gas phase containing reactants, a liquid phase containing reactants, and a solid phase that participates in the reaction catalytically or otherwise. Two examples are:

1. *The Trickle Bed Reactor* The solid is stationary, catalytic, and usually microporous. The liquid flows in a trickling regime where it wets the external surface of the catalyst but leaves substantial voidage available for the flow of gas. The usually industrial design is for cocurrent, downward flow of both liquid and gas. This reactor finds wide use in the hydrogenation and hydrodesulfurization of petroleum fractions.

2. *The Slurry Reactor*   Small solid particles are suspended in a continuous liquid phase. A discontinuous gas phase exists in the form of bubbles. Coal liquefaction is an example where the solid is consumed by the reaction. The three phases are hydrogen, a hydrocarbon solvent–product mixture, and solid coal. Fluidized bioreactors represent the case of a catalytic solid. The liquid phase is water, which contains the organic substrate. The gas phase supplies oxygen and removes carbon dioxide. The solid phase consists of microbial cells grown on the surface of a nonconsumable solid such as activated carbon.

A general model for a gas–liquid–solid reactor would have to consider homogeneous reactions occurring within the various phases and up to three sets of heterogeneous reactions: gas–liquid, gas–solid, liquid–solid. Such a general treatment tends to add notational complexity without providing much additional insight. Specific embellishments can be added as needed for a particular reaction scheme. Here we consider the canonical case of a single set of liquid–solid reactions. The reactions may, or course, involve dissolved gas species. The liquid acts as a transfer medium between the gas phase and the solid phase. The design intent is for the liquid to wet the solid completely since any direct exposure of the solid to the gas phase would not contribute to the reaction.

Consider a trickle bed reactor with cocurrent flow of gas and liquid. Piston flow is a reasonable approximation for both phases. We suppose also that the solid is catalytic and is completely wetted. Then Equations 11.45 and 11.46 can be applied by omitting the gas phase reaction term and by using an effective, pseudohomogeneous reaction rate for the liquid phase:

$$\frac{d(A_g \bar{u}_g a_g)}{dz} = -k_g A_i' (a_g - a_g^*) \tag{11.109}$$

and

$$\frac{d(A_l \bar{u}_l a_l)}{dz} = +k_l A_i' (a_l^* - a_l) + \frac{A_l \eta \rho}{\varepsilon} (\mathcal{R}_A)_{\text{catalyst mass}} \tag{11.110}$$

where η is the effectiveness factor.

The term $(R_A)_{\text{catalyst mass}}$ represents the intrinsic kinetics of the liquid–solid reaction expressed in moles reacted per unit mass of catalyst per unit time. This approach is exactly analogous to the treatment of gas–solid catalytic reactors presented in Chapter 10. The intrinsic kinetics and the effectiveness factor are measured in much the same way as in gas–solid reactions. The ρ/ε term did not arise in Chapter 10 since the catalyst density and void fraction were assumed constant in a packed bed. It arises here since A_1 can vary even though the bed is packed uniformly. (The factor ρ/ε also arose in Section 11.2.1.) Note that ρ and ε should be calculated on the basis of gas-free volume, that is, ρ is the catalyst mass per unit length of reactor divided by $A_l + A_s$. If some of the catalyst is unwetted, it should be excluded in these calculations. Note that

$$A_g + A_l + A_s = A_c \tag{11.111}$$

where A_s is the cross-sectional area associated with the solid phase.

The central difficulty in applying Equations 11.109 and 11.110 lies in measuring the parameters, particularly A_l and k_g. Prediction of packed-bed phase volumes and mass transfer coefficients from first principles is not yet possible, and even empirical correlations are unusually difficult for trickle beds. Vaporization of the liquid phase is common. From a formal viewpoint, this effect can be accounted for through the mass transfer term in Equation 11.96. In practice, results are specific to a particular chemical system and operating mode.

Since the suspending fluid is a liquid rather than a gas, liquid-fluidized beds tend to have a smaller density difference between the particle and the fluid than in gas-fluidized beds. This leads to a special form of fluidization known as *particulate fluidization*. Bubbles do not form above the minimum fluidization velocity. Instead, the bed remains homogeneous and gradually expands as the liquid velocity is increased. Since the liquid phase is the continuous phase in a three-phase slurry reactor, a form of particulate fluidization should still prevail. If the solid particles are catalytic and if the liquid velocity is well below the entrainment velocity, the solid phase can be modeled as well mixed. On the other hand, a slurry reactor with consumable solids will be self-classifying. Particles will stay in the reactor when they are large but will be entrained in the liquid stream as they decrease in size. Careful hydrodynamic design can lead to complete conversion of the solid phase, and an assumption of perfect mixing is clearly inappropriate.

Suggestions for Further Reading

A good discussion of experimental techniques for characterizing fluid–fluid reactors is given in:

K. R. Westerterp, W. P. M. van Swaaij, and A. A. C. M. Beenackers, *Chemical Reactor Design and Operation*, 2nd ed., Wiley, New York, 1984.

This reference also gives tables of typical values for parameters such as k_l and provides references to the research literature.

A recent and comprehensive treatment of reactive gas absorption is given in:

G. Astarita, D. W. Savage, and A. Bisio, *Gas Treating With Chemical Solvents*, Wiley, New York, 1983.

while a classic in the field is:

P. V. Danckwerts, *Gas–Liquid Reactions*, McGraw-Hill, New York, 1970.

Gas–liquid contacting in fermentation reactors is discussed in:

J. E. Bailey and D. F. Ollis, *Biochemical Engineering Fundamentals*, McGraw-Hill, New York, 1977.

The research literature shows that simplifying assumptions—such as perfect mixing in the gas phase—will usually fail under close scrutiny. See for example:

M. Popovic, A. Papalexiou, and M. Reuss, "Gas Residence Time Distribution in Stirred Tank Bioreactors," *Chem. Eng. Sci.*, **38**, 2015–2025 (1983).

A general reference on fluidization is:

D. Kunii and O. Levenspiel, *Fluidization Engineering*, Wiley, New York, 1969.

A more recent, state-of-the-art review is:

J. F. Davidson, D. Harrison, R. C. Darton, and R. D. LaNauze, pp. 583–685 in *Chemical Reaction Theory*: *A Review*, L. Lapidus and N. Amundson, Eds., Prentice-Hall, Englewood Cliffs, NJ, 1977.

For a comprehensive treatment of spouted bed reactors consult:

Y. Arkun, H. Littman and M. H. Morgan, III, "Modeling of Spouted Bed Chemical Reactors," *Encyclopedia of Fluid Mechanics*, Vol. 4, Solids and Gas-Solids, N. P. Cheremisinoff, Ed., 1089–1025, Gulf, Houston, Texas, 1986.

There is yet no comprehensive review of chemical reaction engineering applied to electronics materials processing. For trickle bed reactors, a good guide to the field is provided by:

M. Herskowitz and J. M. Smith, "Trickle-Bed Reactors: A Review," *AIChE J.*, **29**, 1–18 (1983).

Problems

11.1 A reactive gas is slowly bubbled into a column of liquid. The bubbles are small, approximately spherical, and are well separated from each other. Assume Stokes' law and ignore the change in gas density due to elevation. The gas is pure and reacts in the liquid phase with first order kinetics. Derive an expression for the size of the bubbles as a function of height in the column.

11.2 A continuous phenol alkylation is occurring in a stirred tank reactor. The liquid feed is pure phenol and the gas feed is pure alkene. Catalyst selectivity is poor so that five products are possible: two monoalkylates, two dialkylates, and one trialkylate. The following reaction scheme is proposed:

$$A + P \xrightarrow{k_1} M_1$$

$$A + P \xrightarrow{k_2} M_2$$

$$A + M_1 \xrightarrow{k_1/2} D_1$$

$$A + M_1 \xrightarrow{k_2} D_2$$

$$A + M_2 \xrightarrow{k_1} D_2$$

$$A + D_1 \xrightarrow{k_2} T$$

$$A + D_2 \xrightarrow{k_1/2} T$$

(a) Assuming k_1/k_2 is known, determine the product distribution as a function of phenol conversion.

(b) Explain whether or not this reaction scheme, which involves only two rate constants, is plausible from the viewpoint of organic reaction mechanisms.

11.3 The simplified bubble model of Section 11.2.1 predicts zero conversion and 100% bypassing at $Q = Q_{slug}$. This is clearly unreasonable. Suggest a modification to the model that requires only one additional parameter and that allows intermediate levels of bypassing at $Q = Q_{slug}$. Fit your model and predict conversions for $Q_{min} \le Q \le Q_{slug}$ for the parameter values of Equations 11.82 to 11.85. Use the additional fact that there is 15% bypassing at $Q = Q_{slug}$.

11.4 An overly simplified model of fluidized bed combustion treats the solid fuel as spherical particles freely suspended in upward-flowing gas. Suppose the particles react with zero order kinetics and that there is no ash or oxide formation. It is desired that the particles be completely consumed by position $z = L$. This can be done in a column of constant diameter or in a column where the diameter increases with increasing height. Which approach is better with respect to minimizing reactor volume? Develop a model that predicts the position of the particle as a function of time spent in the reactor.

11.5 Determine the position of the reaction front in the diffusion-limited decoking of a spherical cracking catalyst. *Hint*: Use a version of Equation 11.103 but correct for the spherical geometry and replace the convective flux with a diffusive flux.

1.6 Solid rocket motors use a solid–solid reaction (oxidizer–fuel) to generate thrust. The linear burning rate is constant. Thrust is proportional to the mass burning rate. The

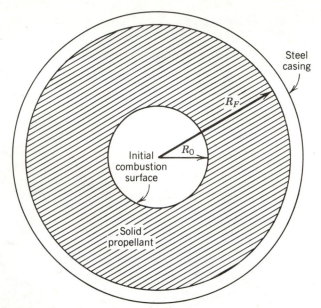

Figure 11.11 Solid rocket motor with thrust that increases with time.

solid propellant is cast inside a cylindrical steel casing as shown in Figure 11.11 and burns from the inside out.

(a) Plot thrust versus time for the simple annular casting shown in Figure 11.11. Suppose $R_0/R_f = 0.8$ and ignore end effects.

(b) Devise a casting geometry that will deliver constant thrust from ignition to burnout.

11.7 The low-pressure chemical vapor deposition of silicon nitride on a silicon substrate involves two gaseous reactants, dichlorosilane and ammonia. The following reactions are believed important under typical conditions of $P = 1$ torr and $T = 1000$ to 1200 K:

$$SiH_2Cl_2 + \tfrac{4}{3}NH_3 \rightarrow \tfrac{1}{3}Si_3N_4 + 2HCl \tag{I}$$

$$SiH_2Cl_2 \rightleftharpoons SiCl_2 + H_2 \tag{II}$$

$$SiCl_2 + \tfrac{4}{3}NH_3 \rightarrow \tfrac{1}{3}Si_3N_4 + 2HCl + H_2 \tag{III}$$

$$NH_3 \rightarrow \tfrac{1}{2}N_2 + \tfrac{3}{2}H_2 \tag{IV}$$

Suppose the reactant gases are supplied continuously in large excess and flow past a single wafer of silicon. By performing multiple experiments, the growth of the nitride layer can be determined as a function of time and reactant concentrations. Develop an experimental program to determine rate expressions for Reactions I through IV. Note that Reactions I, III, and IV are heterogeneous while Reaction II occurs in the gas phase. It is possible to include N_2, H_2, and HCl in the feed if this is useful. Exiting gas phase concentrations cannot be measured.

12

Polymer Reaction Engineering

Polymer reaction engineering is a specialized but very important branch of chemical reaction engineering. The odds strongly favor the involvement of chemical engineers with polymers at some point in their careers. This involvement utilizes the same basic concepts of reaction engineering presented previously in this book. The chemistry is rather more complicated, however, as is the mathematics. The number of chemical species participating in the reaction is potentially infinite, and the mathematical description of a batch polymerization requires an infinite set of differential equations. Analytical and numerical solutions are much more difficult than for the small sets of equations dealt with thus far. Polymerizations also present some interesting mechanical problems in reactor design. Viscosity increases dramatically with molecular weight, and a polymer solution is typically 10^4 to 10^7 times more viscous than an ordinary liquid. Molecular diffusivities in polymer solutions are lowered by a factor of 10 to 10^4. Laminar flow is the rule; pressure drops are high; agitation is difficult; and heat and mass transfer limitations can be very severe.

12.1 Polymerization Reactions

Polymerization reactions are classified as being either **chain growth** or **step growth**. In chain growth polymerizations, a small molecule reacts with a growing polymer chain to form a slightly longer chain:

$$
M + P_n \xrightarrow{k_n} P_{n+1} \qquad n = 1, 2, \ldots . \tag{12.1}
$$

377

where M represents the small molecule which is called the **monomer** and P_n denotes a polymer chain consisting of n monomer units that are chemically bonded. The chain may also contain a residual fragment of an **initiator** molecule that started the chain. Chain growth polymerizations are also called **addition** polymerizations since the monomer adds to the chain one unit at a time.

A quite different polymerization mechanism has two polymer molecules reacting together to form a much larger new molecule:

$$P_m + P_n \xrightarrow{k_{mn}} P_{m+n} \qquad n, m = 1, 2, \ldots . \tag{12.2}$$

This mechanism is called **step growth**, or **condensation**, polymerization. The polymer grows in steps of sizes m and n as the precursor molecules condense together.

12.1.1 Step Growth or Condensation Polymerizations

Condensation polymers are often formed from two distinct monomers, each of which is difunctional. A typical example is

$$n(Cl - \bigcirc - \overset{O}{\underset{O}{\overset{\|}{S}}} - \bigcirc - Cl) + n(Na - O - \bigcirc - \overset{CH_3}{\underset{CH_3}{\overset{|}{C}}} - \bigcirc - O - Na)$$

Dichlorodiphenyl sulfone   Sodium salt of bisphenol A   (12.3)

$$\longrightarrow Cl - \left[- \bigcirc - \overset{O}{\underset{O}{\overset{\|}{S}}} - \bigcirc - O - \bigcirc - \overset{CH_3}{\underset{CH_3}{\overset{|}{C}}} - \bigcirc - O \right]_n - Na + (2n-1)NaCl$$

Polysulfone   Salt

The reaction actually consists of many steps; we write them symbolically as

$$AMA + BNB \rightarrow AMNB + AB$$

$$AMNB + AMA \rightarrow AMNMA + AB$$

$$AMNB + BNB \rightarrow BNMNB + AB \tag{12.4}$$

$$AMNB + AMNB \rightarrow AMNMNB + AB$$

$$\vdots \qquad \vdots \qquad \vdots \qquad \vdots$$

This reaction set has two monomers (AMA and BNB) which will react with each other but not with themselves. It has only one dimer (AMNB) which is self-reactive. This pattern continues indefinitely with two trimers (AMNMA and BNMNB), one self-reactive tetramer (AMNMNB), and so on.

We have used the symbols A and B to denote the *endgroups*. In the polysulfone example of Equation 12.3, the endgroups are chlorine and sodium and the *condensation product* AB is sodium chloride. Condensation products tend to be small molecules such as hydrogen chloride, salts, water, and alcohols. The reactions often are reversible so that the condensation product must be removed for a high degree of polymerization to occur. A high degree of polymerization also depends critically on the stoichiometric ratio:

$$S_{AB} = \frac{[A]_0}{[B]_0} \tag{12.5}$$

Suppose $[A]_0 < [B]_0$ so that $S_{AB} < 1$. Then as the reaction proceeds, molecules containing an A endgroup will be reacted, and the entire population will consist of molecules having the form $B(NM)_{n-1}NB$, $n = 1, 2, \ldots$. If $S_{AB} \approx 1$, the polymerization can go to completion, with the ultimate product being one very large molecule. We suppose that the A endgroups are limiting and denote the conversion of these endgroups as X_A:

$$X_A = \frac{[A]_0 - [A]}{[A]_0} = \frac{[B]_0 - [B]}{[A]_0} = \frac{1}{S_{AB}} - \frac{[B]}{[A]_0} \tag{12.6}$$

The number of molecules present at any time is

$$\frac{[A] + [B]}{2} = \frac{[A]_0(1 - 2X_A S_{AB} + S_{AB})}{2 S_{AB}} \tag{12.7}$$

where the factor of 2 results from each molecule having two endgroups. The number of molecules present initially is found from Equation 12.7 with $X_A = 0$. The *number-average degree of polymerization*, or *number-average chain length*, is defined as

$$\bar{\ell} = \frac{\text{Reactive molecules present initially}}{\text{Reactive molecules present after polymerization}} \tag{12.8}$$

$$\boxed{\bar{\ell} = \frac{1 + S_{AB}}{1 + S_{AB} - 2X_A S_{AB}}} \tag{12.9}$$

Example 12.1   Determine the ultimate molecular weight resulting from a batch polycondensation reaction that has initial stoichiometry S_{AB}.

   Solution   The ultimate molecular weight is approached at long times when $X_A \to 1$. Thus

$$\bar{\ell}(t = \infty) = \frac{1 + S_{AB}}{1 - S_{AB}} \tag{12.10}$$

A few values for this function are

$$
\begin{array}{cc}
S_{AB} & \bar{\ell}\,(t=\infty) \\
0.2 & 1.5 \\
0.5 & 3 \\
0.8 & 9 \\
0.98 & 99 \\
0.998 & 999
\end{array}
$$

When $X_A = 1$, all molecules have the form $B(NM)_n NB$. The **number-average molecular weight** can be found from this chemical formula by setting $n = \bar{\ell}$.

For the polysulfone example in Reaction 12.3, the usual practice is to polymerize with a slight excess of sodium endgroups and then to convert these to methyl endgroups. See Example 12.3. The final structures in the formula $B(NM)_\ell NB$ are

$$B = -CH_3$$

and the number-average molecular weight is

$$\overline{M}_n = 15 + 442\bar{\ell} + 226 + 15 = 256 + 442\bar{\ell}$$

A commercial polymer might have $\bar{\ell} = 150$ so that $\overline{M}_n \approx 66,500$.

The example shows that very accurate control of stoichiometry is needed to achieve high molecular weight polymers. This is one reason why **binary polycondensations** are usually performed in batch vessels with batch weighing systems. The reaction of AMA with BNB is termed a binary polycondensation because two monomers are involved. The repeating unit along the polymer chain is the group MN, and the polymer can be considered a **strictly alternating copolymer** of M and N. **Random copolymers** are possible but unusual in polycondensations. A ternary mixture of AM_1A, AM_2A, and BNB can lead to molecules such as

$$AM_1NM_1NM_2NM_1NM_2NM_2NM_1NM_2NB$$

where the M_1 and M_2 groups strictly alternate with N groups but are distributed randomly with respect to each other. Such a structure is sometimes called a **terpolymer** because three monomers are involved, but the term copolymer is used inclusively for any polymer formed from more than one monomer.

Polycondensations of a self-reactive monomer AMB give molecules of the form AM_nB which can be considered as **homopolymers** of M. Such systems eliminate the need for precise control of stoichiometry in order to achieve high molecular weights. This makes them more amenable to continuous polymerization since accurate metering equipment is unnecessary. The chemistry takes the form

$$2AMB \rightarrow AM_2B + AB$$

$$AM_2B + AMB \rightarrow AM_3B + AB$$

$$2AM_2B \rightarrow AM_4B + AB \tag{12.11}$$

$$AM_3B + AMB \rightarrow AM_4B + AB$$

$$\vdots \qquad \vdots$$

Example 12.2   Poly(ethylene terephthalate), commonly called PET, is nominally the copolymer of terephthalic acid and ethylene glycol:

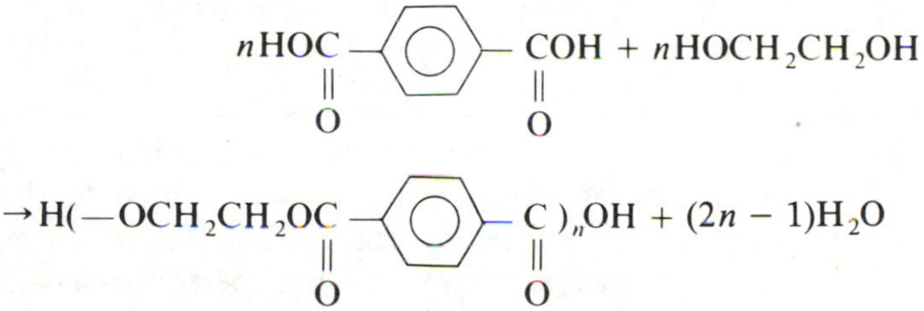

$$\tag{12.12}$$

As written, this reaction requires exacting stoichiometry. However, an alternative chemistry uses the trimer that is formed when terephthalic acid is reacted with a large excess of ethylene glycol:

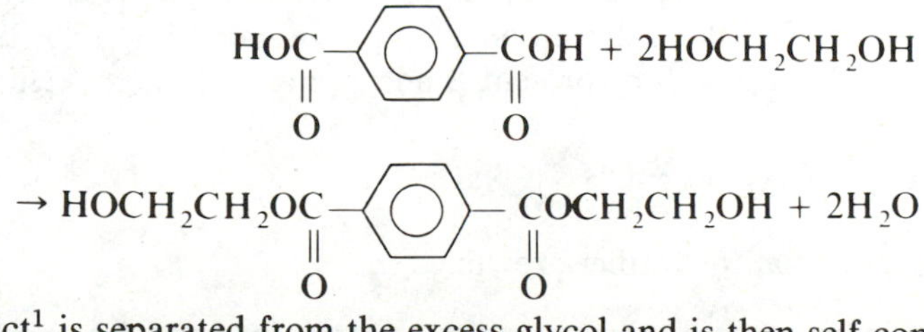

$$\tag{12.13}$$

This product[1] is separated from the excess glycol and is then self-condensed:

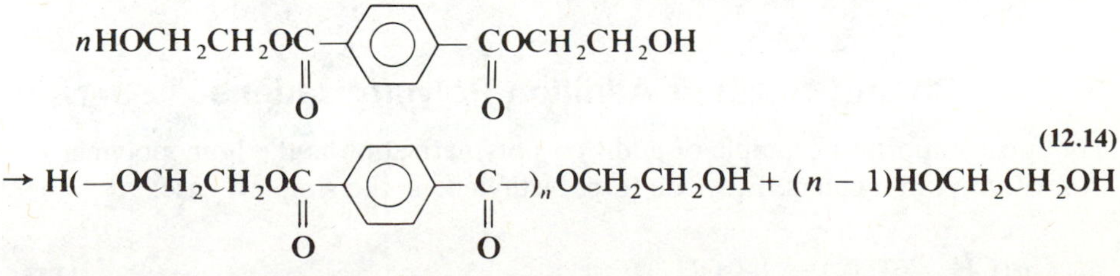

$$\tag{12.14}$$

---

[1]A more common route to this molecule, diglycol terephthalate, is to react ethylene glycol with dimethyl terephthalate and to remove methanol as the condensation product.

The reader may confirm that the overall stoichiometry of Reactions 12.13 and 12.14 is identical to that of Reaction 12.12. The conditions needed for the polymerization in Reaction 12.14 are more severe than those for the polymerization in Reaction 12.12. Also, the condensation product, ethylene glycol, is more difficult to remove than the water formed in Reaction 12.12. However, the need for exact weighing is eliminated, and this enables continuous operation.

The polymerization schemes considered thus far involve difunctional molecules and give **linear polymers**. If a trifunctional molecule is used, branching and crosslinking can occur. A classic example is the phenol–formaldehyde condensation. Phenol is trifunctional in this polymerization since reaction can occur at the para position and both ortho positions. If run anywhere near exact stoichiometry, a three-dimensional network polymer is formed. See Problem 12.7. Commercial phenolic resins are low molecular weight **oligomers**, typically with $\bar{\ell} < 8$, that are crosslinked as part of the final fabrication step.

Condensation polymers remain capable of growth throughout their stay in the reactor. Thus the average lifetime of growing chains is the same as $\bar{t}$. Condensation polymers are an example of **living polymers,** and termination of their potential for growth requires **endcapping**.

Example 12.3   Equation 12.9 suggests that adjustments in stoichiometry can be used to control molecular weight. This method is used for low molecular weight oligomers but is generally impractical for $\bar{\ell} > 100$. When high polymers are desired, the usual approach is to aim for nearly perfect stoichiometry so that the ultimate molecular weight would overshoot the goal. The molecular weight buildup is followed as a function of time (e.g., by measuring viscosity) and the reaction is stopped when the desired $\bar{\ell}$ has been achieved. This leaves live molecules with mixed endgroups which could continue to react during subsequent processing. For the polysulfone example of Equation 12.3, both sodium and chlorine endgroups will be present. Endcapping with methyl chloride,

$$\sim Na + ClCH_3 \rightarrow \sim CH_3 + NaCl$$

(12.15)

removes the potential for further growth.

12.1.2   Chain Growth or Addition Polymerizations

The most important example of addition polymerization is the homopolymerization of vinyl monomers. Ethylene to polyethylene is the simplest case:

$$n(CH_2{=}CH_2) \rightarrow (-CH_2)_{2n}$$

(12.16)

The opening of the double bond can be catalyzed in several ways. Free-radical polymerization is the most common method for styrenic monomers, whereas coordination metal catalysis (Zigler–Natta catalysis) is important for olefin polymerizations. The specific reaction mechanism may generate some catalyst residues, but there are no true by-products. There are no stoichiometry or equilibrium limitations so that quite long chains are formed; $\bar{\ell} > 500$ is typical of addition polymers.

Free-radical polymerization involves initiation, propagation, and termination steps similar to those described in Chapter 2. The initiation step typically employs a peroxide catalyst as the original source of free radicals:

$$I_2 \xrightarrow{k_I} 2I \cdot \tag{12.17}$$

These primary radicals react with monomer to form a propagating radical that contains one monomer unit:

$$I \cdot + M \xrightarrow{k_i} R_1 \cdot \tag{12.18}$$

Example 12.4   *t*-Butyl peroxide is commonly used for the polymerization of styrenic monomers. The initiation step is

$$
\begin{array}{ccc}
CH_3 & & CH_3 \\
| & & | \\
CH_3 - C - O - O - C - CH_3 & \xrightarrow{k_I} \\
| & & | \\
CH_3 & & CH_3
\end{array}
\tag{12.19}
$$

$$
\begin{array}{c}
CH_3 \\
| \\
2CH_3 - C - O \cdot \qquad R_{I\cdot} = 2k_I[I_2] \\
| \\
CH_3
\end{array}
$$

The propagating radical with $\ell = 1$ is

$$
\begin{array}{ccccc}
CH_3 & H & H \\
| & | & | \\
CH_3 - C - O - C - C \cdot \\
| & | & | \\
CH_3 & \bigcirc & H
\end{array}
$$

Most styrenic monomers also exhibit a form of spontaneous initiation called *thermal initiation*. See Example 12.7.

The propagation reaction is

$$R_n \cdot + M \xrightarrow{k_P} R_{n+1} \cdot \tag{12.20}$$

There are two general types of termination reactions. The first, called **chain transfer**, stops chain growth but does not terminate the free radical. Chain transfer to monomer is common:

$$R_n \cdot + M \xrightarrow{k_t} P_n + M \cdot \tag{12.21}$$

Chain transfer to a solvent or to polymer is also possible. The latter terminates one polymer chain but reactivates what was presumably a previously terminated chain which can then begin to grow at the new site. This leads to long-chain branching.

The second type of termination stops chain growth and also terminates the free radical. Two mechanisms are common. **Termination by combination** produces a single molecule of **dead polymer**:

$$R_m \cdot + R_n \cdot \xrightarrow{k_c} P_{m+n} \tag{12.22}$$

The free radicals combine to form a carbon-to-carbon bond and give a saturated polymer molecule. **Termination by disproportionation** produces two polymer molecules, one of which will contain a double bond:

$$R_m \cdot + R_n \cdot \xrightarrow{k_d} P_r + P_s \tag{12.23}$$

where $m + n = r + s$.

In contrast to condensation polymers, the lifetime of a growing chain is very short for free-radical polymers.[2] Typically, the chain lifetime will be $10^{-3} \bar{t}$ or less. Free-radical concentrations—which include all the growing chains—are very low, typically 10^{-3} to 10^{-5} moles/m³. To a reasonable approximation, the system consists of unreacted monomer, unreacted initiator, and dead polymer. The quasi-steady state hypothesis gives

$$\frac{d[R \cdot]}{d\alpha} = 2k_I[I_2] - (k_c + k_d)[R \cdot]^2 \approx 0 \tag{12.24}$$

where $[R \cdot]$ denotes the total concentration of radicals. In writing Equation 12.24, it was assumed that the reaction rates for termination are independent of

---

[2]Some addition polymerization schemes lack termination mechanisms and also give rise to living polymers.

chain length and that the termination rate constant, $k_c + k_d$, incorporates the factor of two that might seem to be missing from Equation 12.24. This is a specific example of the **equal reactivity hypothesis** discussed in Section 12.2. Applying it to the propagation reaction as well gives

$$\mathscr{R}_{\text{propagation}} = k_p[\text{R} \cdot][\text{M}] = k_p[\text{M}]\sqrt{\frac{2k_{\text{I}}[\text{I}_2]}{k_c + k_d}} \qquad (12.25)$$

where we have used Equation 12.24 to eliminate $[\text{R} \cdot]$. The propagation reaction accounts for nearly all consumption of monomer. Thus Equation 12.25 predicts that the polymerization rate will be first order in monomer concentration and half order in initiator concentration. This is confirmed by experiments at low polymer concentrations but is violated when the polymer concentration becomes high. The termination mechanisms require pairwise interactions between large molecules, and these become increasingly difficult at high polymer concentrations due to chain entanglements. The propagation reaction is less affected, and the net rate of polymerization can actually increase. The phenomenon of the rate increasing as $[\text{M}]$ decreases is a form of autoacceleration known as the **gel**, or **Trommsdorff effect**, and is particularly noticeable in the polymerization of methyl methacrylate. See Section 2.3.4 and Figure 2.4.

The ratio of propagation rate to termination rate gives the **dynamic chain length** of the growing polymer:

$$\bar{\ell} = \frac{\mathscr{R}_{\text{propagation}}}{\mathscr{R}_{\text{termination}}} = \frac{k_p[\text{M}]}{\sqrt{2(k_c + k_d)k_{\text{I}}[\text{I}_2]}} \qquad (12.26)$$

As suggested by the notation, this quantity is similar to the condensation polymerization counterpart given by Equation 12.9. It applies to the growing chain before termination. Assuming equal reactivity, the dead polymer will have the same average length if termination is by disproportionation and will have twice this length if termination is by combination. Unlike the condensation case, $\bar{\ell}$ for free-radical polymers does not include any unreacted monomers in the calculations of average chain length.[3] Free-radical polymers are formed in fractions of a second and, once formed, typically remain inert for the remainder of the reaction. In a batch reactor, some high molecular weight polymer is formed from the onset, and the molecular weight of the polymer being formed remains

---

[3] Some free-radical polymerizations have side reactions which lead to small amounts of low molecular weight oligomers. By convention, these are usually excluded from both experimental and theoretical determinations of average molecular weight. In measurements, it is typical to have a **low molecular weight cutoff** at about 2000. Molecules having a molecular weight less than the cutoff are excluded from the measurement.

approximately constant as the reaction proceeds. *Conversion* in a free-radical polymerization means the conversion of monomer to high molecular weight polymer. In contrast, molecular weight increases gradually in a batch poly-condensation, and really high molecular weights are not achieved until near the end of the batch reaction. *Conversion* in a polycondensation refers to consumption of the stoichiometrically limiting endgroup.

12.1.3 Vinyl Copolymerizations

Vinyl monomers are often copolymerized using free-radical or coordination metal catalysis to obtain more-or-less random arrangements of the monomers along the polymer chain. Random copolymers are important items of commerce. They also present fascinating problems in reactor design. Consider the free-radical poly-merization of two vinyl monomers denoted by X and Y. Each addition in the propagation reaction can add either an X or a Y to the growing polymer chain, and it is unrealistic to assume that the monomers have equal reactivities. Furthermore, reaction probabilities can depend on the composition of the polymer chain already formed. We suppose they depend only on the last member added to the chain which also contains the site of the free radical. There are four propagation reactions to consider:

$$X_n \cdot + X \overset{k_{XX}}{\rightarrow} X_{n+1} \cdot$$

$$X_n \cdot + Y \overset{k_{XY}}{\rightarrow} Y_{n+1} \cdot$$

$$Y_n \cdot + X \overset{k_{YX}}{\rightarrow} X_{n+1} \cdot \qquad (12.27)$$

$$Y_n \cdot + Y \overset{k_{YY}}{\rightarrow} Y_{n+1} \cdot$$

The initiation and termination steps may also come in several varieties, but they will have little effect on overall chain composition provided the chains are long. The monomer reaction rates are

$$\mathscr{R}_X = -k_{XX}[X \cdot]x - k_{YX}[Y \cdot]x$$
$$\mathscr{R}_Y = -k_{XY}[X \cdot]y - k_{YY}[Y \cdot]y \qquad (12.28)$$

The ratio of these propagation rates is the ratio in which the two monomers are incorporated into the polymer:

$$\frac{x_P}{y_P} = \frac{\mathscr{R}_X}{\mathscr{R}_Y} = \frac{x}{y}\left(\frac{-k_{XX}[X \cdot] - k_{YX}[Y \cdot]}{-k_{XY}[X \cdot] - k_{YY}[Y \cdot]}\right) \qquad (12.29)$$

This result can be simplified considerably by observing that, except for end effects, the number of transitions from X to Y along the polymer chain (i.e., structures like ⁓XY⁓) must equal the number of transitions from Y to X (i.e., structures like ⁓YX⁓). This requires

$$k_{XY}[X \cdot]y = k_{YX}[Y \cdot]x \tag{12.30}$$

Combination with Equation 12.29 gives

$$\boxed{\frac{x_P}{y_P} = \frac{x}{y}\left(\frac{r_X x + y}{x + r_Y y}\right)} \tag{12.31}$$

where $r_X = k_{XX}/k_{XY}$ and $r_Y = k_{YY}/k_{YX}$ are known as **copolymer reactivity ratios**. If $r_X > 1$, monomer X tends to homopolymerize so that sequences like ~ XX ~ will be favored. If $r_X < 1$, copolymerization is preferred, and sequences like ~ YX ~ will be common. If both $r_X = 0$ and $r_Y = 0$, the monomer is incapable of homopolymerization. Binary polycondensation actually fits this scheme, with $r_A = r_B = 0$.

Equation 12.31 shows that copolymers will generally be different in composition than the monomers from which they were formed. In a batch reactor, the relative monomer concentrations will drift with time. A semibatch polymerization with the more reactive monomer being continuously added to the system is required to obtain a compositionally homogeneous copolymer. Note, however, that

$$\frac{x_P}{y_P} = \frac{x}{y} \quad \text{if} \quad \left(\frac{r_X x + y}{x + r_Y y}\right) = 1 \tag{12.32}$$

Equation 12.32 gives the condition for a **copolymer azeotrope** to exist. The azeotropic composition is

$$\left(\frac{x}{y}\right)_{azeotrope} = \frac{1 - r_Y}{1 - r_X} \tag{12.33}$$

which implies either both $r_X, r_Y > 1$ or $r_X, r_Y < 1$. The situation with $r_X < 1$ and $r_Y < 1$ is quite common. The situation with $r_X > 1$ and $r_Y > 1$ is extremely rare.

Example 12.5 The reactivity ratios for X = styrene, Y = acrylonitrile are $r_X = 0.41$ and $r_Y = 0.04$. Both reactivity ratios are less than 1 so that copolymerization is preferred over homopolymerization. A 50/50 copolymer would tend to have the styrene and acrylonitrile groups alternating along the chain rather than to occur in long sequences.[4] An azeotrope exists at

$$\frac{x_P}{y_P} = \frac{x}{y} = \frac{1 - 0.04}{1 - 0.41} = 1.63$$

This corresponds to 62 mole-%, or 76 weight-% styrene.

Example 12.6 A 50/50 (molar) mixture of styrene and acrylonitrile is to be batch polymerized until 60 percent (molar) conversion of the monomers is achieved. Determine the copolymer composition distribution.

    Solution Perhaps surprisingly, the solution to this problem does not require explicit knowledge of the polymerization rate. For a batch reaction, Equation 12.31 can be written in terms of monomer concentrations as

$$\frac{dx}{dy} = \frac{x}{y}\left(\frac{r_X x + y}{x + r_Y y}\right) \tag{12.34}$$

This equation can be integrated analytically, but a numerical solution is adequate for the present purpose. Set $x_0 = y_0 = 0.5$ and take small steps in Δy until $x + y = 0.4$. Some results are:

Moles Monomer Remaining	Mole Fraction Styrene in Monomer	Mole Fraction Styrene in Polymer	Cummulative Weight Fraction of Polymer
1.000	0.500	0.576	0
0.926	0.494	0.574	0.124
0.852	0.487	0.571	0.248
0.779	0.479	0.569	0.370
0.706	0.470	0.566	0.492
0.634	0.460	0.563	0.612
0.562	0.447	0.559	0.732
0.491	0.431	0.554	0.850
0.421	0.410	0.548	0.966
0.400	0.403	0.547	1.000

The last two columns constitute the cumulative distribution function of copolymer compositions. The polymer composition drifts by 3 mole-% due to the changing monomer composition. In addition to this **macroscopic composition distribution**, there is also a **microscopic composition distribution** due to statistical fluctuations at the molecular level. Two molecules formed from the same monomer mixture will not have the same sequence down the chain nor exactly the

---

[4]Sequence length distributions can be calculated from the reactivity ratios and the overall polymer composition. See, for example, G. E. Ham, *Copolymerization*, Interscience, New York, 1964.

same overall composition. The microscopic distribution is approximately Gaussian and fairly narrow for high molecular weight polymers. See W. H. Stockmayer, *J. Chem. Phys.*, **15**, 199 (1945) for details.

Reactivity ratios are ratios of propagation constants for rather similar reactions which tend to have similar activation energies. Thus, to a first approximation, they are independent of temperature. They can be measured by low-conversion polymerizations (so the monomer composition does not drift appreciably) run at two or more monomer compositions.

12.2 Molecular Weight Distributions

To a first approximation, a free-radical homopolymerization involves only two chemical species: monomer and high molecular weight polymer. Closer examination, say, by an analytical technique such as *size exclusion chromatography* (also known as *gel permeation chromatography*), reveals a more complex situation. The high molecular weight polymer shows a broad spectrum of molecular weights. There are also some low molecular weight species, typically dimers and trimers, which did not exist prior to polymerization. Complete understanding of this situation requires definition of all molecular structures that could result from the polymerization and determination of the concentrations for each structure. A complete description of the products of a polymerization is now feasible. Computational tools are not a serious limitation. However, approximate characterization using the first few moments of the molecular weight distribution is an attractive option for most systems. For free-radical polymers, it is common practice to exclude monomers and low molecular weight oligomers from the moment calculations.

12.2.1 Polymerization Kinetics and Equal Reactivity

For linear polymers with $\bar{\ell}$ of a few hundred or even a few thousand, brute force enumeration of molecular structures and their concentrations is quite possible, at least for polymerizations in one of the ideal reactors.

Example 12.7 A simplified kinetic scheme for the thermally initiated free-radical polymerization of styrene is

$$3M \rightarrow R_1 \cdot + R_2 \cdot \qquad \mathscr{R} = k_I [M]^3$$

$$R_n \cdot + M \xrightarrow{k_n} R_{n+1} \cdot \qquad n = 1, 2, \ldots \qquad \text{(12.35)}$$

$$R_m \cdot + R_n \cdot \xrightarrow{k_{mn}} R_{m+n} \qquad m, n = 1, 2, \ldots$$

Use this mechanism to develop species material balance equations for the isothermal, batch polymerization of styrene.

Solution   The material balance equations are just Equations 1.31 and 2.8 applied to each component. The various components are monomer (M), live polymer ($P_n \cdot$, $n = 1, 2, \dots$), and dead polymer (P_ℓ, $\ell = 2, 3, \dots$). For monomer,

$$\frac{d[M]}{d\alpha} = -3k_1[M]^3 - M \sum_{n=1}^{\infty} k_n[R_n \cdot] \tag{12.36}$$

For live polymer of length $n > 2$,

$$\frac{d[R_n \cdot]}{d\alpha} = k_{n-1}[M][R_{n-1} \cdot] - k_n[M][R_n \cdot]$$

$$\tag{12.37}$$

$$-[R_n \cdot] \sum_{m=1}^{\infty} k_{nm}[R_m \cdot] \qquad n > 2$$

The astute reader will notice that termination by combination allows a n-mer to react with another n-mer, giving a reaction with a stoichiometric coefficient of -2 for this special case of $m = n$ rather than -1 for the usual case of $m \neq n$. We regard the factor of 2 as being incorporated into the rate constant[5]: $k_{nn} = 2k'_{nn}$, where the termination reaction is

$$2R_n \cdot \overset{k'_{nn}}{\to} P_{2n} \tag{12.38}$$

Thus the k_{nn} term in Equation 12.37 is $-2k'_{nn}[P_n \cdot]^2 = -k_{nn}[P_n \cdot]^2$.

Live polymer of length $\ell = 1$ is formed by the initiation reaction rather than by propagation. For it,

$$\frac{d[R_1 \cdot]}{d\alpha} = k_1[M]^3 - k_1[M][R_1 \cdot] - [R_1 \cdot] \sum_{m=1}^{\infty} k_{1,m}[R_m \cdot] \tag{12.39}$$

---

[5] This definition is a mathematical convenience since it makes for neat-looking summations. It also has a physical basis when the reaction probabilities are independent of chain length. See Example 12.9.

Live polymer of length $\ell = 2$ can be formed by either initiation or propagation:

$$\frac{d[R_2 \cdot]}{d\alpha} = k_I[M]^3 + k_1[M][R_1 \cdot] - k_2[M][R_2 \cdot] - [R_2 \cdot] \sum_{m=1}^{\infty} k_{2,m}[R_m \cdot]$$

$$(12.40)$$

Dead polymer of length ℓ is formed by the reaction of m-mer with $(\ell - m)$-mer. For the case of $\ell = 4$, for example,

$$\frac{d[P_4]}{d\alpha} = k'_{2,2}[R_2 \cdot]^2 + k'_{1,3}[R_1 \cdot][R_3 \cdot] + k'_{3,1}[R_3 \cdot][R_1 \cdot]$$

$$(12.41)$$

$$= \frac{k_{2,2}[R_2 \cdot]^2}{2} + k_{1,3}[R_1 \cdot][R_3 \cdot]$$

where $2k'_{1,3} = 2k'_{3,1} = k_{1,3}$. For the general case,

$$\frac{d[P_\ell]}{d\alpha} = \frac{1}{2} \sum_{m=1}^{\ell-1} k_{m,\ell-m}[R_m \cdot][R_{\ell-m} \cdot]$$

$$(12.42)$$

The sum in Equation 12.42 contains terms like $k_{1,3}[R_3 \cdot][R_1 \cdot]$ and $k_{3,1}$ $[R_3 \cdot][R_1 \cdot]$. From a mathematical viewpoint, these terms could be different, but they are the same from the viewpoint of chemical kinetics: $k_{mn} = k_{nm}$. The factor of 2 in Equation 12.42 arises because the sum counts each reaction twice (except for the special case of $\ell = 2m$).

The batch polymerization is governed by Equations 12.36, 12.37, 12.40, and 12.42. Given values for all the rate constants, this set of equations can be numerically integrated by truncating the summations at some large value of n. The set of equations will be very stiff, since the propagation rate will be orders of magnitude larger than the initiation and termination rates. Nevertheless, the computation is entirely feasible using algorithms intended for stiff equations. The marching-ahead technique is not well suited to this integration.

The brute-force method can also be applied to condensation polymers. It turns out to be even simpler, since $\bar{\ell}$ is usually small and since the equations are not stiff. Numerical solutions for $\bar{\ell}$ of a few hundred were feasible in the early 1960s.

Example 12.8 Write the component material balances for an isothermal, batch self-condensation. Do not ignore the possibility of cyclization.

Solution   There are only two types of molecules in the system: AM_nB, which is a living polymer, and M_n, which is a cyclic oligomer. Denote them as P_n and C_n. The reactions are propagation:

$$P_n + P_m \overset{k_{mn}}{\rightarrow} P_{m+n} \qquad \mathscr{R} = k_{mn} p_m p_n \qquad m \neq n$$

$$\qquad\qquad\qquad\qquad = \frac{k_{nn}}{2} p_n^2 \qquad m = n$$

(12.43)

and cyclization:

$$P_n \overset{k_n}{\rightarrow} C_n \qquad \mathscr{R} = k_n p_n \tag{12.44}$$

The batch material balances are

$$\frac{dp_n}{d\alpha} = \frac{1}{2} \sum_{m=1}^{n-1} k_{m,n-m} p_m p_{n-m} - p_n \sum_{m=1}^{\infty} k_{nm} p_m - k_n p_n \tag{12.45}$$

and

$$\frac{dc_n}{d\alpha} = k_n p_n \tag{12.46}$$

The factor of 2 appears in Equation 12.45 for the same reason it appeared in Equation 12.42.

Equations 12.45 and 12.46 are ready to be solved given values for k_{nm}, k_n, and the initial concentrations. An example of such a solution is given in Section 12.2.2.

Example 12.9   Determine *initial* formation rates for the batch self-condensation of a mixture containing 50 mole percent AMB and 50 mole percent AM_2B. Assume all endgroups are equally reactive.

Solution   At $\alpha = 0$ there are two reactants, AMB and AM_2B, and three possible products, AM_2B, AM_3B, and AM_4B. Adopting the notation of Example 12.8, the reactions are

$$P_1 + P_1 \overset{k_{11}'}{\rightarrow} P_2 \qquad \mathscr{R} = \frac{k_{11}}{2} p_1^2$$

$$P_1 + P_2 \overset{k_{12}}{\rightarrow} P_3 \qquad \mathscr{R} = k_{12} p_1 p_2 \tag{12.47}$$

$$P_2 + P_2 \overset{k_{22}'}{\rightarrow} P_4 \qquad \mathscr{R} = \frac{k_{22}}{2} p_2^2$$

The initial formation rates are

$$\frac{dp_1}{d\alpha}\bigg|_0 = -2\left(\frac{k_{11}}{2}p_1^2\right) - k_{12}p_1p_2$$

$$\frac{dp_2}{d\alpha}\bigg|_0 = +\frac{k_{11}}{2}p_1^2 - k_{12}p_1p_2 - 2\left(\frac{k_{22}}{2}\right)p_2^2$$

$$(12.48)$$

$$\frac{dp_3}{d\alpha}\bigg|_0 = k_{12}p_1p_2$$

$$\frac{dp_4}{d\alpha}\bigg|_0 = \frac{k_{22}}{2}p_2^2$$

These equations are consistent with Equation 12.45 when there is no cyclization, $k_n = 0$.

Suppose

$$k_{11} = k_{12} = k_{22} = k \qquad\qquad (12.49)$$

Then

$$\frac{dp_1}{d\alpha}\bigg|_0 = -\frac{k}{2}$$

$$\frac{dp_2}{d\alpha}\bigg|_0 = -\frac{3k}{8}$$

$$(12.50)$$

$$\frac{dp_3}{d\alpha}\bigg|_0 = +\frac{k}{4}$$

$$\frac{dp_4}{d\alpha}\bigg|_0 = +\frac{k}{8}$$

We now show that the assumption of Equation 12.49—where we set all the k's equal, including the specially defined k for P_n reacting with itself—is consistent with the physical notion of equal reactivity. We adopt a probability viewpoint, choosing a pair of molecules at random and letting them react.

Probability that both molecules are P_1 is $\frac{1}{2} \cdot \frac{1}{2} = \frac{1}{4}$.
Probability that both molecules are P_2 is $\frac{1}{2} \cdot \frac{1}{2} = \frac{1}{4}$.
Probability that the molecules are different is $\left(\frac{1}{2} \cdot \frac{1}{2}\right) + \left(\frac{1}{2} \cdot \frac{1}{2}\right) = \frac{1}{2}$.

Suppose the molecules are picked at a rate of k per second. Then the rate at which P_1 molecules are selected and reacted is

$$\left.\frac{dp_1}{d\alpha}\right|_0 = -k\left(\tfrac{1}{2}\right) = -\frac{k}{2} \qquad\qquad\qquad \text{(12.51)}$$

since the P_1 type is selected half the time. The same selection rate applies to P_2 since the two types are equally abundant and equally reactive. Now, however, some P_2 is also being formed. The rate at which P_2 is formed is half the rate at which a P_1 molecule is selected to react with another P_1 molecule, half because it takes two P_1 molecules to make one P_2 molecule. Thus

$$\left.\frac{dp_2}{d\alpha}\right|_0 = -\frac{k}{2} + \frac{1}{2}\left(+\frac{k}{4}\right) = -\frac{3k}{8} \qquad\qquad \text{(12.52)}$$

The formation rates for P_3 and P_4 are also half the rates at which the parent molecules are selected:

$$\left.\frac{dp_3}{d\alpha}\right|_0 = \frac{1}{2}\left(+\frac{k}{2}\right) = \frac{k}{4} \qquad\qquad\qquad \text{(12.53)}$$

$$\left.\frac{dp_4}{d\alpha}\right|_0 = \frac{1}{2}\left(+\frac{k}{4}\right) = \frac{k}{8} \qquad\qquad\qquad \text{(12.54)}$$

Equations 12.51 to 12.54 agree with Equations 12.50. Thus Equation 12.43 is confirmed as the mathematical analog of the equal-reactivity assumption.

Example 12.10 Write the component material balances for a binary poly-condensation occurring in a batch reactor. Assume equal reactivity of all end-groups.

Solution The monomers are AMA and BNB. There are three generic types of polymer molecules: $A(MN)_n MA$, $BN(MN)_n B$, and $A(MN)_n B$. Denote these as A_n, B_n, and C_n. Then

$$a_n = \left[A(MN)_n MA\right] \qquad n = 0, 1, \ldots$$

$$b_n = \left[BN(MN)_n B\right] \qquad n = 0, 1, \ldots \qquad\qquad \text{(12.55)}$$

$$c_n = \left[A(MN)_n B\right] \qquad n = 1, 2, \ldots$$

The reactions can be written as

$$A_m + B_n \rightarrow C_{m+n+1} \qquad \mathscr{R} = k_{AB} a_m b_n$$

$$A_m + C_n \rightarrow A_{m+n} \qquad \mathscr{R} = k_{AC} a_m c_n$$

$$B_m + C_n \rightarrow B_{m+n} \qquad \mathscr{R} = k_{BC} b_m c_n$$

$$C_m + C_n \rightarrow C_{m+n} \qquad \mathscr{R} = k_{CC} c_m c_n \qquad m \neq n \qquad (12.56)$$

$$= \frac{k_{CC}}{2} c_m^2 \qquad m = n$$

where the various rate constants, k_{AB}, k_{AC}, k_{BC}, k_{CC}, have been assumed independent of chain length. They do, of course, depend on the generic types of polymer molecules involved in the reaction. Any endgroup on a type-A molecule will react with any endgroup on a type-B molecule, but a given endgroup on a type-A molecule will react with only half the endgroups on a type-C molecule. Thus, for a binary polycondensation, the equal reactivity assumption takes the form

$$k_{AB} = 2k_{AC} = 2k_{BC} = 2k_{CC} = k \qquad (12.57)$$

The material balance equations are

$$\frac{da_n}{d\alpha} = \mathscr{R}_{A_n} = +\frac{k}{2} \sum_{m=0}^{n-1} a_m c_{n-m} - ka_n \sum_{m=0}^{\infty} b_m - \frac{k}{2} a_n \sum_{m=1}^{\infty} c_m$$

$$\frac{db_n}{d\alpha} = \mathscr{R}_{B_n} = +\frac{k}{2} \sum_{m=0}^{n-1} b_m c_{n-m} - kb_n \sum_{m=0}^{\infty} a_m - \frac{k}{2} b_n \sum_{m=1}^{\infty} c_m$$

$$\qquad (12.58)$$

$$\frac{dc_n}{d\alpha} = \mathscr{R}_{C_n} = +k \sum_{m=0}^{n-1} a_m b_{n-m-1} + \frac{k}{4} \sum_{m=1}^{n-1} c_m c_{n-m} - \frac{k}{2} c_n \sum_{m=0}^{\infty} a_m$$

$$\qquad - \frac{k}{2} c_n \sum_{m=0}^{\infty} b_m - \frac{k}{2} c_n \sum_{m=1}^{\infty} c_m$$

The usual initial conditions associated with Equations 12.58 are

$$a_n = b_n = c_n = 0 \quad \text{for} \quad n \geq 1$$

$$\qquad (12.59)$$

$$(a_0)_0 = S_{AB}(b_0)_0$$

A boundary condition like

$$a_n = b_n = c_n = 0 \quad \text{for} \quad n \neq 1$$

$$a_1 = b_1 = 0 \tag{12.60}$$

$$c_1 = (c_1)_0$$

converts the binary polycondensation to a self-condensation. Thus this example includes self-condensation as a special case.

Example 12.11   Repeat Example 12.10 for a perfect mixer.

Solution   A binary condensation in a batch reactor is governed by a infinite set of differential equations. A binary polycondensation in a perfect mixer is governed by an infinite set of algebraic equations. Suppose

$$(a_n)_{\text{in}} = (b_n)_{\text{in}} = (c_n)_{\text{in}} = 0 \quad \text{for} \quad n \geq 1$$

$$(a_0)_{\text{in}} = S_{\text{AB}}(b_0)_{\text{in}} \tag{12.61}$$

The set of algebraic equations is

$$(a_0)_{\text{out}} = (a_0)_{\text{in}} + \bar{t}\mathcal{R}_{a_0}$$

$$(b_0)_{\text{out}} = (b_0)_{\text{in}} + \bar{t}\mathcal{R}_{b_0}$$

$$(a_n)_{\text{out}} = \bar{t}\mathcal{R}_{a_n} \qquad n = 1, 2, \ldots$$

$$(b_n)_{\text{out}} = \bar{t}\mathcal{R}_{b_n} \qquad n = 1, 2, \ldots \tag{12.62}$$

$$(c_n)_{\text{out}} = \bar{t}\mathcal{R}_{c_n} \qquad n = 1, 2, \ldots$$

where the various reaction rates are the same as in Equations 12.58. Now, of course, the reaction rates are evaluated at the reactor outlet concentrations, and it is these outlet concentrations for which Equations 12.62 are to be solved. A variety of analytical and numerical techniques are available for this solution. One simple approach is to convert Equations 12.62 to the set of differential equations corresponding to an unsteady perfect mixer. See Section 8.1.1 and Equations 8.28. Solving this set for the startup transient gives the various outlet concentrations as the asymptotic solution. This method works equally well for the general case of unequal reactivity and an arbitrary distribution of oliogomers in the inlet stream.

The last few examples introduced the *equal reactivity assumption*. This assumption is widely used in the polymer literature. It states that all endgroups have the same probability of reaction regardless of the length of the molecule to which they are attached. The assumption has the merits of simplicity and plausibility for many polymerizations. It sometimes enables analytical solutions for the molecular weight distribution. It can be faulted in some systems for the first few oligomers but typically becomes an excellent approximation for n greater than 3 or 4. It can be faulted in many systems producing high molecular weight polymers at high conversions. Chain entanglements will lower the mobility of endgroups attached to long chains while having less effect on short chains. This phenomenon occurs in both free-radical and condensation polymerizations. In free-radical polymerizations, it increases molecular weight by slowing the termination reaction. In condensation polymerizations, it retards molecular weight by reducing the probability of long-chain interactions.

12.2.2 Properties of the Molecular Weight Distribution

A molecular weight distribution is described by its frequency function:

$f(\ell)$ = Fraction of polymer molecules that are exactly ℓ units long   (12.63)

This function is similar to the differential distribution of residence times discussed in Chapter 9, except that $f(t)$ is a continuous distribution whereas $f(\ell)$ is discrete, since ℓ takes integer values. The definitions of means and moments use sums rather than integrals. In particular, we suppose $f(\ell)$ to be normalized so that

$$\sum_{\ell=1}^{\infty} f(\ell) = 1 \qquad\qquad (12.64)$$

Normalization is straightforward for step growth polymers. Molecular weight increases gradually with conversion and the starting monomers are included in the sum in Equation 12.64. For addition polymers, the monomer and low molecular weight oligomers are usually excluded from the sum so that $f(1) = 0$ for the "polymer" population. When these low molecular weight species are excluded, the molecular weight distribution of a free-radical polymer becomes approximately independent of conversion.

The mean of $f(\ell)$ is the *number-average degree of polymerization*

$$\boxed{\bar{\ell} = \bar{\ell}_n = \sum_{\ell=1}^{\infty} \ell f(\ell)} \qquad\qquad (12.65)$$

Thus far we have considered distributions in terms of number fractions of the polymer molecules. They can also be defined in terms of weight fractions:

$$g(\ell) = \begin{array}{l} \text{Weight fraction of the polymer molecules} \\ \text{that are exactly } \ell \text{ units long} \end{array} \qquad (12.66)$$

where

$$g(\ell) = \frac{\ell f(\ell)}{\bar{\ell}} \tag{12.67}$$

This definition assumes that the weight of a polymer molecule is proportional to its length, an assumption that ignores the weight of endgroups. The mean of $g(\ell)$ is the **weight-average chain length**:

$$\bar{\ell}_w = \sum_{\ell=1}^{\infty} \ell g(\ell) = \frac{1}{\bar{\ell}} \sum_{\ell=1}^{\infty} \ell^2 f(\ell) \tag{12.68}$$

Both $\bar{\ell}_n$ and $\bar{\ell}_w$ can be multiplied by the weight of a repeating unit to give the **number- and weight-average molecular weights**, $\overline{M}_n$ and $\overline{M}_w$. The ratio,

$$\text{Polydispersity} = \frac{\bar{\ell}_w}{\bar{\ell}_n} = \frac{\overline{M}_w}{\overline{M}_n} \tag{12.69}$$

is called the **polydispersity** and is commonly used to describe the breadth of a molecular weight distribution. The polydispersity has a lower limit of 1.0, which can only be achieved when all molecules have the same length. Such a system is called **monodisperse**. In practice, a polydispersity of 1.01 is very narrow indeed, and a polydispersity of 2 to 4 is typical of most commercial polymers.

The kinetic schemes of the previous section can be solved to find $f(\ell)$. Analytical results usually require the equal reactivity assumption. For a binary polycondensation with perfect initial stoichiometry,

$$f(\ell) = (1 - X)X^{\ell-1} \tag{12.70}$$

This is called the **Flory**, or **most probable, distribution**, and it often arises in polymerizations where reaction rates are independent of chain length. X is the fractional conversion of endgroups; see Equation 12.6. The mean of $f(\ell)$ is

$$\bar{\ell}_n = \frac{1}{1 - X} \tag{12.71}$$

in agreement with Equation 12.9 for $S_{AB} = 1$. The weight-average chain length is

$$\bar{\ell}_w = \frac{1 + X}{1 - X} \tag{12.72}$$

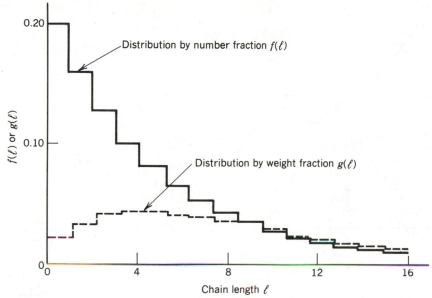

Figure 12.1 Flory distribution for a conversion of $X_A = X_B = 0.8$, $\bar{\ell}_n = 5$; $\bar{\ell}_w = 9$.

and the polydispersity is

$$\text{Polydispersity} = \frac{\bar{l}_w}{\bar{l}_n} = 1 + X \tag{12.73}$$

Figure 12.1 illustrates the Flory distribution for the case $\bar{\ell}_n = 5$, where the discrete nature of the distribution is apparent. Figure 12.2 shows the case of $\bar{\ell}_n = 50$ where the distribution is essentially continuous. A polydispersity of about 2 is typical of high molecular weight condensation polymers. A polydispersity of 1.5 to 2.0 is typical of the **instantaneous molecular weight distribution** of a free-radical polymer, but the composite distribution from a moderate- to high-conversion reactor will often be broader than this due to thermal inhomogeneities. Coordination catalysis produces very broad distributions, while some low-temperature, ionic polymerizations can give nearly monodisperse polymers.

Example 12.12 Determine the molecular weight distribution as a function of conversion for the self-condensation of Equations 12.43 through 12.46. Do the equal reactivity case and compare it to a variable reactivity case with $k_{mn} = k_{11}(mn)^{-0.6}$. Ignore cyclization.

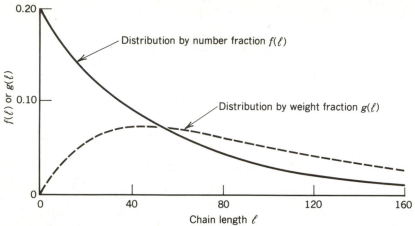

Figure 12.2 Flory distribution for a conversion of $X_A = X_B = 0.98$, $\ell_n = 50$, $\ell_w = 99$.

Solution Let $p_0 = (p_1)_0$ be the initial concentration of monomer. Then Equation 12.45 becomes

$$\frac{d(p_n/p_0)}{d\tau} = \frac{1}{2}\sum_{m=1}^{n-1} m^{-\beta}(n-m)^{-\beta}\frac{p_m}{p_0}\frac{p_{n-m}}{p_0} - \frac{p_n}{p_0}\sum_{n=1}^{\infty}(mn)^{-\beta}\frac{p_m}{p_0}$$

(12.74)

where $\tau = k_{11}p_0\alpha$ is a dimensionless reaction time, and $\beta = 0$ or $\beta = 0.6$ depending on the case. This set of equations is integrated subject to the initial conditions that $p_1/p_0 = 1$, $p_n/p_0 = 0$, $n = 2, 3, \ldots$, at $\tau = 0$. Some results are

	$\beta = 0$			$\beta = 0.6$		
$\tau = k_{11}p_0\alpha$	$1 - X$	$\bar{\ell}_n$	$\bar{\ell}_w$	$1 - X$	$\bar{\ell}_n$	$\bar{\ell}_w$
0	1.000	1.000	1.000	1.000	1.000	1.000
0.125	0.941	1.063	1.125	0.942	1.061	1.120
0.25	0.889	1.125	1.250	0.893	1.120	1.231
0.50	0.800	1.250	1.500	0.812	1.232	1.433
0.75	0.727	1.375	1.750	0.749	1.336	1.612
1.00	0.667	1.500	2.000	0.698	1.434	1.774
1.50	0.571	1.750	2.500	0.619	1.615	2.016
2.00	0.500	2.000	3.000	0.562	1.779	2.311
3.00	0.400	2.500	4.000	0.483	2.027	2.738
4.00	0.333	3.000	5.000	0.430	2.329	3.100
5.00	0.286	3.500	6.000	0.391	2.559	3.418

The molecular weights for the case of $\beta = 0$ follow the Flory distribution. Given a value for X, the $\bar{\ell}_n$ and $\bar{\ell}_w$ values agree with those calculated from Equations 12.71 and 12.72. Further, the conversion follows a simple rate law:

$$\frac{dX}{d\alpha} = \frac{k_{11}a_0}{2}(1 - X)^2 \tag{12.75}$$

This result is not happenchance. It can be deduced from Equation 12.45 for the case of equal reactivity with no cyclization. Confirmation of a "known" result such as this lends credibility to simulations for cases where analytical solutions are not possible.

The simulation for $\beta = 0.6$ gives the expected result of lower conversions and a narrower molecular weight distribution. The results are fairly close to the equal reactivity case, and it would be difficult to distinguish between the two situations experimentally. One reason for this close agreement is that Equation 12.9 must still hold since its derivation did not require an equal reactivity assumption.

Example 12.13   Apply the quasi-steady and equal reactivity assumptions to a batch reaction to determine the instantaneous number- and weight-average chain lengths for dead polymer. Assume a free-radical mechanism with chemical initiation and termination by combination.

Solution   The equal reactivity assumptions give $k_n = \text{constant} = k_p$ and $k_{mn} = \text{constant} = k_c$. The quasi-steady hypothesis gives $d[R_n \cdot]/d\alpha = 0$. Applying these to Equation 12.37 gives

$$0 = k_p[M][R_{n-1} \cdot] - k_p[M][R_n \cdot] - k_c[R_n \cdot]\Psi \tag{12.76}$$

where

$$\Psi = \sum_{m=1}^{\infty} [R_m \cdot] = \text{const} \tag{12.77}$$

Solving for $[R_n \cdot]$ gives

$$[R_n \cdot] = \frac{k_p[M][R_{n-1} \cdot]}{k_p[M] + k_c\Psi} = \omega[R_{n-1} \cdot] \tag{12.78}$$

where ω is a constant proportionality factor. Since the instantaneous distribution is desired, $[M]$ and thus ω are constant.

For the thermal initiation scheme is Section 12.2.1, Equation 12.78 applies for $n > 2$. For chemical initiation it applies for $n = 2$ as well. Define $R_n = [R_n \cdot]/[R_1 \cdot]$. Then the relative concentrations of chain lengths are distributed according to

n	R_n
1	1
2	ω
3	ω^2
⋮	⋮

This is a geometric progression with sum $(1 - \omega)^{-1}$. The frequency function of chain lengths is

$$f_G(n) = (1 - \omega)\omega^{n-1} \tag{12.79}$$

which is identical in form to Equation 12.70. Thus the growing polymer chains have a Flory distribution. The dead polymer has a different and narrower distribution. It is narrower due to the averaging process that results when termination is by combination. An unusually long chain will usually combine with a shorter chain, giving a sum that is not too different from the average sum.

We assume that termination occurs between pairs of growing chains chosen completely at random. Thus a well-known theorem in probability theory states that

$$\sum_{\ell=1}^{\infty} f_d(\ell)e^{s\ell} = \left[\sum_{n=1}^{\infty} f_g(n)e^{sn} \right]^2 \tag{12.80}$$

where $f_d(\ell)$ and $f_g(n)$ are frequency functions for the dead and living polymer, respectively. The exponentially weighted sums are known as **_moment generating functions_**. They have the property that

$$\lim_{s \to 0} \frac{d^N}{ds^N} \sum_{\ell=1}^{\infty} f_d(\ell)e^{s\ell} = \mu_N \tag{12.81}$$

where μ_N is the Nth moment of the distribution. Note that $N = 0$ gives $\mu_0 = 1$, $N = 1$ gives $\mu_1 = \bar{\ell}_N$, and $N = 2$ gives $\mu_2 = \bar{\ell}_N\bar{\ell}_W$.

For the case at hand,

$$\sum_{\ell=1}^{\infty} f_d(\ell)e^{s\ell} = \left[\sum_{n=1}^{\infty} (1 - \omega)\omega^{n-1}e^{sn} \right]^2 \tag{12.82}$$

Applying Equation 12.81 gives

$$\bar{\ell}_N = \frac{2}{1 - \omega} \qquad \bar{\ell}_W = \frac{2 + \omega}{1 - \omega} \tag{12.83}$$

and

$$\text{Polydispersity} = \frac{2 + \omega}{2} \tag{12.84}$$

so that the polydispersity has a limiting value of 1.5 for the assumed kinetic scheme.

12.3   Polymerization Reactors

The properties of a polymer depend not only on its gross chemical composition but also on its molecular weight distribution, copolymer composition distribution, branch length distribution, and so on. The same monomer(s) can be converted to widely differing polymers depending on the polymerization mechanism and reactor type. This is an example of *product by process*, and no single product is best for all applications. Thus there are several commercial varieties each of polyethylene, polystyrene, and poly(vinyl chloride) that are made by distinctly different processes.

Table 12.1 classifies polymerization reactors by the number and type of phases involved in the reaction. The terms *homogeneous* and *heterogeneous* have their usual meanings. However, the most important consideration is not whether the system is homogeneous or heterogeneous but whether the continuous phase contains a significant concentration of high molecular weight polymer. When it does, the reactor design must accommodate the high viscosities and low diffusivities typical of concentrated polymer solutions. It is the design of reactors for operation with a continuous polymer phase that most distinguishes polymer reaction engineering from chemical reaction engineering. When the polymer-rich phase is dispersed, the chemical kinetics and mass transfer steps may be very complex, but the qualitative aspects of the design resemble those of the heterogeneous reactors treated in Chapter 11.

Table 12.1   Classification of Polymerization Reactors

Continuous Phase	Dispersed Phase	Type of Polymerization
Polymer solution	None	Homogeneous bulk or solution polymerization
Polymer solution	Any (e.g., a condensation product)	Heterogeneous bulk on solution polymerization
Water or other nonsolvent	Polymer or polymer solution	Suspension, dispersion, or emulsion polymerization
Liquid monomer	Polymer (swollen with monomer)	Precipitation or slurry polymerization
Gaseous monomer	Polymer	Gas phase polymerization

A polymerization reactor will be heterogeneous whenever the polymer is insoluble in the monomer mixture from which it was formed. This is a fairly common situation and gives rise to the precipitation, slurry, and gas phase polymerizations listed in Table 12.1. If the polymer is soluble in its own monomers, a dispersed phase polymerization requires the addition of a non-solvent (typically water) together with appropriate interfacial agents. These extraneous materials require additional downstream purification and separation steps which may negate any savings in reactor design. Thus *bulk* polymerizations from pure, undiluted monomer are usually preferred for systems where the polymer is soluble in the monomer. For high-volume polymers, like high-volume chemicals, continuous operation is generally preferred over batch. The following sections outline design concepts for two forms of *bulk continuous* polymerizers: the laminar flow tube and the laminar flow stirred tank.

12.3.1   Tubular Polymerizers

Tubular reactors are occasionally used for bulk, continuous polymerizations. A monomer or monomer mixture is introduced at one end of the tube and, if all goes well, a high molecular weight polymer emerges at the other. Practical problems arise from three types of instability:

- *Velocity profile elongation*   Low fluid velocities near the tube wall give rise to high extents of polymerization, high viscosities, and to yet lower velocities. The velocity profile elongates, possibly to the point of hydrodynamic instability.
- *Thermal runaway*   Temperature control in a tubular polymerizer depends on convective diffusion of heat. This becomes difficult in a large-diameter tube, and temperatures may rise to a point where a thermal runaway becomes inevitable.
- *Tube-to-tube interactions*   The problems of velocity profile elongation and thermal runaway can be eliminated by using a multitubular reactor with many small-diameter tubes in parallel. Unfortunately, this may give rise to a new form of instability. Imagine a 1000-tube reactor with 999 of the tubes plugged with solid polymer!

Present design methods are suitable for two situations: independently fed, single tubes or multitubular finishing reactors where the entering polymer concentration is already quite high.

Figure 12.3 illustrates stability regimes for the thermally initiated polymerization of styrene with laminar flow in a single tube. Fluid properties and reaction kinetics are fixed by specifying a bulk polymerization with no polymer in the feed. Design and operating variables are the physical dimensions of the tube, the operating temperatures and pressures, and the flow rate. It turns out that the aspect ratio L/R is unimportant, provided it is reasonably large, and that the operating pressure is unimportant unless it is so high that liquid compressibility becomes significant. Steady state operation can be described by four variables: T_{in}, T_{wall}, R, and $\bar{t}$. Figure 12.3 shows the influence of two of these with the other two held constant at plausible values. Stable operation is easy to achieve in

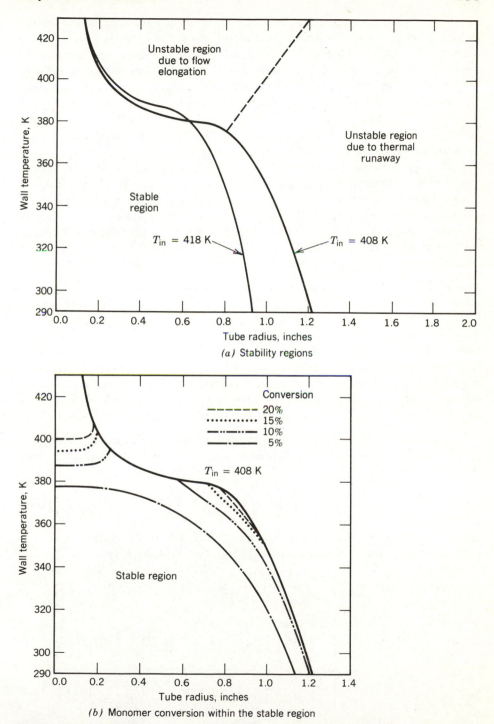

Figure 12.3 Performance of a laminar flow, tubular reactor for the bulk styrene polymerization with $T_{in} = 135°C$ and $\bar{t} = 1$ hr.

small-diameter tubes since mass and thermal diffusion removes gradients in the radial direction. Thus the viscosity is uniform across the tube, and no hot spots develop. As the tube diameter is increased, mass and heat transfer become progressively more difficult. Since $\mathscr{D} \ll \alpha_T$, concentration gradients develop before thermal gradients; and velocity profile elongation emerges as the first limit on conversion. In still larger tubes, operation is limited by thermal stability. For the polystyrene example of Figure 12.3, a thermal runaway will occur if the tube radius exceeds 1.25 in. This is true even if the tube wall were held at 0 K (pretending the model remained valid).

The polystyrene example in Figure 12.3 incorporates a detailed kinetic model of the polymerization including calculation of ℓ_N and ℓ_w. It also includes a reasonably sophisticated description of viscosity as a function of temperature, polymer concentration, and molecular weight. Assembling such a model requires weeks or months. The following example uses simplifying assumptions to reduce the time requirements to hours or days. Hopefully it will still reveal the essence of design problems in tubular polymerizers.

Example 12.14   Model the performance of a laminar flow, tubular reactor for the polymerization of a self-condensing monomer A. The polymerization rate may be taken as $\mathscr{R} = ka^2$, where $a = [A]/[A]_{\text{in}}$ is the dimensionless concentration of monomer. Assume density is constant and viscosity varies with conversion as $\mu = \mu_0(100 - 99a)$. The heat of reaction is small so that isothermal operation is possible. Ignore the radial component of velocity in the axial equation of motion but not in the convective diffusion equation. Assume a reasonably fast reaction with $k\bar{t} = 4$ and a moderately low diffusivity with $\mathscr{D}_A\bar{t}/R^2 = 0.01$.

Solution   The problem requires solution of the convective diffusion equation for mass but not for energy. Rewriting Equation 6.79 in dimensionless form gives

$$\frac{v_z}{\bar{u}}\frac{\partial a}{\partial z} + \frac{v_r}{\bar{u}}\left(\frac{L}{R}\right)\frac{\partial a}{\partial \imath} = \frac{\mathscr{D}_A\bar{t}}{R^2}\left(\frac{1}{\imath}\frac{\partial a}{\partial \imath} + \frac{\partial^2 a}{\partial \imath^2}\right) + \bar{t}\mathscr{R}_A \tag{12.85}$$

Using central differences in the r-direction and a forward difference in the z-direction gives the marching-ahead equations (see Sections 6.4):

$$a(\imath, z + \Delta z) = \left[1 - \frac{2\bar{u}}{v_z}\left(\frac{\mathscr{D}_A\bar{t}}{R^2}\right)\frac{\Delta z}{\Delta \imath^2}\right]a(\imath, z)$$

$$+ \left[\frac{\bar{u}}{v_z}\left(\frac{\mathscr{D}_A\bar{t}}{R^2}\right)\frac{\Delta z}{\Delta \imath^2}\left(1 + \frac{\Delta \imath}{2\imath}\right) - \frac{1}{2}\frac{v_r}{v_z}\left(\frac{L}{R}\right)\frac{\Delta z}{\Delta \imath}\right]a(\imath + \Delta \imath, z)$$

$$+ \left[\frac{\bar{u}}{v_z}\left(\frac{\mathscr{D}_A\bar{t}}{R^2}\right)\frac{\Delta z}{\Delta \imath^2}\left(1 - \frac{\Delta \imath}{2\imath}\right) + \frac{1}{2}\frac{v_r}{v_z}\left(\frac{L}{R}\right)\frac{\Delta z}{\Delta \imath}\right]a(\imath - \Delta \imath, z)$$

$$+ \frac{\bar{u}}{v_z}\bar{t}\mathscr{R}_A \Delta z \tag{12.86}$$

where $\mathscr{R}_A = -ka^2(\imath, z)$. Special forms are needed at the wall and the centerline:

$$a(1, z + \Delta z) = \frac{4a(1 - \Delta \imath, z + \Delta z) - a(1 - 2\Delta \imath, z + \Delta z)}{3} \tag{12.87}$$

and

$$a(0, z + \Delta z) = \left[1 - \frac{4\bar{u}}{v_0}\left(\frac{\mathscr{D}_A \imath}{R^2}\right)\frac{\Delta z}{\Delta \imath^2}\right]a(0, z)$$

$$+ \left[\frac{4\bar{u}}{v_0}\left(\frac{\mathscr{D}_A \imath}{R^2}\right)\frac{\Delta z}{\Delta \imath^2}\right]a(\Delta \imath, z) + \frac{\bar{u}}{v_0}\imath\mathscr{R}_A \Delta z \tag{12.88}$$

The forms recognize that $v_r = 0$ at $\imath = 1$ and at $\imath = 0$. The stability criterion is

$$\Delta z_{max} \le \left[\frac{v_z(1 - \Delta \imath, z)R^2}{2\bar{u}\mathscr{D}_A \imath}\right]\Delta \imath^2 \tag{12.89}$$

The reader may find it helpful to compare Equations 12.86 through 12.89 with Equations 6.28, 6.29, 6.32, and 6.33.

The stability criterion is based on the axial velocity at one radial increment away from the wall. Note, however, that v_z will change in the axial direction and that the stability criterion becomes more severe as the flow elongates. The usual procedure is to pick a value for $\Delta \imath$, to calculate a maximum value for Δz, and then to pick some smaller value for Δz which is the reciprocal of an integer. Since the minimum value for $v_z(1 - \Delta \imath)$ will not be known in advance, the choice of Δz will have to be quite conservative. A better approach is to use a variable step size in the axial direction, changing it according to the dictates of Equation 12.89 as the calculations progress. This method naturally requires a final, variable-length step so that the calculation finishes exactly at $z = 1.00$.

The marching-ahead equations require knowledge of the dimensionless velocity components $v_z/\bar{u}$ and $(v_r/\bar{u})(L/R)$. These are calculated using the methods of Section 6.5. With constant density, Equation 6.66 simplifies to

$$\frac{v_z}{\bar{u}} = \frac{\int_\imath^1 (\imath/\mu)\, d\imath}{\int_0^1 (\imath^3/\mu)\, d\imath} \tag{12.90}$$

These integrals are evaluated using viscosity values estimated at the current axial position.

Equation 6.77 must be converted to dimensionless variables. The result for constant density is

$$\frac{v_r}{\bar{u}}\left(\frac{L}{R}\right) = -\frac{1}{\imath}\int_0^\imath \imath' \frac{\partial(v_z/\bar{u})}{\partial z}\, d\imath' \tag{12.91}$$

The values of $\partial(v_z/\bar{u})/\partial z$ are evaluated using a first order difference approximation:

$$\frac{\partial(v_z/\bar{u})}{\partial z} = \frac{[v_z(r, z + \Delta z)]/\bar{u} - [v_z(r, z)]/\bar{u}}{\Delta z} \tag{12.92}$$

The initial conditions are

$$a(\imath, 0) = 1.0$$

$$v_z(\imath, 0) = 2\bar{u}(1 - \imath^2) \tag{12.93}$$

$$v_r(\imath, 0) = v_r(\imath, \Delta z) = 0$$

The boundary conditions have already been incorporated into the marching-ahead equations.

The above equations allow concentrations and velocities to be calculated in a stepwise manner starting at $z = 0$. The procedure assumes that all variables are known at position z. They are then evaluated at $z + \Delta z$ using the following sequence of calculations:

1. $a(0, z + \Delta z)$ is calculated using Equation 12.88.
2. The $a(\imath, z + \Delta z)$ are calculated for $\Delta\imath \leq \imath \leq 1 - \Delta\imath$ using Equation 12.86.
3. $a(1, z + \Delta z)$ is calculated using Equation 12.87.
4. The $\mu(\imath, z + \Delta z)$ are estimated using the viscosity correlation. Note that the velocity calculations depend only on relative changes in viscosity. Thus the choice of $\mu_0 = 1$ is always possible.

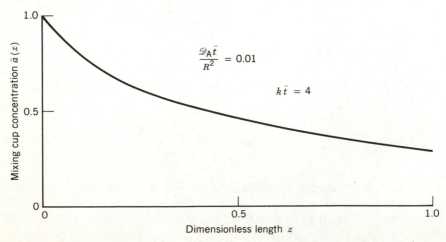

Figure 12.4 Mixing-cup concentration versus length for a tubular polycondensation.

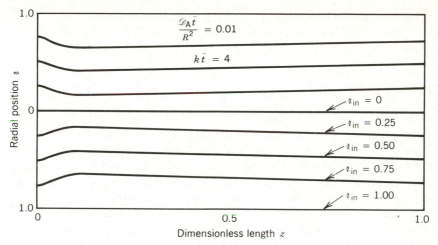

Figure 12.5 Streamlines for polycondensation in a tubular reactor.

5. The integral in the denominator of Equation 12.90 is evaluated. This and all other integrations should use a method (e.g., the trapezoidal rule) that converges at least $0(\Delta \imath^2)$.

6. The $v_z(\imath, z + \Delta z)/\bar{u}$ are calculated using Equation 12.90.

7. The partial derivatives of $v_z/\bar{u}$ are calculated using Equation 12.92.

8. Values for $(v_r(\imath, z + \Delta z)/\bar{u})(L/R)$ are calculated using Equation 12.91.

9. All variables have now been calculated at position $z + \Delta z$. $(\Delta z)_{\max}$ is calculated using Equation 12.89, and the entire procedure is repeated until $z = 1.00$

Figure 12.4 shows the mixing cup average concentration

$$\bar{a}(z) = \frac{\int_0^1 2\pi\imath v_z(\imath, z) a(\imath, z)\, d\imath}{\int_0^1 2\pi\imath v_z(\imath, z)\, d\imath} \tag{12.94}$$

as a function of z for the example reactor, with $k\bar{t} = 4$ and $\mathscr{D}_A\bar{t}/R^2 = 0.01$. Figure 12.5 shows the curved fluid streamlines which are a consequence of this tubular polymerization.

The marching-ahead technique is suitable for simple reactor models and for limited use on fairly complex models such as the polystyrene example of Figure 12.3. More sophisticated numerical methods are needed for design optimization studies and for models that treat individual polymer species. See Suggestions for Further Reading at the end of Chapter 6. For free-radical polymerizations, existing models lump all polymer molecules into a single "polymer" species about which only $\bar{\ell}_n$ and $\bar{\ell}_w$ are known. More recent models for condensation polymerizations treat 20 or so oligomers exactly and also calculate $\bar{\ell}_n$ and $\bar{\ell}_w$ for the entire population. See Suggestions for Further Reading at the end of this chapter.

12.3.2 Stirred Tank Polymerizers

Continuous flow, stirred tank reactors are widely used for bulk, free-radical polymerizations. They have two main advantages: autorefrigeration can be used to remove the heat of polymerization, and fairly narrow molecular weight and copolymer composition distributions can be achieved, at least for the free-radical case.[6] Their principal disadvantage is that high conversions are not possible. The limit on conversion is not due just to the usual problem of a low average reaction rate in a perfect mixer compared to a batch or piston flow reactor. Instead, it is also due to the high viscosities encountered in typical, free-radical polymerizations. The pure polymer, undiluted by residual monomer, will have a viscosity two or more orders of magnitude higher than the design limit for stirred tanks. Practical designs hold polymer concentrations to less than 85 per cent by weight; any polymerization beyond this point is done in a tubular post reactor.

Laminar flow is almost inevitable in a polymer reactor due to the high viscosities of concentrated polymer solutions. Molecular diffusivities are also low. This combination means that segregation is a real possibility in a stirred tank reactor. See Chapter 9. A segregated stirred tank will tend to give a broader molecular weight distribution, a higher conversion, and a broader copolymer composition distribution than a perfect mixer. Given the appropriate rate expressions, calculations are quite feasible for the limiting cases of complete segregation and perfect mixing.

Example 12.15 Poly(styrene–acrylonitrile) is a free-radical copolymer of styrene and acrylonitrile which can be manufactured by bulk polymerization in a stirred tank reactor. Determine the macroscopic copolymer composition distribution for 40 percent conversion in a perfect mixer and in a completely segregated stirred tank. The average acrylonitrile content is adjusted to be 42 mole-% in both cases.

Solution The macroscopic composition distribution in a perfect mixer is a delta distribution. All copolymer molecules are formed from the same monomer mixture, and there is no variation except for probabilistic effects. Thus the first part of the problem is done in a formal sense. As a practical matter, it remains to determine the feed composition to the perfect mixer which will give a copolymer containing 42 mole-% acrylonitrile when operating at 40 percent conversion. Set X = styrene, Y = acrylonitrile and use $r_X = 0.41$ and $r_Y = 0.04$.

---

[6] Binary polycondensations show a much broader molecular weight distribution in a perfect mixer than in a batch or piston flow reactor.

Then Equation 12.31 gives

$$\frac{x_p}{y_p} = \frac{0.58}{0.42} = \frac{x(0.41x + y)}{y(x + 0.04y)} = \frac{F_X(0.41F_X + F_Y)}{F_Y(F_X + 0.04F_Y)} \tag{12.95}$$

where $F_Y = 1 - F_X$ is the mole fraction of acrylonitrile in the monomer mixture (polymer-free basis). Solution of the resulting quadratic equation gives $F_Y = 0.486$. A material balance is needed to find the acrylonitrile content in the feed. Use 1 mole of feed as the basis:

$$(F_Y)_{in} = 0.6(F_Y)_{out} + 0.4(F_Y)_p$$

$$= 0.6(0.486) + 0.4(0.420) = 0.460 \tag{12.96}$$

Thus the feed mixture to the perfect mixer should contain 46 mole-% acrylonitrile. See also Problem 12.2, which adds a second perfect mixer in series with the first.

The case of the completely segregated reactor is more complex. It can be modeled as a large number of piston flow reactors operating in parallel with the reactor lengths being distributed exponentially. Suppose we approximate the segregated reactor using N piston flow elements. Then the residence time in the nth element is

$$t_n = -\bar{t} \ln\left(1 - \frac{2n - 1}{2N}\right) \qquad n = 1, 2, \ldots, N \tag{12.97}$$

The reader may wish to use the concepts of Chapter 9 to confirm that Equation 12.97 indeed approximates an exponential distribution of residence times as $N \to \infty$. Note that each piston flow element receives a fraction Q/N of the total flow to the system.

The methods of Section 12.1.3 allow the compolymer composition distribution to be calculated when the conversion in a batch or piston flow reactor is known. Unfortunately, we know only the average conversion and not its distribution between the various piston flow elements in the model. Thus a rate expression is needed. A suitable one for a free-radical copolymerization is

$$-\frac{d(x + y)}{d\alpha} = \mathscr{R}_p = \frac{\sqrt{\mathscr{R}_I \cdot /(k_c + k_d)} \, k_{XX}(r_X x + 2xy + r_Y y)^2}{r_X x + (k_{XX}/k_{YY})r_Y y} \tag{12.98}$$

where $\mathscr{R}_I \cdot$ denotes the generation rate for primary radicals. See Problem 12.4.

Suppose the various parameters in Equation 12.98 are known. Then the copolymer composition distribution can be calculated using the following proce-

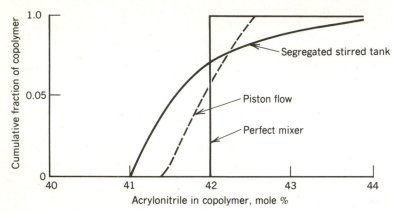

Figure 12.6 Comparison of copolymer composition distributions for ideal reactors.

dure:

1. Assume values for the inlet feed composition and system residence time $\bar{t}$.
2. Integrate the batch rate equation, Equation 12.98, and the copolymer composition equation, Equation 12.34, for each piston flow element.
3. Stop integration for the nth element when $t = t_n$ as given by Equation 12.97.
4. Sum the contributions from the various elements. Are the overall conversion and average polymer compositions correct? If not, return to Step 1 and repeat the calculation.

The simultaneous integration of Equations 12.98 and 12.34 is a situation we have not previously encountered. Using the marching-ahead technique, Equation 12.98 gives $\Delta x + \Delta y = -F_1 \Delta \alpha$, where F_1 represents the right-hand side evaluated at the old values of x and y. Equation 12.34 gives $\Delta x / \Delta y = F_2$. Simultaneous solution gives

$$\Delta x = \frac{-F_1 F_2\, \Delta \alpha}{1 + F_2} \qquad \Delta y = \frac{-F_1\, \Delta \alpha}{1 + F_2} \tag{12.99}$$

so that both X and Y can be incremented to their new values.

Figure 12.6 shows the cummulative composition distributions for a perfect mixer, a batch reactor, and a segregated stirred tank. All three reactors have been devised to give 40 percent conversion of the entering monomers and to give a copolymer containing 42 mole-% acrylonitrile. The results for the segregated stirred tank assume $k_{XX}/k_{YY} = 1$.

The example illustrates the interesting situation where a batch reactor gives a performance which is intermediate between that of a perfect mixer and that of a segregated stirred tank. A real reactor will also lie somewhere between these extremes, but quantitative predictions are not yet possible.

Suggestions for Further Reading

A general reference on polymerization reaction engineering is:

J. A. Biesenberger and D. H. Sebastian, *Principles of Polymerization Engineering*, Wiley, New York, 1983.

A comprehensive treatment of molecular weight distributions is given in:

L. H. Peebles, Jr., *Molecular Weight Distributions in Polymers*, Interscience, New York, 1971.

while a briefer survey of suitable mathematical techniques is given in:

D. C. Chappelear and R. H. Simon, "Polymerization Reaction Engineering," *Adv. in Chem.*, **91**, 1–24 (1969).

Several tubular polymerization models were cited in Chapter 6. An example of a design and optimization study using a fairly sophisticated model for styrene polymerization is given in:

R. Mallikarjun and E. B. Nauman, "A Staged Multitubular Process for Crystal Polystyrene," *Poly. Process Eng.*, **4**, 31–52 (1986).

Another sophisticated model, this time of the high-pressure process for polyethylene, is given by:

S. K. Gupta, A. Kumar, and M. V. E. Krishnamurthy, "Simulation of Tubular Low-Density Polyethylene," *Polym. Eng. Sci.*, **25**, 37–47 (1985).

The state of the art in modeling binary polycondensations in tubular reactors is given by:

E. B. Nauman and H. S. McLaughlin, "Modeling Condensation Polymers in Flow Reactors," Paper 48b, San Francisco AIChE Meeting, 1984.

The effects of mixing on copolymer composition distributions and an approach to measuring segregation in a stirred tank reactor are discussed in

E. B. Nauman, "New Results and Old Problems in Residence Time Theory," *Instn. Chem. Eng. Symp. Ser.*, **87**, 569–581 (1984).

Problems

12.1 A binary polycondensation of AMA and BNB is to be performed in a batch reactor. A number-average chain length of at least 100 is required. What minimum accuracy is required for weighing the two components?

12.2 A continuous polymerization train consisting of two stirred tanks in series is used to copolymerize styrene, $r_X = 0.41$, and acrylonitrile, $r_Y = 0.04$. The flow rate to the first reactor is 5000 pounds per hour and a conversion of 40% is expected. Makeup styrene is fed to the second reactor and a conversion of 30% (based on the 5000 lb/hr initial feed) is expected there. What should be the feed composition and how much styrene should be fed to the second reactor if a compolymer containing 58 wt-% styrene is desired?

12.3 Consider a laminar flow tubular polymerizer with cooling at the tube wall. At what radial position will a hot spot develop: at the tube wall, at the centerline, or at an intermediate radius? Explain your answer. Will the situation change with heating at the wall?

12.4 Use the quasi-steady state hypothesis to show that

$$[X \cdot] + [Y \cdot] = \sqrt{\frac{\mathscr{R}_I \cdot}{k_c + k_d}}$$

for a free-radical copolymerization. Combine this result with Equation 12.30 to eliminate $[X \cdot]$ and $[Y \cdot]$ from the total propagation rate $\mathscr{R}_p = -(\mathscr{R}_x + \mathscr{R}_y)$, thereby obtaining Equation 12.98.

12.5 Many literature results on tubular polymerizers have ignored the effects of radial convection. Delete the appropriate terms in Equation 12.86 and repeat the binary polycondensation example of Section 12.3.1. Compare your results to those in Figure 12.4. Is radial convection important in this situation?

12.6 Example 12.13 used the equal reactivity and quasi-steady state hypotheses to obtain analytical results for $\bar{\ell}_n$ and $\bar{\ell}_w$ in a chemically initiated, free-radical polymerization with termination by combination. Apply the same approach to the thermally initiated polymerization of styrene using the mechanism of Section 12.2.1. Specifically, use Equations 12.39 and 12.40 to obtain expressions for $[R_n \cdot]$ where $n = 1, 2, \cdots$. Determine $\bar{\ell}_n$ and $\bar{\ell}_w$ for dead polymer.

12.7 (a) The phenol–formaldehyde polycondensation will form an infinite network at high formaldehyde concentrations. Use probability or other arguments and the equal reactivity assumption to show that gelation will occur at complete conversion if $S_{AB} > 0.75$ where A denotes formaldehyde and B denotes phenol.[7]

(b) Linear oligimers can also be obtained at very high formaldehyde concentrations, $S_{AB} > 1$. Find the limiting stoichiometry below which complete conversion will lead to gelation.

12.8 Use the quasi-steady state hypothesis and the equal reactivity assumption to confirm Equation 12.24. Note that

$$[R \cdot] = \sum_{n=1}^{\infty} [R_n \cdot]$$

12.9 Example 12.7 treated a simplified model for the thermal initiation of styrene. A more complete mechanism [see F. R. Mayo, *J. Am. Chem. Soc.*, **90**, 1289 (1968) and Y. K. Chong, E. Rizzardo, and P. H. Solomon, *J. Am. Chem. Soc.*, **105**, 7761 (1983)] is believed to be:

$$2M \rightleftharpoons AH \quad \text{(Diels-Alder adduct)}$$

$$2M \rightleftharpoons \cdot M_2 \cdot$$

---

[7]Practical experience shows the gelation point to be at $S_{AB} > 0.8$. This is presumably due to lower reactivity of the para position compared with the ortho positions.

$\cdot M_2 \cdot \rightarrow$ Diphenylcyclobutane   (cyclic dimers)

$M + AH \rightarrow M \cdot + A \cdot$

$M + AH \rightarrow P_3$ (trimers)

$A \cdot + M \rightarrow R_3 \cdot$

$M \cdot + M \rightarrow R_2 \cdot$

$R_n \cdot + M \rightarrow R_{n+1} \cdot$ (propagation)

$R_n \cdot + M \rightarrow P_n + M \cdot$ (chain transfer to monomer)

$R_n \cdot + AH \rightarrow P_n + A \cdot$ (chain transfer to adduct)

$R_n \cdot + R_m \cdot \rightarrow P_{n+m}$ (termination by combination)

(a) Use this mechanism to develop species material balance equations for the isothermal, batch polymerization of styrene.

(b) Apply the quasi-steady state hypothesis to the above mechanism.

Index